# Radionuclides and Heavy Metals in the Environment

This series includes volumes on environmental pollution of air, water and soil, causative toxic agents, including radioactive materials, and remediation strategies. Pollution and its management include interdisciplinary research and applications. Books will cover topics in contamination, toxicology, pollution from heavy metals (water, air, soil), radiobiology, biodegradation, remediation, microbiology, nanotechnology, environment, policy.

**Interested in editing a volume? Please contact Dr. Dharmendra K. Gupta, Series Editor, at** guptadk1971@gmail.com

Pankaj Pathak • Sankar Ganesh Palani
Editors

# Circular Economy in Municipal Solid Waste Landfilling: Biomining & Leachate Treatment

## Sustainable Solid Waste Management: Waste to Wealth

Springer

*Editors*
Pankaj Pathak
Department of Environmental Science
SRM University Andhra Pradesh
Guntur, Andhra Pradesh, India

Sankar Ganesh Palani
Environmental Biotechnology Laboratory
Department of Biological Sciences
Birla Institute of Technology and Science
Pilani, Hyderabad Campus
Hyderabad, Telangana, India

ISSN 2524-7409 ISSN 2524-7417 (electronic)
Radionuclides and Heavy Metals in the Environment
ISBN 978-3-031-07784-5 ISBN 978-3-031-07785-2 (eBook)
https://doi.org/10.1007/978-3-031-07785-2



This Springer imprint is published by the registered company Springer Nature Switzerland AG
The registered company address is: Gewerbestrasse 11, 6330 Cham, Switzerland

# Series Editors' Foreword

This book series will provide recent advancements in the field of pollution management, emphasizing treatment processes and remediation of legacies, waste, and toxic substances including radioactive material of anthropogenic and natural origin in the environment. Pollution, whether of anthropogenic or natural origin (like volcanic eruption), affects the environment and may harm humans. Partly, pollution is a consequence of the increasing use of land and energy resources due to the strongly growing population of our planet (from 2.7 billion in 1955 to 7.6 billion in 2018). In order to preserve our environment, waste must be managed in an appropriate manner, and contaminated areas need to be remediated. This book series will hence focus on technical and biological remediation techniques. These include techniques already widely used but also recent developments and related scientific research.
Pollution is widespread through water, soil, and air. Many cities and rural areas of the world face the challenge of pollution management. Heavy metals (including radioactive materials) and different chemicals (including pesticides, hydrocarbons, and refinery waste) account for partly considerable pollution of the environment. Uncontrolled wastes dumping endangers the environment and may cause large-scale contamination of soil, water, or air. Due to varying interests, different stakeholders like government and industries became proactive in their approach toward pollution prevention and early mitigation to protect the environment. However, classical pollution management strategies like physico-chemical treatments are expensive and suffer process-inherent limitations. However, recent scientific research shows feasibility of large-scale technological developments based on remediation measures which are sustainable for mankind to deal with this multiple problem of pollution.

In the last decades, enormous progress took place in this field. Green technologies, such as phytoremediation or phytofixation, offer solutions to handle the situation. Use of biota for pollution remediation is a natural treatment process.

Varieties of microorganisms (fungi, bacteria, and viruses) are enormously capable of degrading a range of substances. Further, plants are considered green livers, which can remediate polluted components of the environment.

Ministry of Environment, Forest and Climate Change Dharmendra K. Gupta
New Delhi, India

Gottfried Wilhelm Leibniz Universität Hannover, Clemens Walther
Institut für Radioökologie und Strahlenschutz (IRS)
Hannover, Germany

# Preface

This book exclusively focuses on circular economy in municipal solid waste landfilling and provides a highlight of recent advancements in landfill biomining, leachate treatment (organic and inorganic metals), and metal recovery from incinerated landfill waste. With a rapid increase in municipal solid waste generation and concomitant landfilling of the solid waste, investigating new prospects in landfill biomining is imperative. New technologies (physicochemical and biochemical) and processes that are currently used in landfill leachate treatment were discussed in detail. This book will be a collation of invited chapters from experts working in the identified areas of municipal solid waste landfilling.

The landfilling and biomining technologies are the prime focus of this book. The biomining of landfills is among the less used technologies in the landfill remediation process. While few technologies are available for the landfill leachate treatment and heavy metals remediation, most of them are either expensive or inefficient. The biological treatment of landfill leachate remains an unexplored area of research. Hence, this book has focused on identifying crucial processes and technologies for the above-mentioned aspects of landfill operations.

This book will be a valuable addition to the literature by addressing the scarcity of knowledge resources and one of its kind aiming to create a common platform for understanding landfill biomining for resource recycling and utilization and current trends in heavy metals and landfill leachate treatment systems. In this book, Chap. 1 introduces solid waste management and landfill in high-income countries and Chap. 2 describes the significance of landfill mining. Chapters 3 and 4 explore open dumping and associated environmental issues. Moreover, Chap. 5 gives idea of monitoring of the landfill leachate. Chapters 6, 7, 8 and 9 describe the different treatment methods of landfill leachate including operation of bioreactor landfills.

Additionally, Chaps. 10 and 11 explain utilization of inert solid waste as refuse-derived fuel. Whereas Chap. 12 describes metal recovery from incinerated municipal solid waste fly ash.

Guntur, Andhra Pradesh, India Pankaj Pathak

Hyderabad, Telangana, India Sankar Ganesh Palani

# Contents

# Chapter 1
# Solid Waste Management and Landfill in High-Income Countries

**Debananda Roy and Abhrajyoti Tarafdar**

**Abstract** The generation of solid waste (SW) has a direct connection with the monetary status of individuals and shows in terms of gross domestic product (GDP)/gross national income (GNI). The high-income countries (HICs) have shown a lingering generation of solid waste, mainly municipal solid waste (MSW). The MSW of HICs include organic biodegradable, paper, plastic metal, glass, and other inert waste and reported higher for the United States of America (50.70%), Canada (47%), Germany (22%), Canada (13%), Germany (12%), and Hungary (35%). The solid waste generation (kg/capita/day) was reported remarkably higher in Italy (59.83), followed by Australia (8.47), Switzerland (8.08), Hong Kong (7.18), Denmark, Singapore, and Lithuania. The treatment processes such as composting, landfilling or controlled dumping, recycling, and incineration were commonly followed by the HICs. The HICs are spending 82–250 USD/ton, 70–200 USD/ton, 65–150 USD/ton, 40–100 USD/ton, and 35–90 USD/ton for waste collection, waste to energy conversation, anaerobic digestion, sanitary landfill, and composting, respectively. The HICs such as the USA, Canada, Scotland, UK, Northern Ireland, and European Union have strict guidelines and rules for landfilling. The US Environmental Protection Agency has restricted the open dumping process by enforcing laws, administrative actions, and punishments with penalties. The strict restrictions are also effected to dispose bio-hazardous materials and substances generated from demolition/renovation of the building. Other side, adoption of emerging techniques plays significant role in circular economy for high-income countries.

**Keywords** High-income countries · Solid waste management · Landfill · Circular economy approach · Emerging Technologies

---

D. Roy (✉)
Department of Civil and Environmental Engineering, Yonsei University, Seoul, Republic of Korea
e-mail: deba_bappa2003@yahoo.co.in

A. Tarafdar
Department of Environmental Science and Ecological Engineering, Korea University, Seoul, Republic of Korea



P. Pathak, S. G. Palani (eds.), *Circular Economy in Municipal Solid Waste Landfilling: Biomining & Leachate Treatment*, Radionuclides and Heavy Metals in the Environment, https://doi.org/10.1007/978-3-031-07785-2_1

## 1.1 Introduction

Solid waste management (SWM) is one of the key challenges of the twenty-first century. Worldwide three billion people still lack access to controlled waste disposal facilities (Wilson et al. 2015). The wealth of any nation as directly proportional to the rate of urbanization, and consequently, an increase of waste generated. The management of solid waste has direct and indirect impacts on human health, socio-economic, aesthetics, environment, and infrastructure. The sustainable methods for solid waste collection, transportation, and disposal could be solution for solid waste management.

Earlier studies recommend that a nation's financial advancement is emphatically associated with the per capita production of solid waste (SW), with increasing rates of utilization of commercial products and models of development. The production rate of SW or municipal solid waste (MSW) is straightforwardly relative to national gross domestic product (GDP)/gross national income (GNI) and financial status of individuals (Shekdar 2009). The world bank has classified the countries based on per capitagross national income (GNI or GDP), utilizing the World Bank Atlas technique 2011. The nations are: (a) low-income (LICs) with GNI US$1025 or less, (b) lower-middle-income (LMICs) with GNI US$1026–4035, (c) upper-middle-income (UMICs) with GNI US$4036–12,475, and (d) high-income countries (HICs) with GNI US$12,476 or more (World Bank 2015). Around the world, yearly SW production is roughly 17 billion tons, and it is anticipated to arrive at 27 billion tons until 2050 About 1.3 billion tons are MSW which is created by world urban areas and is assessed to arrive at 2.2 billion tons until 2025 (Laurent et al. 2014). The growing amount of MSW requires appropriate administration strategy for removal of waste. But, the management of MSW is now a big challenge because of the inadequate treatment amenities (reusing, fertilizing the soil, and energy from waste) and disposal of enormous measure of waste produced in metropolitan regions.

The MSW consist of recyclable, organic, inflammable, and non-recyclable waste. In LMICs, biodegradable substances signified 46–53% of total MSW load. The amount of reusable waste in LMICs was lower than in HICs (UNEP 2015). The direct disposal of MSW containing high organic matter, act as a source of air pollution by generating green house gases (GHG) and leachate that pollutes groundwater. Open dumping of MSW generates hazards materials harmful to human health and environment. These substances are carried into the air, soil, and water as dioxins and dioxin-like substances (polychlorinated dibenzo-para-dioxins and polychlorinated dibenzofurans) along with toxic heavy metals like nickel and cadmium (Aboyeji and Eigbokhan 2016). The collection of biogas and leachate from MSW treatment process could be economical benefits for the countries. Apart from the economical benefits, the appropriate removal techniques of hazards substances from the MSW could be helpful for the society. Contrastingly, the appropriate waste removal strategies involve bigger venture and operation and maintenance (O & M) costs. Hence, the waste leachate that produced from the landfill are hard to manage due to its long stability. To minimize the adverse impacts, MSW need segregate into reusable, bio-remediable, ignitable, and non-recyclable waste before processing. The waste substances, along with paper, glass and metals, can be reused to bring down the load on original substances. Additionally, biodegradable waste is remediated for stable organic compounds and environmentally friendly power-source as like biogas (UNEP 2015). Compost-treatment of soil changes bio-degradable waste into manures predominantly by utilizing local or decentralized compost. For anaerobic decomposition, biodegradable waste is changed over into methane and utilized as an energy source in emerging nations.

Landfill is applicable for non-recyclable and non-burnable waste. However, it demands immense spaces of land. Thus, the leachate produced by the anaerobic and aerobic deterioration of these wastes contain such elements those are harmful for the environment (Aboyeji and Eigbokhan 2016). At the point when the factors such as ecological effect, health effect, land corruption, and groundwater tainting are considered, landfilling turns into the most exceedingly awful choice. In any case, first-world nations have begun to debilitate landfilling of solid wastes through rigid guidelines, waste decrement and reusing. It is less likely to be the case with high-income economies, where the waste management concept of Reduce, Reuse and Recycle, commonly referred to as the '3R,' is well established and practiced. This distinction is further exemplified in studies emphasizing the philosophy of the "waste management hierarchy". The new emerging technologies have been followed by the HICs to manage the daily solid waste burden. Apart from the conventional SWM techniques (volume reduction and landfilling), vermicomposting method, circular economy approach, and waste to energy system (incinerations, pyrolysis, gasification, anaerobic digestion, biodiesel, biohydrogen, biomethane, bioethanol, and butanol production) being adopted by the HICs. As an example, because of ozone depleting gas discharges, leachates, and land accessibility concerns in excessively populated urban areas, US states are moving from the customary landfill treatments to WTE as a replacement green economical option (Kumar and Samadder 2017).

## 1.2 Solid Waste Types, Compositions and Management

Contingent upon the innate perils related with its physical and compound properties, solid waste can be delegated either hazardous or non-hazardous. Hazardous wastes present considerable or likely dangers to public health or the climate (Fig. 1.1). The corrosive, toxic, and infectious substance that can be generated from hospitals, household, and industries are classed as hazardous. Other side, the substances are not in the list of hazardous materials, despite the fact that they are still harmful to people or the environment. Non-hazardous waste are: farming waste, urban solid waste, and waste from factories. Bio-remediable solid wastes are those that can be degraded (disintegrated) into their components by microbes. The composts and waste generated from crops productions and human eating residues are the fundamental bio-remediable wastes calls food waste.

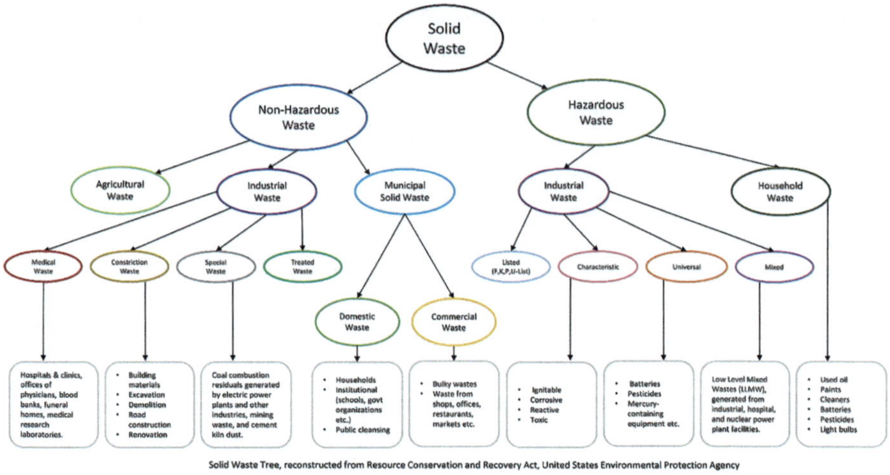


**Fig. 1.1** Classification of solid waste

The correlation of waste produced from different nations with municipal solid waste is displayed in Table 1.1. The waste components are classified to third world nations, second world nations and first world nations. From second and third world nations waste generation was overwhelmed by food wastes, with higher part from Indonesia tailed by Bangladesh, China and Algeria. Poland, Thailand, and Indonesia have a major part in fabric wastes while plastic waste is most influential in Malaysia accompanied by China and Thailand. In correlations with first world nations, like America, Japan and United Kingdom, paper waste is predominating the wastes generated from food waste, fabric, and plastics. In third world nations, the waste from Nigeria, Kenya, Algeria and Ghana accommodates more food wastes than others while wood and concrete waste have higher parts in Tanzania and Uganda. Glass wastes have a major part in Kenya and Nigeria. Looking at the waste arrangement by mean of each class, it can be noticed that food waste is created more in second world nations followed by third world, fabric waste is very little in third world nations while glass waste and papers are a lot higher in first world nations. Wood waste are more in first and third world nations. Overall flammable waste has a normal of 89% at second world nations, 78% in first world nations and 73% in third world nations. Though the third world nations have more waste for energy reproduction, but they can generate less reused energy from the waste than first and third world nations, simply because of less efficient reusing innovation.

In LMICs and UMICs have appropriate perspectives and techniques among nations, consideration should be paid on the qualities and characteristics of waste, along with the amounts of waste produced, composition of wastes, moisture and calorific value. It has been stated that the waste attributes vary essentially among first, second and third world nations (EPA 2012). One of the significant observation from the earlier studies is that the HICs exhibit waste production rates equivalent to or over one kg for per capita daily, while third and second world nations show rates

**Table 1.1** Comparison of waste generated by the countries listed in different income groups (Omari et al. 2014)

| Income groups | Country | Glass | Paper | Textile | Food waste | Debris and construction | Plastic | Metals |
|---|---|---|---|---|---|---|---|---|
| HICs | Germany | 16 | 31 | 17 | 21 | 0 | 10 | 5 |
| | Denmark | 6 | 23 | 21 | 39 | 0 | 7 | 4 |
| | USA | 4.6 | 28.5 | 8.4 | 13.9 | 19.8 | 12.4 | 9 |
| | UK | 9 | 21.4 | 3.3 | 17.3 | 26.6 | 8.8 | 4 |
| | Greece | 5.4 | 21 | 17 | 42 | 0 | 11 | 3.6 |
| | S. Korea | 2.8 | 23.8 | 40.6 | 32.8 | – | 0 | 0 |
| | Japan | 5 | 33 | 12 | 34 | 0 | 13 | 3 |
| | Russia | 13.7 | 15 | 4.8 | 34.9 | 15.1 | 11.3 | 4.7 |
| | France | 14 | 26 | 19 | 24 | 0 | 13 | 4 |
| | **Avg.** | **8.9** | **23** | **13.9** | **29.9** | **6.8** | **11.2** | **4.8** |
| | India | 5 | 30 | 7 | 42 | 1.5 | 10.4 | 4.1 |
| MICs | Turkey | 6.2 | 7.8 | 23 | 43 | 0 | 14.2 | 5.8 |
| | Philippines | 19 | 9 | 49 | – | 17 | 6 | – |
| | Poland | 12 | 18 | 20 | 35 | – | 11 | 4 |
| | China | 5.2 | 8.8 | 4.5 | 67 | – | 13.5 | 0.7 |
| | Malaysia | 1.2 | 16.5 | 1.9 | 61.5 | 0.4 | 15.3 | 0.3 |
| | Bangladesh | 0.7 | 10.7 | 2.2 | 68.3 | – | 4.3 | 2 |
| | Indonesia | 1.7 | 10.9 | 6.2 | 70.2 | – | 8.7 | 1.8 |
| | Nepal | 1.3 | 7.5 | 12 | 60 | 6.7 | 12 | 0.5 |
| | Thailand | 6.6 | 12.1 | 7.3 | 42.7 | 6.9 | 10.9 | 3.5 |
| | Gaza strip | 3 | 10 | 0 | 54 | 18 | 12 | 3 |
| | **Avg.** | **4** | **15.2** | **10.1** | **53** | **6.7** | **10.5** | **2.4** |
| | Uganda | 0.7 | 6.7 | 1.3 | 37.8 | 33.6 | 7.8 | 0.8 |
| LICs | Nigeria | 7 | 6 | 7 | 47 | 18 | 10 | 5 |
| | Ghana | 1.5 | 6.6 | 2.2 | 73 | 11.2 | 3.3 | 2.1 |
| | Algeria | 1 | 9 | 0 | 62 | 0 | 12 | 2 |
| | Tanzania | 4 | 11 | 2 | 37 | 8 | 7 | 1 |
| | Kenya | 6.7 | 17.3 | 5.1 | 52 | 2.4 | 11.8 | 2.8 |
| | **Avg.** | **2.9** | **9.4** | **2.9** | **51.5** | **17.2** | **8.7** | **2.3** |

*HICs* high-income countries, *MICs* middle-income countries, *LICs* low-income countries

under one kg per day. Additionally, the financial status of the neighborhood and at the domestic level influences waste profiles: higher literacy increases the paper content of the waste produced, and wealthier groups often choose to discard durable items instead of repairing them. Parrot et al. (2009) noticed that due their financial growth, LICs and MICs incline towards production of waste with higher biodegradability, with practically no economic benefits can be obtained from it. For sure, it has been noticed that in third world nations like Bangladesh or Pakistan, the organic content is higher than 65%, while in first world nations, like Japan, South Korea, and America the rate is under 20–30% (UNEP 2015). This organic composition presents the potential for composting or biogas treatment to produce compost or domestic energy as a sustainable waste management option in these countries. The waste classification is well defined for the HICs than in LICs, for example, Afghanistan, and MICs in Asia, China, Gaza strip, Kenya, and Ghana. It is noticed that this is a hindrance in management towards a sustainable waste reclamation measure. It is, therefore, necessary to carry out waste characterization to plan and develop recycling initiatives in low-income and middle-income countries. At the municipality level, waste isolation for LICs and MICs are constrained by an insufficiency of information on accessible technologies and better habits for managing the waste, deficiency of suitable equipment for the assortment of arranged substances and absence of decision-makers' attention in environmental affairs.

### *1.2.1 Generation of MSW*

The HICs, UMICs, LMICs, and LICs produce around 1.2–1.8 kg/capita/day, 0.7–1.2 kg/capita/day, 0.5–0.75 kg/capita/day, and 0.3–0.6 kg/capita/day of solid waste, respectively (Hoornweg and Bhada-Tata 2012). The per capita SW generation rates were reported maximum and minimum in Hong Kong and Myanmar with 2.25 kg/capita/day and 0.28 kg/capita/day, respectively (Yadav and Samadder 2017). The chemical compositions and structure of the SW are different for different nations with diverse income groups. The composition of MSW for different income group countries has been presented in Fig. 1.2. The nations with high per capita pay (first world) create more paper, plastic, and glass wastes, while the nations with low per capita pay (second or third world) produce more biodegradable wastes. Moreover, the LICs is producing MSW with high moisture contains because of the MSW of the nations with low per capita pay contain more humidity because of high level of organic content and has a low calorific value, that makes the solid

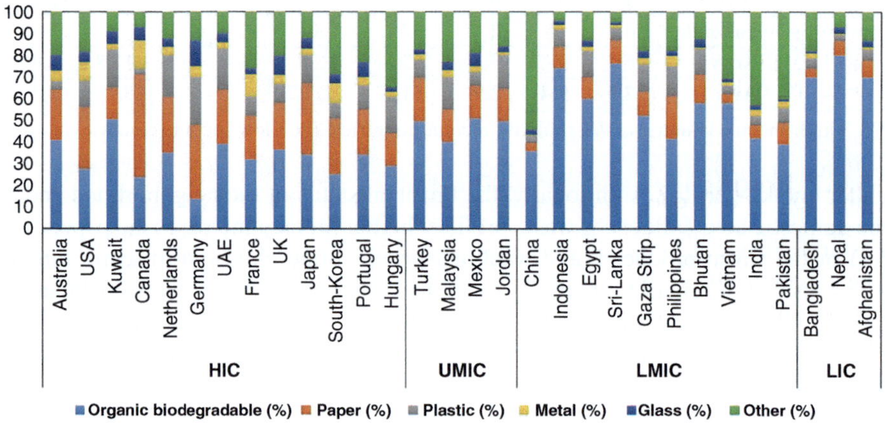


**Fig. 1.2** The composition (in %) of municipal solid waste varies considerably in high, middle and low per capita income countries. (Yadav and Samadder 2017)

wastes incompatible for thermal treatment; however reasonable for biological processes like, fertilizing the soil and anaerobic degradation (Pandyaswargo et al. 2012).

### 1.2.2 Solid Waste Management

The open-air dumping is the most widely recognized technique for disposal of solid wastes in the LICs (Sri Lanka, Pakistan, and Cameroon) and MICs. Open discarding and combustion are roots of air, surface and ground water contamination. It contains toxic compounds existing in landfill gases and leachates. The open dumping connects to numerous unsafe wellbeing, including infections in dermal layer and eye irritations, respiratory issues, and vector-borne infections like diarrhea, typhoid, hepatitis, cholera, malaria, yellow fever, and hypertension (UNEP 2015). In LIC and MIC (third and second world) countries have inadequate information contains quality, characteristic, collection types (bring or kerb collection), transportation of waste (fuel utilization), and disposal methods of SW (Jha et al. 2011). In the absence of accurate data, it is difficult to design better strategies or to make judicious budget estimations and decisions on waste management facilities.

The urban communities in third and second world nations utilize just 4–16% of their entire metropolitan financial plan on MSWM, while the first world nations have bigger financial ability for MSWM. Around 80–90% of solid waste management financial plan is spent on waste collection alone in USA (UN-Habitat United Nations Center for Human Settlement 2010; LICs Yadav and Samadder 2017). Regardless of such big expenditure, waste assortment efficiency is just 41% in LICs and 90% for HICs, which utilizes under 10% financial plan on waste collection (World Bank 2012). The treatment methods of solid waste dependent on pay level of different nations are depicted in Fig. 1.3.

For the HICs, processes such as incineration (Denmark, 54.04%), recycling (Singapore and Hong Kong, 45%), and composting (Austria, 44.72%) is highly recognized. On the other side, landfilling is not in common practice except in Italy (54%) (Fig. 1.3). Controlled dumping (highest in Turkey, 66%) and open dumping (highest in Jordan, 50%) are common choice for the MICs, while composting and reusing are in less choices. Burning is the most un-favored alternative in first world nations. Second and third world nations depend on landfilling (most extreme in China, 55%) and open dumping (highest in Cameroon, 95%) (Fig. 1.3) of total waste. Incineration, composting, and recycling are in common practice for HICs because of high level of metal, plastics, paper, and high calorific value waste (Table 1.1). Skillful discarding is widely recognized waste management practice in the majority of the second and third world nations, yet it isn't the favored choice in first world nations. The majority of the first world nations follow source decrement and reusing (Fig. 1.3). The current traditional strategies for the management of solid waste in first world nations has explored in Table 1.2.

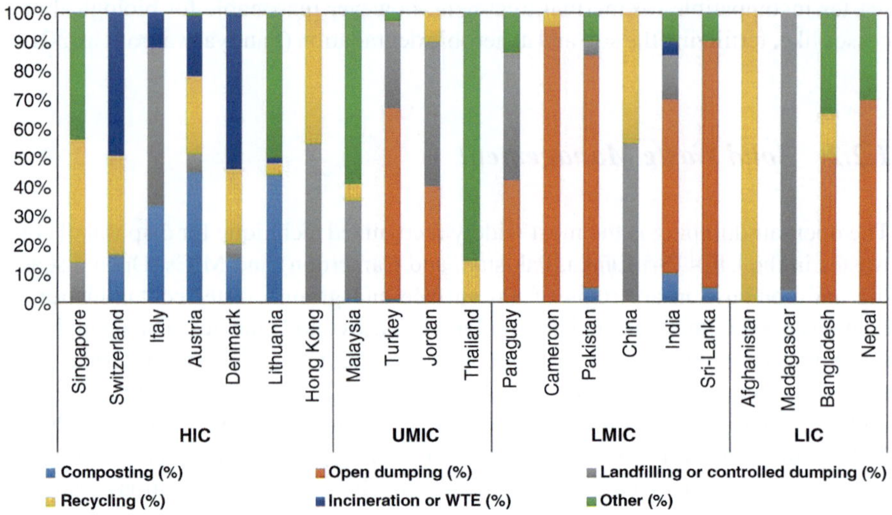


**Fig. 1.3** Solid waste treatment processes for different income group countries

**Table 1.2** Solid waste management methods for high-income groups

| | Methods (M) | | | |
|---|---|---|---|---|
| Country | M1 | M2 | M3 | M4 |
| Greece | Recycling | Anaerobic Digestion | Landfilling | |
| Singapore | Recycling | Composting | incineration | Landfilling |
| Austria | Recycling | Non-recycling (Mechanical & biological) | | |
| Italy | Mechanical pretreatment+ composting + incineration + landfill | Incineration in plant | | |
| Denmark | Standard bioreactor | Semi-aerobic landfill | landfill | Open |
| Spain | Biological treatment + landfilling | | | |
| Sweden | combinations of incineration | materials recycling | biological treatment | Landfilling |
| Portugal | Recycling | mechanical/ biological treatment + aerobic treatment + Landfill | | |
| Kuwait | incineration + Energy recovery | Anaerobic digestion + Landfilling | | |
| UK | Recycling + composting + incineration | Waste recovery | Landfill | Incineration |
| Lithuania | Recycling + composting + incineration | Recycling + incineration | Recycling + composting + landfill | landfilling |

## 1.3 Solid Waste Management by Landfilling with Biogas Generation

Landfilling is considered as the controlled dumping of waste on land to decrease the adverse consequence on the ecosystem through biogas generation and managing leachates (Fig. 1.4). Nonetheless, unhealthy landfilling is an easier and moderate answer for dumping of the expanding waste amount and is the most widely recognized practice in developing nations, that represents a significant warning to the ecosystem (Yadav and Samadder 2017). Earlier publications exhibited that landfilling brings about the most significant ecological effect contrasted with other waste administration choices. It has been accounted for that the majority of the urban developing nations, the waste is dumped on low lying regions situated at the edges of the city. At the point when the variables like effects on ecosystem, influence on wellbeing, land erosion, and groundwater pollution are reviewed, landfilling turns into the most noticeably terrible alternative. Specialists on the matter agree, just 10–15% of the entire waste produced ought to go for landfilling and it needs to be the last alternative for urban communities with limited land area. Although, developed nations have begun to debilitate landfilling of waste through strict protocols, decrease in waste load, and reusing. The leachate generated from landfill is a significant anthropogenic contaminant discharged from dumpsites that contaminates the neighboring surface water and groundwater aquifers. The public guidelines and norms ordinarily control the plan of the leachate collection framework. The administrative laws and guidelines permit the development of the assortment framework that can hold a specific measure of leachate before it is separated through pumping. For USA, Maximum acceptable leachate collection above the liner before extraction is 0.3 m.

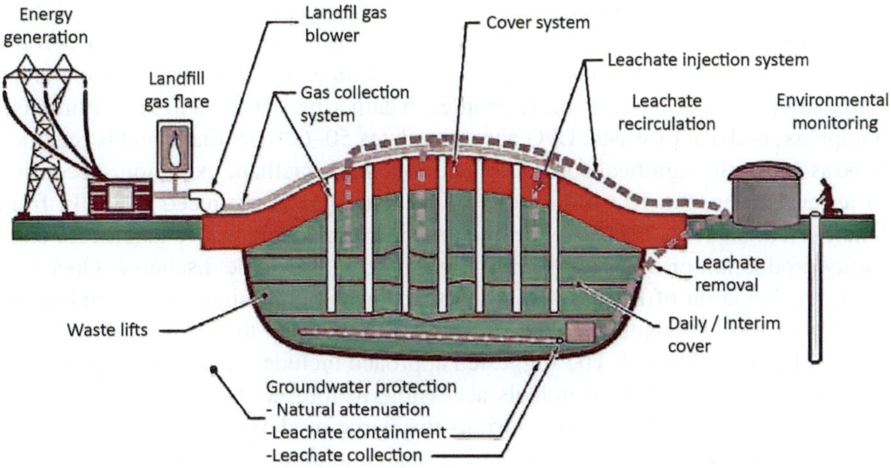


**Fig. 1.4** Typical engineered landfill with biogas recovery system. (Adapted from Kumar and Samadder 2017)

The area of landfills amenities varies with the category of wastes, e.g., demolition, building, and renovation substances, waste from household, toxic and hazards substances, biomedical and biohazards trashes, as well as radioactive substances. The landfills can classify in six Class based on the refuse waste contain. The Class 1–Class 6 denote as soil disposal; building, demolition and renovation substances; MSW dumping; industrial and commercial trashes disposal; hazardous substance dumping; and dumping of hazardous wastes at underground, chronologically. Hence, based on the functionality types, landfill can classify as bioreactor landfill, monofill landfill, reusable landfill, and secure landfill. These techniques have been introduced to bury the losses to defer the unfriendly environmental impacts, accommodation of incinerated wastes, and asset recuperation and fertilizing the soil. Moreover, excavation to recover plastics, glass, metals, and compost cab be done from reusable landfills. Furthermore, bioreactor landfills which is designed as a sanitary landfill along with leachate collection, liners and recirculation arrangements to curtailing the environmental threats introducing microorganisms to degrades the organic wastes.

The organic substance content of the dumped wastes in landfills goes through complicated microbial and chemical disintegration that outcomes in the creation of landfill gas (LFG). The deuteriation of organic substances into LFG happens in five unique stages. The primary stage is aerobic degradation, where the aerobic microbes disintegrate complex organic compounds into $CO_2$ and $H_2O$. The subsequent stage is hydrolysis and fermentation, where the dissolvable organic compounds are decayed into $CO_2$, $H_2$, $NH_3$ and organic acids in presence of facultative microorganisms. The third stage is acidogenesis/acetogenesis, where the organic acids generated during second stage transformed over into acetic acid, formic acid, alcohols, $H_2$ and $CO_2$ by anaerobic microorganisms. During the fourth stage (methanogenesis), the methanogenic microscopic organisms utilize the byproducts of the third stage and generates $CH_4$, $CO_2$, just as other trace gases in more modest sum. The last stage is oxidation, where $CH_4$ gets transformed over into $CO_2$ and $H_2O$ under aerobic condition. The LFG generation rate in a landfill relies upon different components, for example, sort of landfill, waste profiles, weathering (temperature and rainfalls), dampness, and age of waste. LFG accommodates 50–60% methane and is regarded to be as one of the significant origins of anthropogenic methane excretion. According to an evaluation, 30–70 million tons of methane gas is produced annually from landfills (Johari et al. 2012). Hence, reclamation of methane from a landfill for electricity production or other use is important to bring down the discharge. Once in a while, reclamation of LFG is actually not achievable, all things considered onsite ignition of LFG is conducted. Yet, for this, it is important to get the evaluations of trapped LFG in a landfill. The suggested approach includes displaying of LFG production. There are different models accessible to foresee the methane discharges from landfills. Notwithstanding, various discharge models give various outcomes for a solitary landfill and the models give exact outcomes for the locale it has been high-income, as the waste components vary across the nations. Among the seven models, six models depend on the European situation and one on the American situation. These models have enormously diminished the monotonous estimation strategies traditionally applied for methane assessment from landfills. Despite the fact that, TNO model was high-income for the waste attributes of Netherland, however this can be utilized for the assessment of LFG for different nations likewise as it possesses lower relative error (22%) among perceived and determined estimates. In an investigation, it was assessed that 1 ton of MSW creates 80 $m^3$ of LFG and China single handedly may contribute 10 billion $m^3$ LFG to the worldwide LFG emanations in 2020.

The HICs such as USA, Canada, Scotland, UK, Northern Ireland, and European Union has their own strict guidelines and rules on landfilling process. These countries are following state environmental agencies and the US Environmental protection Agency, provincial environmental agencies and Environmental Protection Acts, Scottish Environmental Protection Agency, Landfill Allowance Trading Scheme, Northern Ireland Environmental Agency, and Landfill Directive policies chronologically. Administrative laws in many HICs prohibit the open discarding as banned. While in Canada 79% of MSW goes to landfilled, the Ministry of Environment and Climate Change execute severe guidelines on garbage discarding and unlawful unloading which might bring about punishments, legal examination and indictment, As per US Environmental Protection Agency, restricts open unloading of whip by law alongside the issuance of Administrative action and punishment of US $1500–3000 for littering, scavenging, open burning and proscribed disposal of (bio) hazardous substances as well as demolition, building, and renovation waste on open land or flowing water (https://www.ecos.org/members/illinois/).

## 1.4 History Transition and Adaption of Circular Economy Approach in MSWM for Selected High-Income Countries

Numerous alternatives such as biological /non-thermal/thermal for handling MSW are available for producing energy in the HICs. Moreover, a substantial space in adopting WTE strategy is dominating because of a significant expense of the development of new amenities, monetary threats, and minimal financial advantages towards adopting a circular economic approach.

### *1.4.1 New Zealand*

New Zealand produces around 4 million tons of MSW annually. This sum has a total energy capability of 0.5–1.5 Mtoe annually, given MSW's net energy capability of around 0.13–0.38 toe per ton (Munir et al. 2021). Therefore, fitting MSW-to-energy innovations in New Zealand could counterbalance 10–25% of its absolute energy shortage, around 6.8, 8.5, and 11 Mtoe in 2018, 2030, and 2050 sequentially (Munir et al. 2021). An entire energy capability of 0.35 toe per ton of MSW and roughly 90% waste decrease after treatment was utilized (waste to-energy advancements like plasma gasification or hydrothermal procedures). Conveniently, waste to-energy handling can lessen the energy shortfall by roughly 20% with huge decrease in waste landfilling (Munir et al. 2021). In any case, right now landfilling is the principle MSW removal strategy (72%) in New Zealand with outstanding MSW (28%) reused. As such, MSW in New Zealand is generally either unloaded in landfills or reused without critical utilization of waste to-energy innovations rehearsed in other first world or OECD nations as displayed in Fig. 1.5 (Munir et al. 2021). In New Zealand, there have been limited scale waste to-energy projects utilizing anaerobic assimilation, combustion, pyrolysis, and gasification. Be that as it may, their commitment is almost immaterial when contrasted with other first world nations. Further significant truth is that the quantity of accessible landfill destinations in New Zealand is diminishing. Their new measure is less than 50–in view of more noteworthy interest for land as the populace increments, higher land cost and related difficulties (Munir et al. 2021).

The circular economy activity plan means to reuse substances where conceivable, to utilize traditional and nonconventional advances for waste to-energy handling, and to limit MSW landfills. Besides, this arrangement proposes to isolate organics and inorganics after material reusing as displayed in Fig. 1.6. Organics are shipped off classical facilities, for example, fertilizing the soil, and anaerobic degradation, and inorganics that can't be reused are shipped off non-classical methods like gasification and hydrothermal treatment. For a reliable stock of MSW, maybe more modest circulated waste to-energy plants can be thought about as well. Helpful items and energy from waste to-energy advances can be reused back to domestic and commercial purposes. Accordingly, materials utilized can be kept in the economy for a more extended time frame, energy delivered from MSW can limit energy shortage in New Zealand, MSW can be securely overseen, and altogether decreased volume of MSW would be shipped off landfills as the landfill choice becomes unreasonable. In New Zealand, expenses of non-classical treatments for waste to-energy handling can be diminished with proficient factory installations creating extra income from their items, for example, pyrolysis oil, syngas, and organic acids. High-esteemed finished items incorporate fuels, synthetic substances, and hydrogen. Different methodologies to lessen cost incorporate bringing the tipping expense

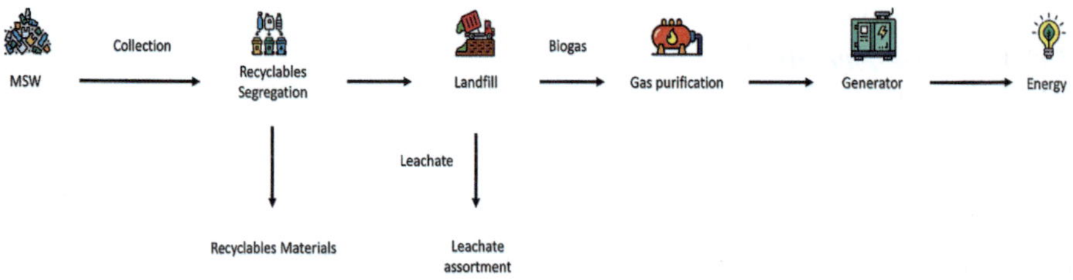


**Fig. 1.5** Typical landfilling process of MSW, New Zealand

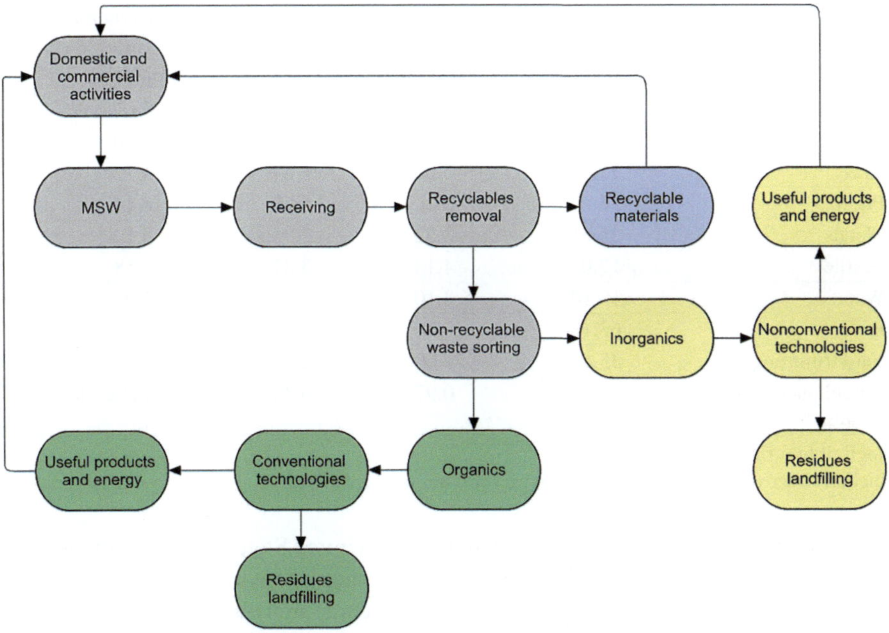


**Fig. 1.6** Closed-loop action plan for MSWM in New Zealand

down to draw in MSW, expanding factory limit, using products like vitrified slag in plasma gasification, utilizing lean production processes like 5S technique, and coordinating energy to bring down costs for energy requirement.

## *1.4.2 United States*

Every year more than 136 and 33 million tones of MSW is landfill and incinerated In the USA (USEPA 1976). But, land filling of MSW turns downed from 89% in 1980 to below 53% in 2014 in the USA because of the technical improvement in impregnate, recycling, incineration and energy recovery techniques. Before 1880s, American cities were unorganized in WMS, and this brings in periodic epidemics. This outbreak aids the desire to better public health and the environment. The burden of WM was on the local authority, and municipal dumps served as disposal locations. In the next many years, mechanical developments, for example, assortment trucks, incineration, motorized street sweepers and sanitary landfills, were high-income. Also, the Resource Conservation and Recovery Act (RCRA) of 1976 forced the closure of municipal dumps. Consequently, this resulted in a garbage crisis in the late 1980s and 1990s. However, information of regional waste management facilities resolved the crisis. Since the 1980s, the United States has invested $7 billion in the construction of 90 waste incineration plants. Table 1.3 shows the

**Table 1.3** Generation, Recycling and Energy Recovery of Materials in MSW, 2018 (in millions of tons and percent of generation of each material) (USEPA 2018)

| Materials | Weight generated | Weight recycling | Energy recovery | Energy recovery (%) |
|---|---|---|---|---|
| Paper and paperboard | 67.40 | 45.98 | 4.21 | 5.90 |
| Glass | 12.25 | 3.06 | 1.64 | 14.25 |
| Plastics | 35.67 | 3.09 | 5.63 | 18.14 |
| Metals | 25.61 | 8.72 | 2.95 | 13.19 |
| Textiles | 17.05 | 4.18 | 3.22 | 24.58 |
| Wood | 18.10 | 3.10 | 2.84 | 17.91 |
| Food[a] | 63.13 | 20.00 | 7.56 | 21.75 |
| Yard trimmings | 35.41 | 22.60 | 2.57 | 7.71 |
| Miscellaneous inorganic wastes | 4.06 | 0.97 | 0.80 | Negligible |

Negligible = Less than 0.05%
[a]Includes recovery of other MSW organics for composting

history of waste timeline in the United States. The United States Pollution Prevention Act, passed in 1990, was to establish the waste management hierarchy. Several considerations like incentive programs, education campaign, training programs and technical assistance were triggered (Sakai et al. 1996). The historical transition and progression of MSWM in the above table is an indication of its significance in middle-income strategic frameworks for sustainable MSWM. Table 1.3 presents the generation and recovery of materials in MSW for USA.

In North America, methane gas recovery from landfill is well known solid waste disposal technique. Landfills are progressively turning out to be full close to significant urban areas and arriving at their maximum limit. In San Diego, California, Miramar landfill is the only regional active dumping site occupying 607 ha of land. This landfill will arrive at its most extreme limit by 2022. Efforts are being made to expand the life expectancy of this landfill to 2030 applying a new waste volume reduction technique and following the regulations, incentives and fee hikes to raise the recycling rates (https://www.sandiegouniontribune.com/news/politics/sdut-miramar-landfill-lifespan-zero-waste-compaction-2015aug05-story.html). Fortistar Methane Group LLC invested in utilizing Miramar Landfill's biogas for generating 51,224 MWh net powers for the local communities (https://www.energy.ca.gov/almanac/renewables_data/biomass/). Landfilling is often more economical than burning waste in incinerators (Assamoi and Lawryshyn 2012). Traditional landfilling involved dumping MSW into pits and then burying it to decompose naturally over years. The US-EPA regulations on landfills, based on the Resource Conservation and Recovery Act of 1976, made construction and operation of most local dumps illegal. The requirements for current operating landfills are to minimize odor, eliminate any seepages of leachate, and lessen greenhouse gas emissions by burning them off. Today's controlled and sanitary landfills are technologically more advanced with leachate and landfill gas (LFG) management. Landfill managers are employing new efficient solutions, such as aeration of landfill to accelerate MSW

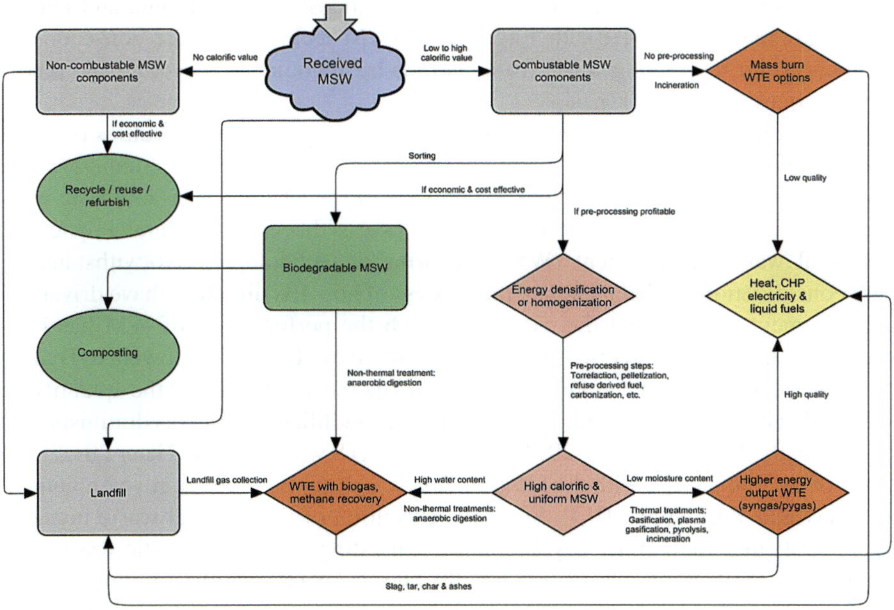


**Fig. 1.7** Closed loop action plan for MSWM in USA. (Assamoi and Lawryshyn 2012)

stabilization and control methane and nitrous oxide emissions. Current technological advancements in the waste to energy process techniques is shown in Fig. 1.7.

## *1.4.3 England*

Landfill is reflected as the most un-ideal disposal technique. The European Union (EU) Landfill Directive (ELD) proclaimed in 1999 needs associate countries to minimize the volume of landfilled by segregation of decomposable substances. The implementation of ELD started the evolution of MSW management system in UK. To combat the difficulties of dealing with the expanding measure of waste and related unfavorable effects on human health, environment, and the climate change from landfills, the EU Landfill Directive (EU Directive 99/31/EC) (ELD) was announced in 1999 (Burnley 2001). ELD seats specific boundaries on the amount of decomposable municipal waste shipped off landfills. EU Member States were needed to bring into power the laws, guidelines and regulatory arrangements to obey with ELD within 2 years of its entry into force. Then, the EU Waste Framework Directive (EU Directive 2008/98/EC) established a "waste management hierarchy", which puts the subsequent techniques in descending order of priority: counteraction, reuse, recycling, recovery and landfill. The EU commands have been moved into national legislations in EU member states as part of European waste management strategy development, to encourage separate collection and waste pre-treatment, as well as upgrading disposal methods (Stanic-Maruna and Fellner 2012; Brennan et al. 2016). In England, MSWM techniques were consecutively announced for redirecting waste from landfills by announcing recycling and recovery practices (Fisher 2006).

Numerous studies have been accompanied to recognize the difficulties of meeting the objectives set in the EU directives (Pires 2007; Lasaridi 2009; Stanic-Maruna and Fellner 2012) to investigations the impacts of the EU directives on waste management legislations and practices, and to assess the environmental impacts of potential waste management (WM) scenarios or technologies. Notwithstanding, less consideration has been paid on the process of how EU directives have driven the advancement of WM and the extent to which the performance of WM has been enhanced under the direction of the EU directives. Landfill Allowance Trading Scheme (LATS) was announced in 2005 to increasingly reduce the quantity of BMW that could be landfilled. Accordingly, the landfilled BMW was diminished by 7% annually during 2005/06–2011/12, though LATS was postponed later 2012/13 in view of its conjunction with the Landfill Tax, which applies comparative implementation (Calaf-Forn et al. 2014). Since the execution of the national waste management strategies, the public recycling and composting rates of domestic waste have been consistently improved, while landfill rate has been steadily diminished. This caused in an enhancement in waste recycling and a reduction in landfill size, particularly the landfilled BMW portion by isolating green garden waste. As results, the recycling and a composting portion of domestic waste in England increased from around 10% in 2001 to 44% in 2015, the landfill share of MSW reduced from 84% in 1996/7 to 44% in 2015, and the landfilled BMW in 2016 reduced to 21% of that in 1995. The improvement of recycling and recovery, also prevention, potentially lead to a remarkable reduction of LCR in Nottingham from 54.7% in 2001/02 to 35.3% in 2006/07 and further to as low as 7.3% in 2016/17. In the S3, only the residual waste from residual MRF that cannot be recycled or processed to RDF was landfilled. It is believed that with continued improvement of separated source collection to prevent cross-contamination, the LCR can be further reduced to approach the zero landfill target set by the Nottingham Waste Strategy 2010–2030. Wang et al. (2020) compared three historical situations (S1–S3) and an alternative scenario (S4) of MSW management to assess the transition of MSW management and to facilitate the future improvement for meeting the targets set in waste management regulations (Fig. 1.8).

### *1.4.4 Australia*

Australia is an island continent with a population of nearly 25 million people occupying 7.692 million km$^2$ of land mass and a waste production of 2.2 tonnes per capita in 2010–2011 alone. 60% of this generated waste was either recovered or recycled, with the rest 40% sent for disposal to landfills. The Details of the selective landfill facilities in Australia is presented in Table 1.4.

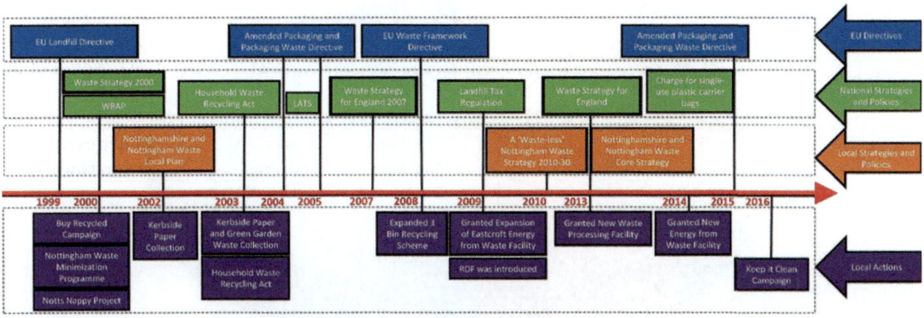


**Fig. 1.8** Historical states in MSWM in England. (Adapted from Wang et al. 2020)

**Table 1.4** Details of the selective landfill facilities in Australia

| Landfill number | State | Operator type | Name of the company (or operator) | Name and address of the landfill facility |
|---|---|---|---|---|
| L1 | New South Wales | Public | Newcastle City Council | Summerhill Waste Management Centre, 141 Minmi Rd., Wallsend NSW 2287 |
| L2 | New South Wales | Private | SUEZ Australia | Lucas Heights Landfill (Sydney), New Illawarra Rd., Lucas Heights NSW 2234 |
| L3 | Victoria | Public | City of Greater Geelong | Drysdale landfill, Becks Rd., Drysdale VIC 3222 |
| L4 | Victoria | Private | SUEZ Australia | Hallam Road Landfill, 274 Hallam Rd., Hampton Park VIC 3976 |
| L5 | Western Australia | Public | City of Cockburn | Henderson Waste Recovery Park, 920 Rockingham Rd. Henderson WA 6166 |
| L6 | Western Australia | Public | City of Armadale | Armadale Landfill and Recycling Facility, Lot 600 Hopkinson Road Hilbert WA 6112 |
| L7 | Western Australia | Public | Eastern Metropolitan Regional Council | Red Hill Waste Management Facility, 1094 Toodyay Road, Red Hill WA 6056 |
| L8 | Western Australia | Private | SUEZ Recycling and Recovery | North Bannister Resource Recovery Park – 6364 Albany Hwy, North Bannister WA 6390 |

Different types of waste management facilities in Australia is shown in Fig. 1.9. For the municipal solid waste (MSW) management, most of these facilities follow the pattern of separation of waste, recycling, aerobic composting and landfilling. It can be observed that the majority of the waste ends up in landfilling facilities. Figure 1.10 shows the distribution of landfilling facilities in Australia by state (Geoscience Australia 2017). It can be noticed that New South Wales (NSW) has the maximum number of landfill sites, followed by Western Australia (WA),

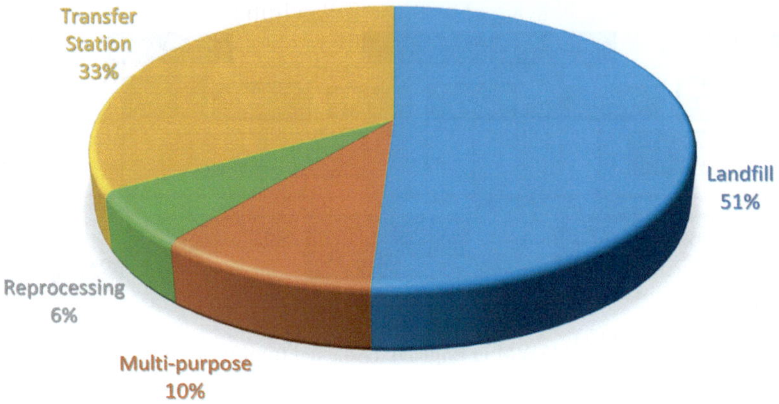


**Fig. 1.9** Different types of waste management facilities in Australia

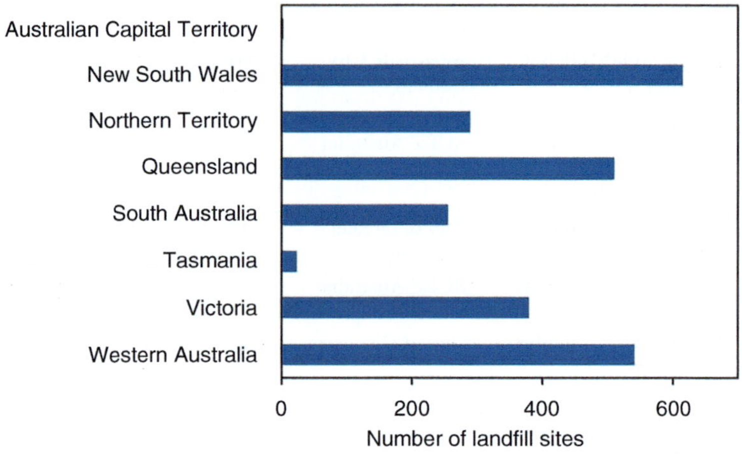


**Fig. 1.10** Distribution of landfilling facilities in Australia by state. (Adapted from Pandey and Shukla 2019)

Queensland (QLD) and Victoria (VIC) at 2nd, 3rd and 4th position, respectively. The Australian Capital Territory has the lowest number of landfill sites, specifically only two. The Norfolk Island has three landfilling facilities, while the Christmas Island and Cocos (Keeling) Islands has one landfilling facility each. As a result, most of the waste production is focused in the major urban areas.

Construction of new landfill sites is time-consuming due to several procedures such as siting, design and construction, planning and appeal processes, and so on. In addition, commencing new landfills is generally not supported by the local population. Therefore, Australian landfills are often developed in old quarries as it is a cost-effective method of rehabilitating used quarries. Additionally, more material is removed from quarries than the amount of waste generated to fill it. Hence, sufficient quarry space is available for use in landfilling. However, owing to geographical and geological constraints, difficulty in obtaining approval for a new site, etc., there is a scarcity of available space. Furthermore, if the landfills are situated far off from the metropolitan areas, the cost of transportation and disposal of wastes becomes extremely high. Hence, although Australia has a huge land mass, the siting, design, operation, and proper maintenance of landfilling facilities have major concerns.

In Australia, waste management can be dated back to the eighteenth century when the European settlers arrived. Before the arrival of the Europeans, the Aboriginal people who lived for over 60,000 years produced very little rubbish. Not only was their waste mostly organic and decomposed naturally, piles of debris consisting of bones, shells, and stone tools, called middens, were managed. Post colonization recorded small amounts of waste generation in Australia until the mid-1900s. The 1970s marked the increase in economic growth, which resulted in the generation of more waste. The waste management transition made by Australia was not void of challenges. Although, increase in population has resulted in an increase in waste generation, there is a reflection on the government's commitment to achieving environmental objectives. Many high-income economies are transitioning from the challenges MSW faced in the earlier centuries, towards a zero-waste society. The application of issue-driven analytical framework and environmental laws are some of the many drivers responsible for the transition. Impressively, Australia recorded a 16% increase in the amount of waste diverted from landfills to recycling facilities between 1999 and 2007 (Pandey and Shukla 2019).

## 1.5 Emerging Technologies for Solid Waste Management Adopted by the High-Income Group Countries

The objectives for any waste management framework are material and energy reuse, trailed by removal of the residues. However, an ideal decision for a waste handling system isn't simply dependent upon monetary prerequisites, energy reclamation, or waste removal capacity, yet additionally to search for ecological administrative compliance necessities of the region of interest. Subsequently, it is important to choose the best accessible methodology for waste processing, which satisfies entire of the necessary standards for a fruitful activity. An assortment of waste transformation measures is accessible, in which the three most broadly utilized methods are (Sharholy et al. 2008; Kalyani and Pandey 2014): (I) thermal conversion [(incineration, pyrolysis, gasification, energy generation from refuse derived fuel (RDF)], (ii) microbial transformation (anaerobic digestion/biomethanation and fertilizing the soil), and (iii) landfilling with gas reuse. The MSW processing methods alongside the common reaction byproducts are displayed in Fig. 1.11.

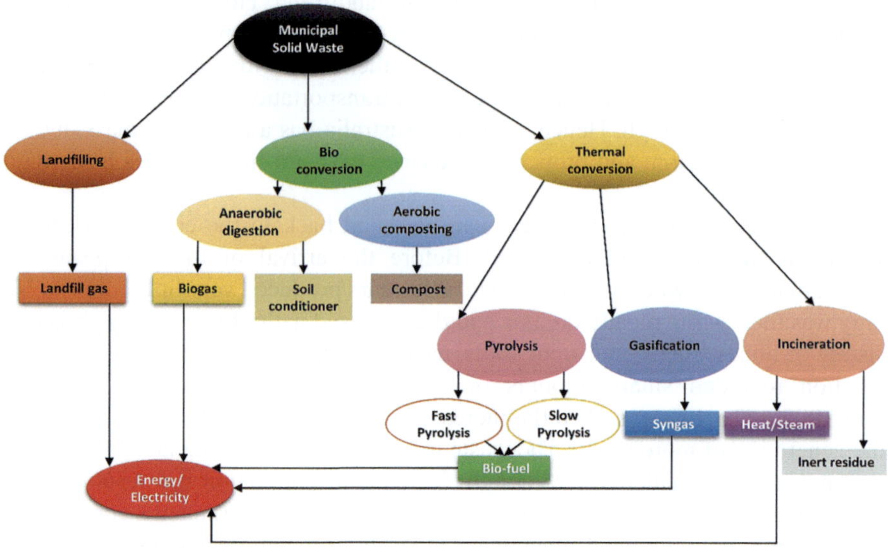


**Fig. 1.11** Solid waste management techniques

Thermal transformation includes heat treatment of organic compounds existing in MSW to generate heat energy, fuel oil or gas. Thermal transformation methods are for the most part helpful for dry waste (low humidity content) with high level of non-biodegradable organic content. At times, thermal transformation method is applied to RDF, which is a flammable material with high calorific value. For generation of RDF, the reusable and noncombustible components are taken out from MSW followed by shredding and additionally palletization of the excess waste. Incineration, which is a controlled burning of wastes at high temperature is the greatest utilized technique in thermal transformation techniques (Shi et al. 2016). The other thermal transformation advancements (pyrolysis and gasification) are as yet in research stage and they are not achievable for business purpose scale-up, might be because of absence of legitimate MSW characterization information, helpless feedstock quality and improper plan of the plant. There are not many economically working pyrolysis/gasification plants across the world for treating MSW. These plants work for treating MSW alongside another sort of wastes like mechanical waste, biomedical waste, biomass, and so on. The traditional reaction conditions and items from thermal treatment measures are displayed in Table 1.4. The fundamental contrasts among these three thermal treatment measures are the environmental condition (i.e., presence of oxygen) and the working temperature. The nature of the end products and the valuable intermediate items relies principally upon these two variables. Working temperature of heat cycles generally relies upon the method plan and feedstock's. For incineration measure, pre-treatment of MSW usually do not undergo in second world nations; crude MSW is straightforwardly utilized as a feedstock material. Nations Calorific value (kcal/kg).

Biological conversion technology is based on microbial decomposition of the organic content of MSW. Many researchers reported this technology as environmentally suitable for energy recovery from wastes. It is generally preferred for wastes with a high percentage of biodegradable organic matter (putrescible) and high moisture content. The main technological option for energy recovery under this category is anaerobic digestion or bio-methanation. The biological conversion technologies are (1) Anaerobic digestion, (2) Biodiesel production system, (3) Biohydrogen production system, (4) Bioethanol production system, (5) Butanol production system, (6) Vermicomposting, and (7) Bioavailability of heavy metals.

## 1.6 Conclusions

Landfilling is characterized as the administered dumping of waste on land to diminish the adverse consequence on the ecosystem through biogas retrieval and managing leachate. The circular economy activity plan means to reuse substances where conceivable, to utilize traditional and nonconventional advances for waste to-energy handling, and to limit MSW landfills. Land filling is one of the common waste treatment processes for the high-income countries, although it is strictly restricted from the regulatory authorities. The high-income countries such as USA, Canada, Scotland, UK, Northern Ireland, and European Union have their own strict guidelines and rules on landfilling. Resource Conservation and Recovery Act of 1976 (RCRA) forced the closure of municipal dumps. Moreover, US Environmental Protection Agency, restricts open unloading of whip by law alongside the issuance of Administrative action and punishment of US $1500–3000 for littering, scavenging, open burning and unlawful disposal of (bio)hazardous substances as well as demolition, building, and renovation waste on open land or flowing water. Adaption of circular economy approach in municipal solid waste management with introducing emerging techniques to generate waste to energy is now in common practices in high-income countries.

## References

Aboyeji OS, Eigbokhan SF (2016) Evaluations of groundwater contamination by leachates around Olusosun open dumpsite in Lagos metropolis, southwest Nigeria. J Environ Manag 183(1):333–341. https://doi.org/10.1016/j.jenvman.2016.09.002

Assamoi B, Lawryshyn Y (2012) The environmental comparison of landfilling vs. incineration of MSW accounting for waste diversion. Waste Manag 32:1019–1030

Brennan R, Healy M, Morrison L, Hynes S, Norton D, Clifford E (2016) Management of landfill leachate: the legacy of European Union directives. Waste Manag 55:355–363

Burnley S (2001) The impact of the European landfill directive on waste management in the United Kingdom. Resour Conserv Recycl 32:349–358

Calaf-Forn M, Roca J, Puig-Ventosa I (2014) Cap and trade schemes on waste management: a case study of the Landfill allowance trading scheme (LATS) in England. Waste Manag 34:919–928

EPA (2012) Environmental Protection Agency. MSW characterization methodology. [Cited 2021 August 30]

Fisher K (2006) Impact of energy from waste and recycling policy on UK greenhouse gas emissions. Environment Resource Management for Department of Environment, Food and Rural Affairs. Prieiga per interneta: http://randd.defra.gov.uk/Document.aspx

Geoscience Australia (2017) Waste management facilities. Geoscience Australia, Australian Government. ACT, Australia. Available via DIALOG, https://data.gov.au/dataset/waste-management-facilities. Accessed 10 Nov 2020

https://www.ecos.org/members/illinois/. Accessed 28 Mar 2020

https://www.energy.ca.gov/almanac/renewables_data/biomass/. Accessed 28 Mar 2019

https://www.sandiegouniontribune.com/news/politics/sdut-miramar-landfill-lifespan-zero-waste-compaction-2015aug05-story.html. Accessed 28 Mar 2021

Jha A, Singh G, Gupta P (2011) Sustainable municipal solid waste management in low-income group of cities: a review. Trop Ecol 52:123–131

Johari A, Ahmed SI, Hashim H, Alkali H, Ramli M (2012) Economic and environmental benefits of landfill gas from municipal solid waste in Malaysia. Renew Sust Energ Rev 16(5):2907–2912

Kalyani KA, Pandey KK (2014) Waste to energy status in India: a short review. Renew Sust Energ Rev 31:113–120

Kumar A, Samadder SR (2017) A review on technological options of waste to energy for effective management of municipal solid waste. Waste Manag 69(2017):407–422

Laurent A, Bakas I, Clavreul BA, Niero M, Gentile E, Hauschilda MZ, Christensen TH (2014) Review of LCA studies of solid waste management systems – part I: lessons learned and perspectives. Waste Manag 34:573–588

Lasaridi K (2009) Implementing the landfill directive in Greece: problems, perspectives and lessons to be learned. Geogr J 175:261–273

Munir MT, Ahmad M, Nasr AT, Carter S (2021) Municipal solid waste-to-energy processing for a circular economy in New Zealand. Renew Sust Energ Rev 145:111080. https://doi.org/10.1016/j.rser.2021.111080

Omari AM, Kichonge BN, John GR, Njau KN, Mtui PL (2014) Potential of municipal solid waste, as renewable energy source – a case study of Arusha, Tanzania. Int J Renew Energ Technol Res 3(6):1–9

Pandey LMS, Shukla SK (2019) An insight into waste management in Australia with a focus on landfill technology and liner leak detection. J Clean Prod 225:1147–1154

Pandyaswargo AH, Onoda H, Nagata K (2012) Energy recovery potential and life cycle impact assessment of municipal solid waste management technologies in Asian countries using ELP model. Energy Environ Eng 3:1–11

Parrot L, Sotamenou J, Dia BK (2009) Municipal solid waste management in Africa: strategies and livelihoods in Yaounde, Cameroon. Waste Manag 29:986–995

Pires A, Martinho MG, Silveira A (2007) Could MBT plants be the solution of fulfill landfill directive targets in Portugal. Proceedings of International Symposium MBT 2007. 22–24 May, Cuvillier Verlag, 63–72

Sakai S, Sawell SE, Chandler AJ et al (1996) World trends in municipal solid waste management. Waste Manag 16(5–6):341–350

Sharholy M, Ahmad K, Mahmood G, Trivedi RC (2008) Municipal solid waste management in Indian cities – a review. Waste Manag 28(2):459–467

Shekdar AV (2009) Sustainable solid waste management: an integrated approach for Asian countries. Waste Manag 29:1438–1448. https://doi.org/10.1016/j.wasman.2008.08.025

Stanic-Maruna I, Fellner J (2012) Solid waste management in Croatia in response to the European landfill directive. Waste Manag Res 30(8):825–838. https://doi.org/10.1177/0734242X12444897. Epub 2012 May 21. PMID: 22615201

Shi H, Mahinpey N, Aqsha A, Silbermann R (2016) Characterization, thermochemical conversion studies, and heating value modeling of municipal solid waste. Waste Manag 48:34–47

UNEP (2015) United Nations Environment Programme. Global waste management outlook. [Cited 2021Augt 24]

UN-Habitat United Nations Centre for Human Settlement (2010) Solid waste in the world's cities–third global report on water and sanitation in the world's cities. Earthscan Publications, London

US-EPA (1976) Summary of the Resource Conservation and Recovery Act. 42 U.S.C.x6901 et seq. https://www.epa.gov/laws-regulations/summary-resource-conservation-and-recovery-act. Accessed 29 Mar 2021

USEPA (2018) Landfill methane outreach program. www.epa.gov/lmop.

Wang D, Tang YT, Long G, Higgitt D, He J, Robinson D (2020) Future improvements on performance of an EU landfill directive driven municipal solid waste management for a city in England. Waste Manag 102:452–463

Wilson DC, Rodic L, Modak P, Soos R, Carpintero A, Velis K et al (2015) Global waste management outlook

World Bank (2012) What a waste a global review of solid waste management. Urban development series knowledge paper no. 15, Washington, DC

World Bank (2015) Country and lending groups. http://data.worldbank.org/about/country-and-lending-groups. 16 Apr 2021

Yadav P, Samadder SR (2017) A global prospective of income distribution and its effect on life-cycle assessment of municipal solid waste management: a review. Environ Sci Pollut Res 24:9123–9141

# Chapter 2
# Landfill Mining: Significance, Operation and Global Perspectives

**Ali Zoungrana, MD Tanvir Hasnine, and Qiuyan Yuan**

**Abstract** Landfilling has gained worldwide notoriety in solid wastes management and disposal, however, in addition to occupying valuable spaces, traditional old landfills are responsible for greenhouse gases (GHG) emissions, water sources pollution by landfill leachate, and major environmental issues. Landfill mining (LFM) can be a solution to the environmental problems caused by landfilling, furthermore, it consists of the recovery of valuable scarce raw materials such as copper, aluminum, gold, or other valuable metals and combustibles, resulting in the conversion of waste to wealth. This book chapter points out the significance and benefits of LFM, mining techniques, and the importance of feasibility studies. LFM practices are growing, especially in European countries, the USA, China, and India, however, some countries are still unfamiliar with the practice. Though LFM as a reuse or reclamation technique, is a promising solid waste management process, some key factors need to be addressed for the acceptance and adequate integration of the process.

**Keywords** Environmental · Landfill mining · Resource recovery · Solid waste

## 2.1 Introduction

Landfills are commonly used techniques for the disposal of solid waste, mainly due to advantages barely found elsewhere, including the ease of landfill installation and operation, low construction and operating costs. The thousands of landfills available

A. Zoungrana
Department of Environmental Engineering, Yildiz Technical University, Istanbul, Turkey

M. T. Hasnine · Q. Yuan (✉)
Environmental Engineering Program, Department of Civil Engineering, University of Manitoba, Winnipeg, MB, Canada
e-mail: hasninmt@myumanitoba.ca; Qiuyan.yuan@umanitoba.ca



P. Pathak, S. G. Palani (eds.), *Circular Economy in Municipal Solid Waste Landfilling: Biomining & Leachate Treatment*, Radionuclides and Heavy Metals in the Environment, https://doi.org/10.1007/978-3-031-07785-2_2

worldwide have accumulated wastes but also valuable materials. Two main reasons rise the interest in traditional landfill plants; they contain an abundant number of valuable materials (metals, plastics,) and resources (combustibles) which are scarce, they are also a source of environmental pollution by GHG ($CH_4$, $CO_2$) emission, leachate infiltration into groundwater and occupy spaces which disturbs the cities in expansion. Reducing solid waste generation is vital to achieving sustainable development goals through the promotion of the concepts of prevention, reduction, reuse, and reclamation. Given the world's tremendous environmental concerns and resource scarcity, a shift to a resource-efficient, low-carbon circular economy is necessary (Jones et al. 2013). Landfill mining (LFM), a process that consists of recovering valuable resources from landfill plants including the landfill area, might be a suitable solution for sustainable development and waste management. Mining and reuse of resources later in their life cycle can contribute to achieving social, economic, and environmental benefits.

Since the first LFM plant was operated in 1953, many LFM operations emerged worldwide, with more significant numbers in Europe, the US, China and India (Dhar 2017). Their goals include preserving waste volumes in landfills, eliminating potential sources of contamination, reducing site pollution, energy recovery, reuse of recoverable fraction of landfill materials, and contributing to the funding of the site management and renovation costs (Masi et al. 2014). Environmental benefits and the recovery of valuable materials are the main drivers of LFM. They depict great environmental benefits such as controlling $CO_2$ and $CH_4$ emission, preventing leachate pollution of surface and groundwater, the rehabilitation of landfill sites area for the construction of social projects (parks for instance) or used as new landfills (Burlakovs et al. 2017). The reclamation of valuable raw materials (metals, woodlands, combustibles, minerals) is increasingly vital against the global scarcity of these raw commodities (Franke et al. 2010). As a result, the amount of recyclable and hazardous materials, as well as the quantity and quality of valuable resources in the landfill site, are determining criteria in LFM (Frändegård et al. 2013).

LFM processes generally happen with drilling or excavation followed by shredding (volume reduction) and screening. In simple words, the wastes are dug out, sizes are reduced, then necessary screening and separation processes are conducted and the materials recycled (Hogland et al. 2019). For a successful LFM project to be implanted, economic, technical, regulatory, environmental, health and social feasibility studies should be performed and all the factors and role-playing actors addressed before any action is taken (Dodson and Parker 2013; Danthurebandara et al. 2014; Hermann et al. 2014). Even though previous feasibility studies supported positive outcomes of LFM projects, feasibility studies should be specific to particular conditions proper to each landfill site (Mandpe et al. 2021).

The present chapter intends to provide some knowledge regarding landfill mining significance and mining techniques used around the globe. The insight of the different feasibility studies will be examined and global perspectives of LFM will be addressed, and future directions will be discussed.

## 2.2 Landfill Mining

LFM is the process that consists of excavating a landfill using standard surface mining machinery in order to recover valuable items and the land resource itself. The main purpose of LFM, through excavation and treatment, are landfill space conservation, landfill area reduction, elimination and mitigation of existing contamination sources, reuse of recovered materials and recovery of energy from mined wastes, cost-effective waste management system and site re-development (Cossu et al. 1996). Enhanced Landfill Mining (ELFM) is an advanced LFM process that consists of mining and converting deposited waste into resources and energy while taking into account social and environmental considerations. The fundamental goal of ELFM is to use innovative technologies to create value-added goods (waste-to-material and energy) from urban waste streams. It's a holistic approach not only with (geo)strategic considerations, but also potential sociological, environmental, and economic benefits (Einhäupl et al. 2019).

### *2.2.1 Landfill Mining Concept*

With the increased output of total Municipal Solid Waste (MSW), which contains a considerable amount of valuable items transferred to the landfill for disposal, a major portion of items such as organic matter, metals, glass, and other materials are going to waste. Due to the continued global demand for raw materials, particularly metals, increased reliance on mining ores has a negative impact on the geopolitical and socio-environmental aspects, meanwhile, it has been identified that a major portion of these important resources is deposited at low depths in the landfills. Thus, a substantial portion of recyclable materials is present in a large proportion of wastes in landfills. At the same time, mining landfills for material recovery while disposing of the same high-value materials as waste in landfills is an unsustainable approach. As a result, the possibility of recovering these already dumped wastes can help to ease the burden of extracting critical natural resources. Therefore, the possibility of mining existing landfills for materials recovery has sparked interest around the globe (Wagner and Raymond 2015). The obtained volume from landfill mining can later be used for dumping new waste by adopting standard regulations. As reported by (Jones et al. 2013), LFM should be "integrated into a wide resource management paradigm," with resource recovery from the landfill waste such as energy production from biogas prior to the excavation. Five distinct LFM concepts were distinguished, which are ELFM, enhanced biodegradation, sustainable landfill, natural cap/catch, and temporary storage. They are described in Table 2.1. The significance of the various concepts is determined by the intrinsic and extrinsic parameters of the landfill. The Intrinsic parameters generally include the size, location, age, type, composition and available documentation level of the targeted landfill, while the extrinsic parameters include the availability of suitable technologies and societal

**Table 2.1** Concept of a variety of landfill mining

| Concept | Definition | References |
|---|---|---|
| ELFM | Using landfills as a source of materials and energy and gain access to their resources and value. | Jones et al. (2013) |
| Enhanced biodegradation & bioreactor | Cost-effective technology for remediating contaminated soil and water. | Hoekstra et al. (2005) |
| Sustainable landfill | Methods for reducing the possibility of landfill pollution. Pollutants are broken down, removed, or immobilized in landfills using the bioreactor idea and concept. | Raga and Cossu (2014) |
| Natural cap & Natural catch | A natural layer of living organic material serves as an alternative to the usual landfill cover (soil and synthetic foil) known as natural cap. Natural catch is a concept that involves creating wetlands to facilitate natural attenuation of leakage from waste dumps. | Jones et al. (2013) |
| Temporary storage place | Provides an environmentally and structurally secure place to store MSW for future materials recovery, linking the gap between ELFM and resource recovery approaches. | Jones et al. (2013) |

and economic boundary conditions. This helps in decision making, for instance, an industrial waste LFM will be more suitable with an ELFM concept, while an MSW landfill will be more appropriate with a bioreactor concept.

## 2.2.2 *Types of Landfill Mining*

Landfills can be categorized into three different types such as (1) MSW landfills, (2) Mono waste landfills, and (3) Inert waste landfills.

### 2.2.2.1 Municipal Solid Waste (MSW) Landfills

MSW landfills are made up of a variety of materials, including (20%–50%) combustible papers, textiles, wood and plastics, soil-like materials (40%–60%), inorganic matter (10%–20%) that includes construction-demolition wastes and glass, and a little proportion of metals. All these materials can be recycled with a considerable potential for resource recovery as well as energy (Quaghebeur et al. 2013). Though the quantity of metal is less in MSW, there is a vast possibility for energy recovery from such a waste, making LFM a suitable option for MSW management. Landfilling of solid waste remains the most frequently used technique around the world because of its cost-effectiveness and convenience of use, and the application of LFM extends the life of landfills while lowering their environmental impact (Calderón Márquez et al. 2019).

#### 2.2.2.2 The Mono Waste Landfills

Mono waste landfills are used to store homogeneous materials such as dredging debris, fly ash, and other industrial waste. In general, the mining of mono waste landfills is determined by the waste composition in the landfill. It can be utilized for the reclamation of land area because mining these landfills does not provide significant energy or resource recovery benefits and therefore is not viable economically (Quaghebeur et al. 2013). Mono waste landfills can also be mined to protect the environment from various contaminations.

#### 2.2.2.3 The Inert Landfills

On the contrary to the mono waste landfills, inert landfills are a heterogeneous mixture of materials with low value. Glass, inert building materials, and bonded asbestos are examples of items found in these types of landfills. As a result, due to their waste content, these types of landfills are not suitable for LFM. Other types include urban mining and biomining. The first one refers to the process of extracting materials from anthropogenic stocks. These stocks include industries, infrastructure, buildings, and so on, and such an extraction process should be cost-effective (Cossu and Williams 2015). The second one, "biomining" refers to a method of processing concentrates and metal ores that employs microorganisms. Biomining can be used to replace more traditional physicochemical methods of mineral processing. It is commonly utilized to improve gold extraction from ores, concentrate precious metals from sulphide minerals and extract copper from secondary ores.

### *2.2.3 Significance of Landfill Mining*

#### 2.2.3.1 Resource Recovery and Recycling of Waste

The idea of recovering and recycling raw materials, such as metals, woodlands, fuels, and minerals, is becoming increasingly vital as the globe faces a rising scarcity of these raw commodities. Landfill mining's primary purpose is to reduce the generation of waste through recycling (and reuse) in order to generate financial assets, minimize natural resource consumption, and support local businesses while also creating jobs.

The possibility of the recovery of resources from LFM is determined by the fractional waste mix in the landfill, as well as the quality of the various elements. However, in many circumstances, the inefficiency of diversion procedures results in the disposal of a considerable amount of materials in new MSW landfills, which materials could be reusable or recoverable (Wagner and Raymond 2015). Material recovery success is controlled by the age and depth of the waste materials, as well as their moisture content and level of deterioration. As a result, the equipment must be developed specifically for each type of landfill mining (Ford et al. 2013b). According to (Franke et al. 2010), there are more than 16 million Mega-grams (Mg) of metallurgical slag deposited in landfills in Germany alone. Slag is a substance produced during the manufacturing of iron or steel, as well as other nonferrous metals. Slag is a valuable Supplementary Cementitious Material (SCM) that is utilized in concrete mixtures. The recovery of minerals includes the following benefits such as conservation of natural resources. For example, cement is made from a variety of elements, thus recovering materials that might replace or minimize the use of cement is a significant benefit in terms of conserving natural resources. Other benefit includes low-cost raw material. When compared to the cost of cement, recovering a material that was previously regarded as waste from landfills is far less expensive. Also, the elimination of wastes is another benefit through resource recovery. Consuming slag and fly ash from landfills will result in lowering the amounts of waste in landfills and creating a space for other waste that cannot be recycled (Cheerarot and Jaturapitakkul 2004). Other repayable materials such as plastics, glasses, metals, and paper can be reclaimed from landfills and it is reported that in 2010, only in Canada, 25 million tons of waste were sent to landfills (Giroux 2014).

#### 2.2.3.2 Environmental Benefits

LFM initiatives are assumed to have worldwide environmental benefits, such as reducing $CH_4$ emissions, therefore will contribute to the reduction of global warming. The environmental impact of LFM is determined by the following factors: (a) the waste sorting technology's efficiency, the conditions of the weather, electricity consumed for operation, and the composition and the quality of the reclaimed materials; (b) the distance of transportation between the various facilities, which accounts for a significant portion of the photochemical oxidation effects; and (c) the intended final use of the recovered items as a resource (Frändegård et al. 2013). In other words, the 'Closing the Circle (CtC)' idea in LFM has a great potential for reducing GHG emissions, with the majority of the decrease coming from emission savings via material recovery (Van Passel et al. 2013). In addition to the recovery of raw materials, landfills contain a large amount of calorific waste that can be used to generate energy and replace traditional fossil fuels. As emphasized by (Marella and Raga 2014), LFM has social and environmental benefits which include the restoration of nature and the reduction of energy and raw materials importation. It is, however important to mention that the rehabilitation of old landfill sites should undergo an appropriate cost-benefit analysis. For example, the 'Dumpsite Restoration Manual' (Burlakovs et al. 2017) suggests that partial closure of landfill sites by scientific rehabilitation and good use of the site can enhance MSW management in India. LFM has a significant contribution to protecting the environment and ensuring sustainability. The relationship between LFM and sustainable development goals (SDG) is depicted in Table 2.2.

**Table 2.2** Relationship between LFM and sustainable development goals (Calderón Márquez et al. 2019; United Nations 2015)

| SDG | Individual target | Benefits of LFM | Objective |
|---|---|---|---|
| SDG 3: Good health and well-being | 3.2. End preventable deaths of newborns and children below 5 years of age<br>3.3. Combat water-borne diseases<br>3.9. Reduce the number of deaths and illnesses from hazardous chemicals, air, water and soil pollution | 1. Surface water and groundwater protection<br>2. Reduction of waste management environmental impact | Environmental protection |
| SDG 6: Clean water and sanitation | 6.3. Improve water quality by reducing pollution, eliminating dumping, and minimizing the release of hazardous chemicals and materials | Surface water and groundwater protection | Environmental protection |
| SDG 7: Affordable & clean energy | 7.2. Increase substantially the share of renewable energy in the global energy mix | Energy recovery | Recovery of excavated waste |
| SDG 8: Decent work and economic growth | 8.3. Promote development-oriented policies that support productive activities, decent job creation, entrepreneurship, creativity and innovation, and encourage the formalization and growth of micro-, small- and medium-sized enterprises, including through access to financial services | Inclusion of WPs | LFM as a tool for social inclusion |
| SDG 11: Sustainable cities | 11.6. Reduce the adverse per capita environmental impact of cities, special attention to waste management | Reduction of environmental impact related to the waste management | Environmental protection and urbanization |
| SDG 13: Climate action | 13.1. Strengthen resilience and adaptive capacity to climate-related hazards in all countries | Landfill emission and energy recovery | Environmental protection |

## 2.3 Landfill Mining Unit Operations

In general, there are two types of LFM activities. The first is in-situ landfill mining, which refers to on-site mining operations. In-situ operations are materials recovery activities that take place on-site with little or no excavation, such as methane gas extraction, contaminant removal, and soil and water remediation. Ex-situ landfill mining is the second type. This sort of operation entails recovering items from landfills using various excavating methods, then transporting those materials off-site to be treated and recycled (Jones et al. 2013). To further clarify this concept, ex-situ operations are frequently used to approach landfills that include vast amounts of industrial waste such as metals, fly ash, and slag. On the other hand, MSW landfills are approached by in-situ operations like bioreactors. The following figure explains the closed-loop flow of in-situ and ex-situ landfill operations, the flowchart is an actual representation of the waste flow in the REMO landfill, Belgium (Fig. 2.1).

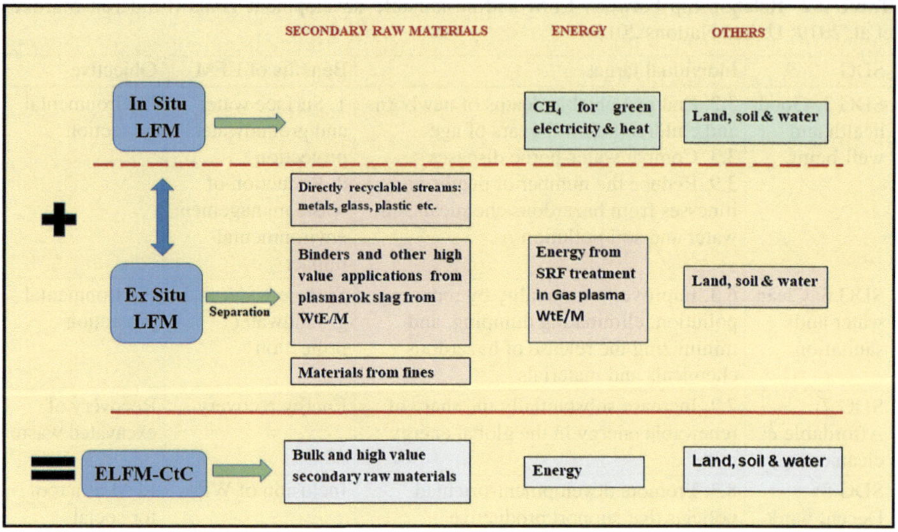


**Fig. 2.1** Resource recovery from in situ and ex-situ landfill at Remo Landfill plant. (Jones et al. 2013)

Here, the double arrow shows that an in situ LFM phase can be combined with an ex-situ LFM for the CtC project. However, the detailed steps for LFM are described in the following sections.

### *2.3.1 Drilling or Excavation*

Modified geological techniques by drilling or excavation can be used for LFM (Denafas and Bučinskas 2017). Mostly, the feasibility of LFM is dependent on the depth of the waste and the waste to soil ratio (Hogland et al. 2019). Drilling is a technique for installing a landfill gas collecting system that involves digging and removing a layer from a vertical waste wall of a specific length (typically 1 m) and size (usually 1 $m^3$). In the excavation process, excavators and/or front-end loaders are used to retrieve waste and soil fraction, which is subsequently kept in stockpiles for further processing. It is recommended to conduct a preliminary in situ aeration by airflow technology with leachate extraction during excavation to ensure suitable working conditions and avoid leachate or biogas emissions in the excavation area (Raga and Cossu 2014). Secondary materials and the fuel fraction can be sent to a processing facility off-site. Non-combustibles and leftover residues are re-landfilled (Savage et al. 1993). Hazardous material must be removed and disposed of properly. Material that can be recycled is removed – prior to the screening, bulk metals such as white goods are removed immediately by the excavator. Paper, glass, and plastic that can be recycled must be physically separated from other debris using a sieve (40 or 50 mm mesh size). After separation, small fractions are characterized by a high calorific value, 7–8 MJ/kg (Hogland et al. 2019), 11 MJ/kg (Pula 1996), allowing energy to be used without any need of auxiliary fuels. The main disadvantages of excavation are the high operating cost and the generation of mined wastes with smaller particle sizes (Kaartinen et al. 2013). Another drawback of this method is the limited sampling depth up to 5 m (Quaghebeur et al. 2013).

### 2.3.2 *Shredding/Volume Reduction*

Landfilled materials are generally diverse and compacted; therefore, the reduction of waste materials size is very important before any other step such as screening and the separation of materials and recyclables items been performed. Shredders and grinders are used for this purpose. Low-speed shredders and multiple shaft machines are usually preferred, and the excavated waste's conditions and composition can affect the device's life and separation efficiency (Ford et al. 2013b). The advantages of shredding must be balanced against the higher expense. Figure 2.2 shows the Process design for an LFM Plant.

### 2.3.3 *Separation & Screening*

Screens are used to sort waste into different sizes. Trommel screens and vibrating screens are two of the most common types of screens used in LFM projects. Trommel screens can be used at various stages of the LFM process based on the

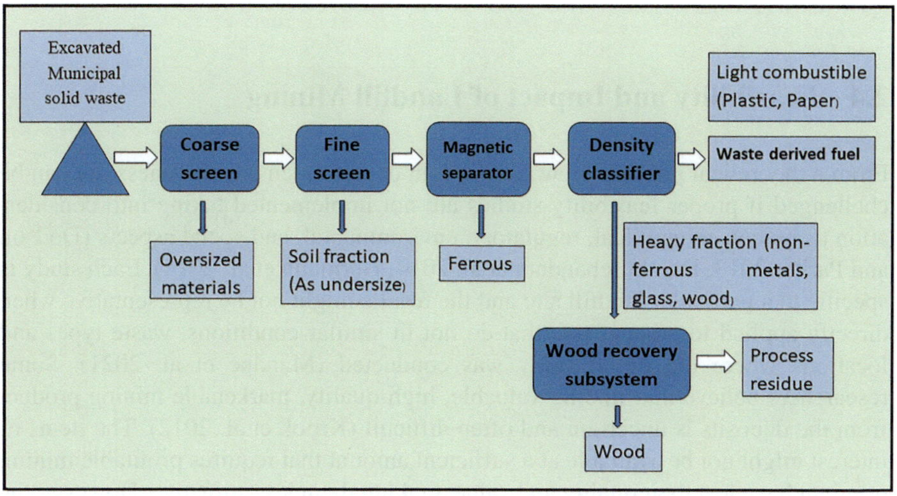


**Fig. 2.2** Process design for a landfill mining plant. (Modified from Hogland et al. (2019))

waste type and reclaimed items requirements, but they are commonly used to separate debris from soil and fine organic materials (Ford et al. 2013b). Light materials such as plastics are removed through air technologies. The next phase is metal separation which involves removing ferrous materials using drum magnets or overbend magnets. Eddy current separators are used in waste management plants to separate non-ferrous metals, such as aluminum and copper, but their performance is likely to be reduced for compacted landfills debris (Ford et al. 2013b). The remaining waste after the reclamation of valuable items, are sent to a new landfill plant, which is usually built in the same area (Townsend et al. 2015). Furthermore, the key requirements for solid recovered fuel (SRF) are related to its calorific value, chlorine, and mercury content, which frequently need additional processing (bio drying for example) before use. The fine fraction, often separated as under sieve 10–20 mm, includes higher amounts of organic compounds and has been suggested as a viable additive to compost, though this option is usually hampered by agricultural high-quality law requirements. In a case study in Thailand, the fine fraction from LFM (under sieve 25 mm) was effectively evaluated as an addition in soils for producing energy crops. The amount of recovered soil/compost mix going through the screens is the most important variable in reclaimed landfills. This changes depending on the amount of cover dirt originally placed as a cover material, the degree of waste degradation, and the screen gauge utilized. The percentage of recovery is determined by the physical and chemical qualities of the ore, as well as the efficiency of the mining equipment used. Soil recovery rates could range from 85% to 95%, ferrous materials from 70% to 90%, and polymers from 50% to 75%. Soil purity could range from 90% to 95%, ferrous materials from 80% to 95%, and polymers from 70% to 90% (Savage et al. 1993). At landfill reclamation sites, the degree of waste decomposition varies depending on the type and depth of waste, type and amount of cover soil used, compaction efforts, local climate, and age. Different landfill mining technologies are presented in Table 2.3.

## 2.4 Feasibility and Impact of Landfill Mining

From a theoretical point of view LFM might depict much attractiveness but can be challenged if proper feasibility studies are not implemented taking into consideration technical, economical, regulatory, environmental, and social aspects (Dodson and Parker 2013; Danthurebandara et al. 2014; Hermann et al. 2014). Each study is specific to a particular landfill site and the results might not be representative when directly applied to other plants that do not fit similar conditions, waste types and locations where the initial study was conducted (Mandpe et al. 2021). Some researchers believe that finding valuable, high-quality, marketable mining product from the deposits is uncertain and often difficult (Krook et al. 2012). The items of interest might not be available at a sufficient amount that requires profitable mining or might have lost their quality and value, making them less attractive. However, the economical aspect based on the value of the reclaimed end product alone is not

**Table 2.3** Various technologies in landfill mining (Ford et al. 2013a)

| Process | Technologies | Description |
|---|---|---|
| Removal of moisture from the waste mass, the collection and extraction of landfill gas | Bioreactor | In-situ. Stabilization of biological waste activity in a short time. |
| Excavation/waste removal | Excavators<br>Bulldozers<br>Grappling Hoes | Ex-situ. Waste is removed from the landfill in order to be processed. |
| Size reduction | Hammermill's shaft<br>Shear shredder<br>Rotary, scissors shears<br>Grinders & mills<br>Flail mill<br>Wet pulper<br>Knife mill | Ex-situ. Facilitation of subsequent material handling and sorting. |
| Screening | Trommel<br>Vibrating<br>Disc/star | Ex-situ. Waste is separated by size by passing through one or more screens |
| Air technologies | Wind shifter<br>Drum separators<br>Air classifiers<br>Air knife | Ex-situ. Separation of waste. |
| Metal separation | Over-band magnets<br>Drum magnets<br>Head pulley magnets<br>Eddy current separators | Ex-situ. Removal of ferrous and non-ferrous metals |

sufficient for a decision to be taken regarding the feasibility of LFM. It is also imperative to consider the recovery of the landfill area, which can have a tremendous socio-economic impact. Although LFM is a very attractive strategy for diverse resources recovery, it is a growing technology with many challenges to overcome. The most obvious challenges include the low quality and value of recovered materials which reduce their application in the market, low separation efficiency of the materials due to poor technology, landfill gas and odors, health risk, poor logistics at the excavation and sorting area (Dhar 2017).

### *2.4.1 Technical Feasibility*

LFM is resource recovery operations from landfills which are mainly performed by excavation, generally using conventional surface mining technologies followed by materials processing intending to recover resources and the land itself. It is necessary to perform a site characterization study to identify potential difficulties for the landfill being considered for mining operations. Quantitative and qualitative waste analyses are required to estimate the landfill mining and recycling potential (Prechthai et al. 2008). Quantification approaches are important techniques for determining the stock, the type of material and quantity of each valuable material contained in the landfill reservoir should be carefully considered, even though it is a time-consuming approach that requires effort (Kapur and Graedel 2006). Excavators, magnets, odor control sprayers, trammel screens, conveyor belts to transport the excavated materials, and front end loaders are commonly used types of machinery in LFM. Conventional excavators are used for digging out the waste which can be segregated in situ or off-site and the resource recovery operations executed (Krook et al. 2012). The conventional LFM technology is well adapted for the reduction of the volume of landfilled waste, landfill area recovery, removal of some fractions of hazardous material, reorganization for further waste to be landfilled (Kieckhäfer et al. 2017), but such technology fails to recover some valuable resources and resource recovery efficiency might be low due to poor segregation performance and the mixed and compacted nature of most landfilled waste. The application of advanced waste separator technologies can improve the separation performance and separated materials quality at low equipment maintenance costs (Zanetti and Godio 2006; Krook et al. 2012; Mandpe et al. 2021). This is known as ELFM which uses innovative technologies for the recovery of many resources and combustible as possible while respecting socio-ecological criteria (Kieckhäfer et al. 2017). From a technical point of view, the techniques for separating excavated waste materials into several waste streams and valuable products to be recovered are available and practical, which results in the conclusion that landfill mining is technically possible and viable. However, generalizing one technical feasibility to all landfill sites might result in mistakes, since, based on the landfill composition, age, nature, geographical position, and main mining goal, each landfill plant requires unique technical applicability.

### 2.4.2 Economic Feasibility and Impact

The implementation of LFM projects largely relies on the investment cost and different outcomes from the site. Integrated cost-benefit considerations, by considering the relation between the LFM costs, revenues from resources and land recovery are among many other factors to follow for proper decision making (Hermann et al. 2014). The economic benefits of LFM can be apprehended in terms of valuable materials recovery (metals, glass, combustible materials, plastics…) or a way to solve the traditional management problems resulting from waste deposits (Prechthai et al. 2008). Van Vossen and Prent investigated the feasibility of a 500,000 tones waste landfill plant metal recovery and reported that the extracted metal could generate sufficient revenue to offset by 8.2% the LFM costs with a full separation of the waste and by 18% for only ferrous metal excavation and separation. However, by considering the landfill area reclamation, the total cost could be compensated, with the generation of a return on investment of 10–20% (Vossen and Prent 2011).

LFM is perceived economical and viable under some conditions, including being an alternative solution to waste management and remediation, for making more space available for landfilling or other use (Krook et al. 2012). Jain et al. investigated a landfill recovery project in Florida covering the excavation of 6.8 ha of unlined cells, resulting in saving about 230,600 $m^3$ of lined airspace for USD 3.09 million, which generated a gross monetary benefit of about USD 6 million (Jain et al. 2013). In terms of valuable materials recovery, LFM has proven beneficial with various materials. It has been estimated that around 393 million metric tons of copper are buried in landfills worldwide compared to the 330 million metric tons of the present copper stock in use within the technosphere (Krook et al. 2012). A benefic LFM should take into account the various combination of benefices including valuable resource recovery, landfill reclamation, combustible (waste to energy) recovery and pollution minimization. A study conducted by Zhou et al. on an old LFM project in China under various scenarios provided a net positive benefit between USD 1.92 and USD 16.63 million (van Vossen and Prent 2011). These economic aspects of LFM prove that it can be profitable even without government subsidies. But it is worth mentioning that LFM projects are not and should not economically be justified only based on valuable materials and energy resources recovery, but land reclamation, waste management, non-monetary factors such as greenhouse gas ($CO_2$) emissions, pollution mitigation (infiltration of leachate into groundwater) should be considered and mainly contribute in improving the economic results of the feasibility (Ford et al. 2013b; Damigos et al. 2016). However, most of the studies in favor of LFM in terms of its economical feasibility are decision-oriented, intending to produce knowledge that supports the decisions of investment on regional levels of the LFM project. Investigation results which highly integrate the information on the economic constraints of LFM are necessary, and research on the acceptance without reluctance of the recovered materials and energy sources in existing markets is important (Esguerra et al. 2019).

### *2.4.3 Environmental, Health and Sociological Feasibility and Impact*

Understanding the environmental and health impact of LFM is as crucial as the technical and economical aspects for global feasibility decisions. LFM can be complex and therefore requires long-term ecological considerations. The environmental and health benefits of LFM is undoubtedly the eradication of different pollution sources including the waste itself and its component, landfill leachate (preventing surface and groundwater contamination), and hazardous gas emission (reducing a source of global warming and health risk and odor problems) (Mandpe et al. 2021). It was found that the risks for occupational hazards are generally negligible, but the generation of landfill gases at the bottom layers of the landfill should be considered when feasibility studies are being performed. Sufficient information is not available concerning the pollutant emissions which occur during the excavation of the landfills and the long-term effects of LFM are not well studied (Krook et al. 2012). But for sure LFM has the potential of occasioning significant environmental impact, health and safety risk, and nuisance risk. During the LFM operations including excavation, transportation, and separation processes can cause the pollution of air, soils and water due to the ease in the emission of $CH_4$ and $CO_2$ gases and volatile organic compounds (VOCs). The cyanogenic materials and medical waste present in the landfills can also be a source of risk to human health and can cause environmental and atmospheric pollution. The collapses into the excavated area, releases of explosives or flammable gases, emissions and increased dispersal of particulate matter, heavy metals, odor, are important threats to humans and the environment (Damigos et al. 2016).

The various environmental, health and sociological issues should be considered during the installation of the LFM project, and measures must be taken to mitigate their effects, which on the other hand might contribute to an increase in the investment cost, resulting in a negative economical consideration. Mining operations for instance have huge negative impacts on the environment and reclamation of mine waste, in addition to providing high valuable minerals can reduce the need for new mining sites and their negative exploitation effects (Agboola et al. 2020). Danthurebandara reported that the total environmental impact of LFM is determined by the waste type and composition, as well as the process technologies used, but also, it is highly dependent on the output products' quantity and quality (Danthurebandara et al. 2014). LFM itself however has a huge positive environmental and social impact. Positive benefits may result from the removal of pollutants and a potential source of pollution. Dangerous solid waste will be dug out, leachate will be managed and gas emissions controlled, with benefit to the surrounding environment (Ford et al. 2013b; Damigos et al. 2016). Sabour et al. studied LFM in Tehran as a solution to leaching of hazardous substances, the generation of $CH_4$ gas, loss of land covered by waste materials, loss and contamination of natural resources (groundwater, soil, minerals) as the result of the do-nothing scenario (landfill's current status) of landfilled materials. Their results proved that LFM could result in tremendous environmental benefits in comparison with the do-nothing scenario of the landfill's current status. With potential profitability estimated at around USD 370 million, compared to the do-nothing scenario, the LFM project lowered the impact of global warming by 1,759,790 tons $CO_2$ eq, a reduction of 134% (Sabour et al. 2020).

LFM might also be a green job opportunity for the local community (Ford et al. 2013b). Such projects require engineers as well as laborers for the excavation and separation processes and treatment of the valuable materials. In all aspects of the LFM process, local job opportunities will thrive (Einhäupl et al. 2021).

### 2.4.4 Regulatory

Most LFM projects are initiated not only for valuables materials recovery but to solve a specific regional problem including the lack of landfill space. They are generally funded and the operations are conducted by local authorities (public or private landfill owners) (Van Der Zee et al. 2004). The influence of political factors can be determining elements in the success of LFM projects. These actions might for instance include regulatory and actions against odor, noise, and dust pollution. Some actions to support measures or bureaucratic efforts to implement such LFM projects have a significant influence (Hermann et al. 2014). There is an extensive legal network in most countries that controls the licensing, operation, closure, and aftercare of landfills, but for many countries, there is no substantial legislation for landfill mining (Geysen et al. 2009), and this could lead to slow decision making and confusion in political and legal framework regarding materials recovery from landfills, the Austria LFM regulatory can be considered for instance (Hermann et al. 2014). Some countries however took some steps in encouraging LFM through their legislation. In 2005 a law was passed in Germany forbidding the relocation of old landfills without resources reclamation. In Bavaria, a public support scheme subsidizes efforts to explore old landfills and the materials stored there. The result of such policies and incentives led to an increase in LFM activities in the years 2007 and 2008 (Bockreis and Knapp 2011; Dhar 2017).

## 2.5 Global Perspectives of Landfill Mining

### 2.5.1 Perspectives

LFM was first introduced in 1953 in Israel (the landfill of Tel Aviv) in order to obtain materials for the generation of fertilizers for orchards and the recovery of ferrous materials (Savage et al. 1993). No LFM project was reported else where till after the 1980s (Calderón Márquez et al. 2019). The first LFM in Europe took place in the 1980s in the United Kingdom where solid waste from the old dockyard tip had to be removed to allow urban development in the Chatham Dockyard area (Ford et al. 2013b). First Asia's LFM project was conducted in Deonar city, India in 1989, to recover fine materials for commercialization as mixed fertilizer (Calderón Márquez et al. 2019). Even though the interest in the United States (US) LFM peaked in the 1990s (Krook et al. 2012), the US had the most LFM projects in the world (Ford et al. 2013b) by the year 2013. LFM in Africa is at an embryonic stage, not well known, and at the stage of characterization of landfill wastes (Owusu-Nimo et al. 2019). Calderón Márquez et al. investigated LFM as a strategic tool for global sustainable development and reported that over 40% of the LFM projects they analyzed were in Europe, 31% in North America, 21% in Asia 4% in the Middle East, while no LFM projects were reported in Latin America, Africa and Australia

(Calderón Márquez et al. 2019). This report might not include all LFM available worldwide, especially some traditional LFM which are not well documented, but it gives an idea about the percentage repartition of LFM practices over the globe.

#### 2.5.1.1 Asia and the Middle East

Since the introduction of LFM in Israel in 1953, many other Asian countries have investigated LFM techniques after the 1980s including India, Thailand, China, Taiwan, South Korea, Sri Lanka, and Pakistan. After India in 1983, other LFM initiatives in this region of the world were conducted between 1997 and 2013, and in the years ahead. The prime objective of these LFM was to prevent any environmental harm recovery of landfill areas (Ranjan et al. 2014; Calderón Márquez et al. 2019). Most LFM projects in Asia are located in India and China due to their high population which requires unique solid waste and landfill management (Gaitanarou et al. 2014), welcoming LFM as a golden alternative. LFM projects have been planned or implemented in Bangkok, Thailand, Seoul (Korea) (Dhar 2017).

China for instance disposes of hundreds of urban landfills for MSW disposal which poses some risks (environmental and health) to cities, therefore LFM is welcome as a solution for urgent old landfills management (Singh et al. 2010) and the removal of landfill plants surrounded by cities due to their rapid expansion (Krook et al. 2012). This pushed some Chinese national waste management institutions to provide a budget for the funding of landfill remediation projects in the country from the year 2011 to 2015 (Zhou et al. 2015).

#### 2.5.1.2 Europe

Between 1995 and 2015, about 5.25 billion tons of trash, including up to 5–25 wt% of plastic waste were landfilled in Europe. These wastes have a significant potential to be reclaimed and reintroduced into the circular economy (Canopoli et al. 2018). In 1993 in Germany, the first European LFM pilot project was conducted to free space in the landfill site and confirm the technical and economic feasibility of the process. Many other projects were also conducted later on in Germany for hazard prevention in landfills (Dhar 2017). The first LFM projects were conducted in Italy and Sweden in 1994 to reduce local risks from poor landfilling and recover landfill areas. Most projects were, however at a pilot scale and mainly located in Germany, the Netherlands, Finland, United Kingdom, with only fewer full-scale installations due to strict environmental legislation (Gaitanarou et al. 2014; Dhar 2017).

Investigations on LFM continue to find a proper solution and better LFM alternatives to landfill sites in Europe. The landfill waste composition and characteristics were investigated for 12 years at the MSW landfill plant in Debrecen, Hungary, and the total wet and dry mass of the valuable materials in the plant was evaluated (Faitli et al. 2019). The NEW-MINE project at the landfill site in Halbenrain (Austria) is part of the same dynamic by studying the efficiency of various sorting technologies for valuable materials recovery from the old landfill plant (López et al. 2018).

#### 2.5.1.3 America

A tremendous amount of waste is produced in the USA over the years and a large proportion of these solid wastes are managed in landfills. Interest in LFM practices in the US grew in the 1990s mainly due to stricter environmental laws and the need for space for new waste, with the US EPA regulations strengthening the minimum criteria for MSW landfills (Krook et al. 2012). The first US LFM projects started in Naples (Florida) from 1986 to 1992, and Edinburgh (New York) in 1988. They were motivated by strict environmental legislation to recover landfill space, waste energy, and the reclamation of valuable materials (Dhar 2017). The most common aim of LFM in the USA has been the relocation of old waste to make space for new landfills. By the year 2009, the USA was home to 32 LFM projects which focused on the conservation of landfill space, remediation of groundwater contamination, reducing the cost of closure, and recovery of energy (Wagner and Raymond 2015).

#### 2.5.1.4 Africa

With a young population attracted by urban life, urbanization in Africa contributed to a signifcant increase of waste mountains, generally municipal waste but industrial waste as well. This raises landfilling practices in the continent, however formal LFM processes are not frequent in Africa. Traditional valuable materials collection from landfill sites are however courant in some areas and is a form of LFM, as an example of Hay el- Zabaleen in central Cairo (Egypt), locally known as Garbage City, where some of the poorest communities collect plastics, metal, and textiles for reuse or recovery purposes (EL-SHERBINY 2019). Very little information is available about the contents of landfills in Africa, resulting in no actual projects in the sector, but this could however change if the wealth in landfills in Africa catches the attention of investors. Information about landfill waste characteristics is a key element in defining the LFM performance. To answer that, Owusu-Nimo et al. investigated the waste components and characteristics of a landfill site in Kumasi (Ghana), by digging samples from various locations and depths of the landfill site, representing various ages of waste deposition. The results pointed out the presence of combustibles (plastics, wood, textiles) and other valuable materials such as metals, leather, glass, which can be profitable in LFM (Owusu-Nimo et al. 2019). Such studies are scarce in Africa and would be helpful as the cornerstone in introducing LFM processes.

### 2.5.2 Future Directions for Landfill Mining

LFM is an old technology used for the first time 70 years back, but the process is still not well developed, and emerging researchers are still uncovering the value of the resources of landfills and proper mining (Zhou et al. 2015). There are hundreds of thousands of old landfills in Europe, and millions worldwide, the ones which are economically and environmentally feasible in terms of LFM are yet to be determined (Geysen et al. 2009). With the use of proper modeling techniques, changes in the composition of landfilled waste can be estimated to produce scenarios for future LFM and management (Bučinskas et al. 2018). Many types of research enlighten us on the potential of LFM, but more investigations are needed and should focus on LFM challenges and potentials instead of only obtaining some accurate results for a given specific scenario (Krook et al. 2019). The technical feasibility in extracting high-quality valuable materials, the efficiency, capacity, and practical suitability should be studied and demonstrated to justify such projects (Krook et al. 2012).

The future of LFM lies in the so-called ELFM process which can compensate the LFM cost by combining the remediation with resource recovery in a cost-effective and environmentally friendly manner. It involves producing energy and recyclable materials to generate cash to offset the exorbitant expenditures. ELFM prime targets are landfills known to contain higher-value materials, such as industrial landfills which conceal aluminum, copper, zinc, or steel like valuable minerals. The process consists of separation, further treatment and recycling, energetic valorization, mineral carbonation, carbon sequestration and storage (CCS), the use of recycled materials in or as building materials (Jones et al. 2013). ELFM target the valorization of historic waste streams, both materials and energy, with the ratio depending on the type of waste streams and advanced technology for material recuperation and energy production. Another relevant goal is to prevent GHG ($CO_2$, $CH_4$) emissions from the energy valorization process. The ideas generated by ELFM become directly relevant for the future development of waste management in relation to mining (Geysen et al. 2009). As reported by Sabour et al. ELFM could lead to huge environmental benefits by optimizing the recovery potential of valuable materials, increasing waste to energy generation efficiency, and reducing the impact of global warming (Sabour et al. 2020). Therefore, an ELFM practice is urgently required.

## 2.6 Conclusions

Valuable resources buried in landfills around the globe are undoubtedly enormous. LFM is a solution to resource recovery, but also a way to recover valuable lands in the cities and prevent environmental pollution caused by traditional landfills. LFM processes have been practiced for around 70 years but the techniques are still under development to reach the balance between economic and environmental implications which are influenced by a variety of site-specific, technical, regulatory, market factors and conditions. The potential economic and environmental benefits of an LFM project must consider carefully the site conditions, existing information and data, to support project planners with adequate knowledge about the landfill characteristics and decide the excavation strategy (and tools), and the marketing and economic strategies to be used. The factors that influence the economic feasibility of an LFM plant vary depending on the site characteristics and the mining goal. With the concept of ELFM, the process will undoubtedly be more attractive in the future for economically and environmentally safe management of landfills.

## References

Agboola O, Babatunde DE, Fayomi OSI, Sadiku ER, Popoola P, Moropeng L, Yahaya A, Mamudu OA (2020) A review on the impact of mining operation: monitoring, assessment and management. Results Eng 8(July):100181. https://doi.org/10.1016/j.rineng.2020.100181

Bockreis A, Knapp J (2011) Landfill mining – landfill sites serving as raw materials sources. Osterreichische Wasser- und Abfallwirtschaft 63(3):70–75. https://doi.org/10.1007/S00506-010-0275-1

Bučinskas A, Kriipsalu M, Denafas G (2018) Proposal for feasibility assessment model for landfill mining and its implementation for energy generation scenarios. Sustain (Switzerland) 10(8). https://doi.org/10.3390/su10082882

Burlakovs J et al (2017) Paradigms on landfill mining: from dump site scavenging to ecosystem services revitalization. Resour Conserv Recycl 123:73–84. https://doi.org/10.1016/j.resconrec.2016.07.007

Calderón Márquez AJ et al (2019) Landfill mining as a strategic tool towards global sustainable development. J Clean Prod 226:1102–1115. https://doi.org/10.1016/j.jclepro.2019.04.057

Canopoli L et al (2018) Physico-chemical properties of excavated plastic from landfill mining and current recycling routes. Waste Manag 76:55–67. https://doi.org/10.1016/j.wasman.2018.03.043

Cheerarot R, Jaturapitakkul C (2004) A study of disposed fly ash from landfill to replace Portland cement. Waste Manag 24(7):701–709. https://doi.org/10.1016/j.wasman.2004.02.003

Cossu R, Williams ID (2015) Urban mining: concepts, terminology, challenges. Waste Manag 45:1–3. https://doi.org/10.1016/j.wasman.2015.09.040

Cossu R, Hogland W, Salerni E (1996) Landfill mining in Europe and USA. ISWA yearbook

Damigos D, Menegaki M, Kaliampakos D (2016) Monetizing the social benefits of landfill mining: evidence from a Contingent Valuation survey in a rural area in Greece. Waste Manag 51:119–129. https://doi.org/10.1016/j.wasman.2015.12.012

Danthurebandara M et al (2014) Assessment of environmental and economic feasibility of Enhanced Landfill Mining. Waste Manag 45:434–447. https://doi.org/10.1016/j.wasman.2015.01.041

Denafas G, Bučinskas A (2017) Feasibilities and perspectives for landfill mining in Lithuania. Linnaeus Eco-Tech. https://doi.org/10.15626/eco-tech.2014.012

Dhar A (2017) Landfill mining – a comprehensive review. https://doi.org/10.13140/RG.2.2.31092.55687

Dodson J, Parker HL (2013) Mining municipal waste: prospective for elemental recovery. RSC Green Chem:220–257

Einhäupl P et al (2019) Eliciting stakeholder needs – an anticipatory approach assessing enhanced landfill mining. Waste Manag 98:113–125. https://doi.org/10.1016/j.wasman.2019.08.009

Einhäupl P et al (2021) The conceptualization of societal impacts of landfill mining – a system dynamics approach. J Clean Prod 296. https://doi.org/10.1016/j.jclepro.2021.126351

EL-SHERBINY E (2019) Digging for gold in Africa's garbage. Africa Science Magazine

Esguerra JL et al (2019) Assessing the economic potential of landfill mining: review and recommendations. Detritus 8(December):125–140. https://doi.org/10.31025/2611-4135/2019.13883
Faitli J et al (2019) Assessment of a residual municipal solid waste landfill for prospective "landfill mining". https://doi.org/10.1177/0734242X19881197
Ford S, Warren K, Read A (2013a) An evaluation of the feasibility and viability of Landfill Mining and Reclamation (Lfmr) from a UK perspective. Sardinia 2013, …, (October 2013)
Ford S et al (2013b) Feasibility and viability of landfill mining and reclamation in Scotland
Frändegård P et al (2013) A novel approach for environmental evaluation of landfill mining. J Clean Prod 55:24–34. https://doi.org/10.1016/j.jclepro.2012.05.045
Franke M, Mocker M, Faulstich M (2010) Resource potential of landfill mining–a national and regional evaluation. ISWA World Congress 49(0)
Gaitanarou Z et al (2014) Landfill mining: an empirical review on past and state-of-the-art applications
Geysen D et al (2009) Enhanced landfill mining – a future perspective for landfilling, (March)
Giroux L (2014) State of Waste Management in Canada. Canadian Council of Ministers of Environment
Hermann R et al (2014) Landfill mining in Austria: foundations for an integrated ecological and economic assessment. Waste Manag Res 32(July):48–58. https://doi.org/10.1177/0734242X14541168
Hoekstra N et al (2005) Enhanced microbial degradation of chloroethenes and chloroethanes in a bioscreen. IAHS-AISH Publication 298:75–81
Hogland W, Gomes MM, Thörneby L (2019) Landfill mining: space saving, material recovery and energy use. Linnaeus Eco-Tech. https://doi.org/10.15626/eco-tech.1997.041
Jain P, Townsend TG, Johnson P (2013) Case study of landfill reclamation at a Florida landfill site. Waste Manag 33(1):109–116. https://doi.org/10.1016/j.wasman.2012.09.011
Jones PT et al (2013) Enhanced landfill mining in view of multiple resource recovery: a critical review. J Clean Prod 55:45–55. https://doi.org/10.1016/j.jclepro.2012.05.021
Kaartinen T, Sormunen K, Rintala J (2013) Case study on sampling, processing and characterization of landfilled municipal solid waste in the view of landfill mining. J Clean Prod 55:56–66. https://doi.org/10.1016/j.jclepro.2013.02.036
Kapur A, Graedel TE (2006) Copper mines above and below the ground. Environ Sci Technol 40(10):3135–3141. https://doi.org/10.1021/es0626887
Kieckhäfer K, Breitenstein A, Spengler TS (2017) Material flow-based economic assessment of landfill mining processes. Waste Manag 60:748–764. https://doi.org/10.1016/j.wasman.2016.06.012
Krook J, Svensson N, Eklund M (2012) Landfill mining: a critical review of two decades of research. Waste Manag 32(3):513–520. https://doi.org/10.1016/j.wasman.2011.10.015
Krook J et al (2019) What do recent assessments tell us about the potential and challenges of landfill mining? In: Sustainable resource recovery and zero waste approaches. Elsevier B.V., Amsterdam, Netherlands. https://doi.org/10.1016/B978-0-444-64200-4.00018-9
López CG et al (2018) Landfill mining: a case study regarding sampling, processing and characterization of excavated waste from an Austrian landfill. Detritus 2(June):29–45. https://doi.org/10.31025/2611-4135/2018.13664
Mandpe A, Pratap V, Lakshmikanthan P (2021) Landfill mining. In: Current developments in biotechnology and bioengineering. BV, Amsterdam, Netherlands, Elsevier. https://doi.org/10.1016/b978-0-12-821009-3.00010-5
Marella G, Raga R (2014) Use of the Contingent Valuation Method in the assessment of a landfill mining project. Waste Manag 34(7):1199–1205. https://doi.org/10.1016/j.wasman.2014.03.018
Masi S et al (2014) Assessment of the possible reuse of MSW coming from landfill mining of old open dumpsites. Waste Manag 34(3):702–710. https://doi.org/10.1016/j.wasman.2013.12.013
Owusu-Nimo F et al (2019) Characteristics and management of landfill solid waste in Kumasi, Ghana. Sci African 3:e00052. https://doi.org/10.1016/j.sciaf.2019.e00052

Prechthai T, Padmasri M, Visvanathan C (2008) Quality assessment of mined MSW from an open dumpsite for recycling potential. Resour Conserv Recycl 53(1–2). https://doi.org/10.1016/j.resconrec.2008.09.002
Pula SM (1996) Sardinia'95 – fifth international landfill symposium. Waste Management and Research. https://doi.org/10.1007/bf02986760
Quaghebeur M et al (2013) Characterization of landfilled materials: screening of the enhanced landfill mining potential. J Clean Prod 55:72–83. https://doi.org/10.1016/j.jclepro.2012.06.012
Raga R, Cossu R (2014) Landfill aeration in the framework of a reclamation project in Northern Italy. Waste Manag 34(3):683–691. https://doi.org/10.1016/j.wasman.2013.12.011
Ranjan MR et al (2014) Landfill mining: a case study from Ghazipur landfill area of Delhi. Int J Environ Sci 4(5):919–925. https://doi.org/10.6088/IJES.2014040404530
Sabour MR, Alam E, Mostafa Hatami A (2020) Environmental and economic assessment of Enhanced Landfill Mining in Tehran. Environ Sci Pollut Res 27(27):34469–34483. https://doi.org/10.1007/s11356-020-09458-w
Savage GM, Golueke CG, von Stein EL (1993) Landfill mining: past and present. Biocycle 34(5):58–61
Singh RK, Datta M, Nema AK (2010) Review of groundwater contamination hazard rating systems for old landfills. Waste Manag Res 28(2):97–108. https://doi.org/10.1177/0734242X09103816
Townsend TG et al (2015) Sustainable practices for landfill design and operation. Springer Nature, Switzerland. https://doi.org/10.1007/978-1-4939-2662-6
United Nations (2015) Transforming our World: the 2030 agenda for sustainable development A/RES/70/1, United Nations
Van Der Zee DJ, Achterkamp MC, De Visser BJ (2004) Assessing the market opportunities of landfill mining. Waste Manag 24(8):795–804. https://doi.org/10.1016/j.wasman.2004.05.004
Van Passel S et al (2013) The economics of enhanced landfill mining: private and societal performance drivers. J Clean Prod 55:92–102. https://doi.org/10.1016/j.jclepro.2012.03.024
Vossen W van, Prent OJ (2011) Feasibility study sustainable material and energy recovery from landfills in Europe. In: Proceedings of the thirteenth international waste management and landfill symposium, Sardinia 2011
Wagner TP, Raymond T (2015) Landfill mining: case study of a successful metals recovery project. Waste Manag 45:448–457. https://doi.org/10.1016/j.wasman.2015.06.034
Zanetti M, Godio A (2006) Recovery of foundry sands and iron fractions from an industrial waste landfill. Resour Conserv Recycl 48(4):396–411. https://doi.org/10.1016/j.resconrec.2006.01.008
Zhou C et al (2015) A cost-benefit analysis of landfill mining and material recycling in China. Waste Manag 35:191–198. https://doi.org/10.1016/j.wasman.2014.09.029

# Chapter 3
# Open Dumps and Circular Economy: Assessing Landfill Mining Potential, Benefits and Challenges for Indian Dumpsites

**Ayush Singh, Roshan Vilasrao Mankhair, and Munish Kumar Chandel**

**Abstract** Increasing economic growth and rapid resource consumption have led us to think about concepts like closing the loop, circular economy and sustainable growth. Landfill mining (LFM) is one of the approaches to achieve sustainability in waste management. LFM entails mining of the dumped waste to clear landfill with potential material and energy recovery. This study analyses the LFM objectives, potentials, benefits, and challenges of LFM by reviewing published literature in the Indian context. The results show that interest in the LFM has increased in the recent decade with the municipal solid waste rules 2016 asserting to investigate old landfills for possible reclamation. Research on LFM in India is mainly upon the waste characterization, with the recent interest in waste valorization. The common conclusion from the published literature is that the fine fraction (soil-like material) is main constituent of landfills comprising nearly half of the landfilled waste and needs to be utilized for the success of LFM. In India, LFM can be seen as the method to solve urban environmental concerns by clearing landfill space for future waste management while reclaiming dumped resources. However, for LFM to be successful, local government participation, subsidy and incentives, strategic planning, and public participation are required.

**Keywords** Reclamation · Remediation · Open dumpsites · Resource recovery · India · Landfill mining · Legacy waste

---

A. Singh · R. V. Mankhair · M. K. Chandel (✉)
Environmental Science and Engineering Department, Indian Institute of Technology Bombay, Mumbai, Maharashtra, India
e-mail: munish.chandel@iitb.ac.in



P. Pathak, S. G. Palani (eds.), *Circular Economy in Municipal Solid Waste Landfilling: Biomining & Leachate Treatment*, Radionuclides and Heavy Metals in the Environment, https://doi.org/10.1007/978-3-031-07785-2_3

## Abbreviations

| | |
|---|---|
| C&D | Construction and Demolition |
| MSW | Municipal Solid Waste |
| CSO | Central Statistics Office |
| CPHEEO | Central Public Health and Environmental Engineering Organization |
| TPD | Tonnes per day |
| MT | Metric Tonnes |
| MoHUA | Ministry of Housing and Urban Affairs |
| RDF | Refuse Derived Fuel |
| VSR | Void Space Recovery |

## 3.1 Introduction

In the past, landfills were seen as the most promising solutions for managing municipal solid waste (MSW). However, with the developing technologies in waste management and the increasing government awareness on waste management issues, landfills have become the least preferable option. However, for developing countries like India, where more than 63% of collected MSW still ends up in the dumpsites, the need for landfill space will not go for the near future (CSO 2021). Furthermore, the increasing population of India demands more land, which is instigating a reduction in space available for MSW disposal. Most old landfills in India have already exhausted their capacity, but stockpiling of waste is still going on such sites due to lack of land availability. Hence a need arises for reclaiming the old landfill space.

The landfilled waste has a negative impact on societal and environmental levels. Gases generated (methane and carbon dioxide) during waste decomposition contribute to greenhouse gas emissions in the environment. Further landfill fire is another issue that arises due to the burning of methane in the landfill. Landfills like Deonar, Ghazipur, Autonagar and many others have reported fire incidents in the past few years owing to air pollution in nearby areas. Further, gases like hydrogen sulfide generated due to waste decay emit a foul odour which is a nuisance to people living nearby. The leachate generated from waste contains various heavy metals and organic pollutants that contaminate the water resources if not handled carefully (Kanmani and Gandhimathi 2013; Guleria and Chakma 2021). A Nagpur landfill study showed that the leachate traversed the aquifer situated below the landfill, polluting the aquifer (Pujari and Deshpande 2005).

India has brought the concept of smart cities similar to the model city programme of the United States of America or the national model city plan of China. To plan a smart city, there would be a need to clear these large historic dumps as they add a grotesque picture to the scenery. Hence to deal with the dumped legacy waste, an integrated waste management approach is needed. A sustainable option to achieve such goals can be landfill mining (LFM). LFM entails the total recovery of waste from a landfill. Through LFM, the old landfills can be excavated and equipped with sanitary measures to control environmental threats. Also, the recovered space could be used to develop an integrated solid waste management facility to deal with the old and incoming fresh waste. The excavated waste from dumpsite could be recycled to monetize some benefits.

The focus of the study is to assess the potential of LFM and apply the concept of LFM as a crucial part of sustainable landfill management for the rehabilitation of the landfill in India. Hence, a literature review was carried out to identify the key objectives, findings, and challenges pertaining to LFM in India. We carried out a meta-analysis using keywords "landfill mining," "biomining," "dumpsite remediation," "dumpsite reclamation" in combination with "India," using different scientific databases to select research papers for the review. Total 34 relevant research papers (articles in research journal, book chapters and conference proceedings) were selected and further reviewed for the study (Table 3.1). There have been trials/pilot projects for landfill reclamation in India which have not been reported in the literature. However, the information regarding the same is minimal.

## 3.2 Indian Scenario

### *3.2.1 Literature Studies*

The literature available for LFM in India is very limited and had a sharp increase in research activities in the last 2 years (Fig. 3.1). The first study for LFM in India was carried out by (Manfred and Bhattacharyya 1997) to valorize fine fraction as compost. In the last decade (2000–2010), only 7 research studies are available, with the rest 25 studies carried out in the current decade (2014–2021). The research papers were divided into research themes for systematic review, i.e., characterization, environmental, economic, modeling, valorization, and others. Many papers deal with multiple research themes. However, to simplify the understandings, only one central theme was attributed to each paper. It is evident from the literature that studies carried out in India are on waste characterization and valorization (Fig. 3.1).

#### 3.2.1.1 Characterisation

For the success of LFM, waste characteristics are a primary prerequisite. Since the dumped waste characteristic varies from country to country and landfill to landfill, the experience of a developed country cannot be directly adapted to other developing countries. The waste characteristics depend on the dumped waste composition, which depends on the lifestyle of people, governmental policies in the region, geographical location of the dumpsite, climatic conditions, and many other factors. Characterization of legacy waste is the second most studied subject within LFM in

**Table 3.1** Literature studies on landfill mining in the Indian context

| Year | Site | Main field | Objective | References |
|---|---|---|---|---|
| 1997 | Deonar (Mumbai) | Valorization | Valorization of fine fraction as compost | Manfred and Bhattacharyya (1997) |
| 2003 | Kodungaiyur (Chennai), Perungudi (Chennai) | Waste characterization | Physical and chemical characterization of excavated waste to study the feasibility of landfill mining | Kurian et al. (2003) |
| 2005 | N.A. | Conceptual/ modelling | Decision-making tool for dumpsite rehabilitation | Kurian et al. (2005) |
| 2005 | Kodungaiyur (Chennai) | Environmental | Heavy metal speciation in decomposed MSW | Esakku et al. (2005) |
| 2006 | Ghazipur (Delhi) | Environmental | Water composition and methane emission was studied | Mor et al. (2006) |
| 2007 | N.A. | Valorization | Concept of compost mining from stabilized waste and bioreactor landfill studied | Joseph et al. (2007) |
| 2007 | Kodungaiyur (Chennai) | Conceptual/ modelling | Anaerobic treatment for stabilization of mined waste | Karthikeyan et al. (2007) |
| 2007 | Chennai | Economical | Cost-benefit analysis for dumpsite rehabilitation | Joseph and Nagendran (2007) |
| 2014 | Ghazipur (Delhi) | Valorization | Emission from landfills studied for landfill gas recovery | Ranjan et al. (2014) |
| 2016 | 32 cities | Economical | Suitability of landfill mining as a remediation technique for dumpsites studied | Dubey et al. (2016) |
| 2017 | Ghazipur (Delhi) | Characterization and valorization | Waste characterization and application of segregated MSW as embankment material | Havanagi et al. (2017) |
| 2018 | Okhla (Delhi), Jawaharnagar (Hyderabad), Ukkayyapalli (Kadapa) | Characterization | Physical and chemical characterization of soil-like fraction (fine fraction) | Somani et al. (2018) |
| 2019 | Chennai | Characterization and environmental | Characterization of waste and pollution from buried waste studied | Peter et al. (2019) |

(continued)

**Table 3.1** (continued)

| Year | Site | Main field | Objective | References |
|---|---|---|---|---|
| 2019 | Delhi, Hyderabad, Kadapa | Valorization | Leaching behaviour of soil-like fraction studied for application as earth fill | Somani et al. (2019) |
| 2019 | N.A. | Valorization | Potential application of legacy waste studied | Singh et al. (2019) |
| 2019 | Nagpur | Characterization and economics | Water composition and economic analysis | Mandpe et al. (2019) |
| 2019 | N.A. | Valorization and modelling | Model analysis for soil-like material as backfill material | Rawat and Mohanty (2019) |
| 2019 | Okhla, Ghazipur, and Bhalswa landfill sites of Delhi, Ariyamangalam (Tiruchirappalli) open dumpsite | Environmental | Health risk due to groundwater contamination in dumpsites | Guleria and Chakma (2021) |
| 2020 | Okhla (Delhi), Bhalswa (Delhi), Noida (Delhi-NCR), Hyderabad and Kadapa. | Valorization | Physical and chemical characterization of soil-like fraction (fine fraction) to assess contamination | Somani et al. (2020) |
| 2020 | Mulund (Mumbai) | Characterization | Effect of age on waste characteristics studied | Singh and Chandel (2020a) |
| 2020 | Deonar (Mumbai) | Characterization | Physical characterization of Deonar dumpsite | Singh and Chandel (2020b) |
| 2020 | Barshi landfill (Barshi) | Economical | Economic feasibility of landfill mining assessed | Gaikwad and Patil (2020) |
| 2020 | Okhla (Delhi) | Characterization and valorization | Composition analysis and potential use of stabilized waste studied | Saluja et al. (2020) |
| 2020 | Vishakhapatnam | Characterization and valorization | Physicochemical characterization of aged waste to investigate RDF potential | Cheela et al. (2021) |
| 2020 | N.A. | Conceptual/ modelling | Model for landfill site ranking for remedial strategies | Naveen and Apoorva (2018) |
| 2020 | Chennai | Conceptual/ modelling | Framework for biomining of MSW dumpsite | Mohan and Joseph (2020) |
| 2021 | N.A. | Economics | Financial and sustainable details of landfill mining in global and Indian context | Mandpe et al. (2021) |

(continued)

**Table 3.1** (continued)

| Year | Site | Main field | Objective | References |
|---|---|---|---|---|
| 2021 | Mulund (Mumbai) | Valorization | Characterization and valorization of fine fraction | Singh and Chandel (2021) |
| 2021 | Okhla (Delhi), Bhalswa (Delhi), Hyderabad and Kadapa. | Valorization | Feasibility of using soil-like material as earth fill and compost | Datta et al. (2021a) |
| 2021 | Delhi | Valorization | Feasibility of local soil substitution by MSW residues (aged MSW and bottom ash) | Datta et al. (2021b) |
| 2021 | Ramana (Varanasi) | Valorization | Potential use of fine fraction in geotechnical application | Rawat and Mohanty (2021) |
| 2021 | Jawharnagar (Hyderabad) | Environmental | Techno-commercial assessment between scientific capping and biomining | Choudhury et al. (2021) |
| 2021 | N.A. | Conceptual/ modelling | Approach towards remediation of the dumpsite | Sughosh et al. (2021) |
| 2021 | N.A. | Valorization | Seismic response of soil-like material studied | Shubham et al. (2022) |

*NA* Not applicable

India. In most studies, excavated waste from the dumpsite was sieved into different size fractions and then manually sorted into waste fractions such as paper/cardboard, textile, metal, glass, wood, inert and fine fraction. The research on characterization provides vital knowledge to assess the viability of LFM in India. The studies have shown that the waste composition varies from landfill to landfill and within the landfill (Table 3.2). It adds to uncertainties in LFM based projects and therefore necessitates site-specific characterization, a fundamental requirement of LFM. However, some common conclusions regarding the waste composition can be drawn from the studies. The dumped waste mainly comprises 44–75% of soil-like material, typically referred to as fine fraction, which is composed of degraded organic matter, soil, dirt and pebbles (Havanagi et al. 2017; Somani et al. 2018; Mandpe et al. 2019; Peter et al. 2019; Cheela et al. 2021; Saluja et al. 2020; Singh and Chandel 2020a, b).

The fine fraction tends to rise with the age of the waste in the dumpsite (Singh and Chandel 2020a). The other significant material in the dumpsite is inorganic material (C&D waste and glass) 8–33%. The inorganic material is mainly comprised of C&D waste, with glass only 1–4% of total dumped waste. The metal content in the dumpsites was 1% on average and absent in many dumpsites. The combustible fraction ranged from 4% to 70% in the landfill. In most of the studies, the major part of the combustible fraction was plastics. The studies on waste characterization also dealt with the potential valorization of recovered fractions and the

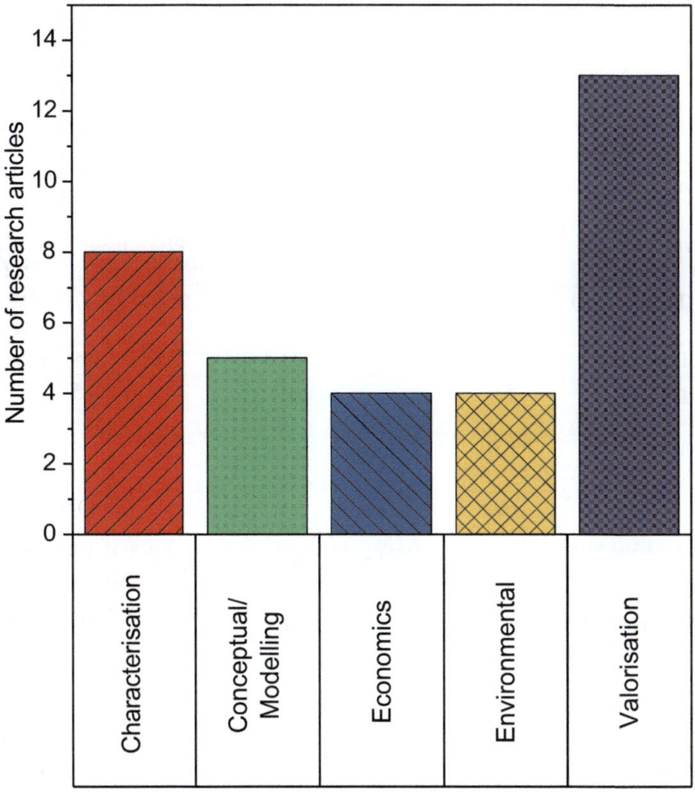


**Fig. 3.1** Major themes addressed by the research papers based upon its primary objective for landfill mining in India. For each paper, one theme is attributed

environmental burden posed by such sites (Peter et al. 2019; Singh and Chandel 2020a).

#### 3.2.1.2 Economics

The studies carried in India for the economics of LFM are very limited. Joseph and Nagendran (2007) carried out a cost-benefit analysis for LFM to recover land for landfilling. The study reported a project cost of Rs. 147 million with a benefit of Rs.162 million arising from the sale of reclaimed soil as compost, landfill space and sale of recyclables. The study further added that the environmental benefits also needed to be considered while evaluating LFM projects. Dubey et al. (2016) also carried out a cost-benefit analysis for open dumpsites in metro cities of India and had a similar conclusion to that of Joseph and Nagendran (2007). Mandpe et al. (2019) carried out a cost assessment for two scenarios for LFM; one included the MSW transfer to sanitary landfill and other dumpsite mining for a period of 50 years.

**Table 3.2** Excavated waste composition from different landfills in India

| Dumpsite | Age (Year) | Combustibles (%) | Metal (%) | Inorganics (%) | Fine fraction (%) | Others (%) | References |
|---|---|---|---|---|---|---|---|
| Perungudi (Chennai) | ~10 | 39.4 | 0.2 | 19.3 | 40.1 | – | Kurien et al. (2003) |
| Kodungaiyur (Chennai) | ~10 | 3.5 | 0.1 | 28.7 | 67.8 | – | |
| Kodungaiyur (Chennai) | – | 26 | 1 | 24 | 49 | – | Karthikeyan et al. (2007) |
| Ghazipur (Delhi) | 10 | 26.45 | 1.7 | 0 | 71.5 | – | Havanagi et al. (2017)[1] |
| Okhla (Delhi) | – | 4.3 | – | 23.6 | 71.9 | 0.5 | Somani et al. (2018) |
| Jawaharnagar (Hyderabad) | – | 5.6 | – | 18 | 73.1 | 0.5 | |
| Ukkayyapalli (Kadapa) | – | 6.01 | – | 17.93 | 75.2 | 1.5 | |
| Chennai | – | 69.5 | – | 0 | | 30.5 | Peter et al. (2019)[2] |
| | – | 22 | 2 | 0 | 72.5 | | Mandpe et al. (2019)[3] |
| Mulund (Mumbai) | 2–10 | 20.8 | 0.6 | 31.9 | 44.8 | 1.9 | Singh and Chandel (2020a) |
| Deonar (Mumbai) | 1–32 | 22.5 | 0.2 | 32.8 | 44.4 | – | Singh and Chandel (2020b) |
| Okhla (Delhi) | – | 1.871 | – | 13.85 | 73.6 | 1.43 | Saluja et al. (2020)[4] |
| Visakhapatnam | – | 26.6 | 4 | 8.1 | 52.1 | 9.3 | Cheela et al. (2021)[5] |

[1]Weighted average of >80 mm and <80 mm fraction, fine fraction is the combination of C&D waste and soil-like material
[2]Data is the average of both sites in the study
[3]Average of waste at 2 m and 4 m depth, fine fraction is the sum of inert fractions like soil, stone and C&D waste
[4]<80 mm (9.21%) sorting not provided
[5]Average of 5–15 years old waste

The study showed that dumpsite mining is a more economical solution than transferring MSW to a new sanitary landfill. Gaikwad and Patil (2020) studied the economics of Barshi LFM using cost-benefit analysis. The benefit to cost ratio in their study was 1.28 suggesting the project would be feasible. Mandpe et al. (2021) discussed success stories of LFM scenarios around the globe. The study also discussed the policies, feasibility and economics of LFM in a global context and the need for LFM in India.

The studies carried out for the economics of LFM are preliminary in nature, with no detailed costing provided for LFM operations. Further, the benefits arising from LFM need to be cross-checked with the recycling market for LFM products. The significant revenue in all cost-benefit studies was from selling the fine fraction as compost/fertilizer. Analysis of all direct and indirect costs/benefits need to be carried out. The contamination in recovered resources from landfills may affect the benefits arising from their direct sale. Hence new valorization options which can put a check on the contamination needs to be evaluated. The economics of LFM needs to be understood in much more depth and is currently lacking in the Indian context.

#### 3.2.1.3 Environment

The studies on the environmental aspect of LFM mainly covered the pollution arising from landfills. No study has been carried out to understand the environmental performance of LFM in India. However, the studies carried out to show the pollution aspect make a strong case in favour of LFM as it can help reduce those emissions. Mor et al. (2006) observed elevated concentrations of electrical conductivity, dissolved solids, sulphates, nitrate in groundwater near the Ghazipur landfill site. Guleria and Chakma (2021) observed data for groundwater samples near dumping sites and calculated the human health risk arising from groundwater contamination due to leachate. The study observed that most of the groundwater samples were within safe limits for dumping sites except in the case of Okhla dumpsite. Study by Choudhury et al. (2021) analyzed the environmental aspect of open dumping capping efficiency. The study argued that the refuse-derived fuel produced from the excavated waste is of low quality, making LFM less desirable. Further, the partial capping had reduced the pollutant concentration in groundwater. Based on the above observations, the study suggested capping as the preferred alternative. However, the choice between capping and mining is debatable and can vary from site to site. To follow sustainable development goals and achieve a circular economy approach, the resources present in the dumpsite cannot be overlooked. The environmental burden due to dumpsites is well understood by the research community, creating a push towards research to resolve the issues. LFM is one solution, but the pollution arising from LFM activities also needs to be understood to weigh the pros and cons of LFM.

#### 3.2.1.4 Conceptual/Modelling

This category included papers that showed a concept related to LFM or some assessment tool for LFM. Two research papers had covered a decision-making model for ranking dumpsites for rehabilitation and reclamation based on their risk to the surroundings. Kurian et al. (2005) developed a decision-making tool using an integrated risk-based approach to guide local government in prioritizing the actions related to dumpsite rehabilitation. They applied the framework to the Kodungaiyur and Perungudi dumpsite of Chennai, India. They indicated that both dumpsites had moderate hazard potential and needed to be rehabilitated immediately. Naveen and Apoorva (2018) proposed a hazard decision model based on source-pathway and receptor relationships. For all the analyzed landfills in their study, Deonar dumpsite had a very high hazard score. The framework can be used to prioritize the sites for LFM based on their repercussions on the environment.

Karthikeyan et al. (2007) carried out anaerobic digestion of mined waste for waste stabilization. The study observed that bioreactor landfills could increase organic degradation in the landfill. Similar observations were observed by Kurian et al. (2007). Mohan and Joseph (2020) developed an integrated approach of biomining, which includes the first stabilization of waste using bioreactor landfill treatment followed by mining waste. The concept of waste stabilization after mining will further help in waste valorization.

#### 3.2.1.5 Valorization

Dumped waste valorization is the most covered topic in literature and has gained much interest in the last few years. The characterization papers have shown that fine fraction is present in the highest amount in the landfills and is the most covered fraction for valorization potential. The primary application of fine fraction is suggested as compost or geotechnical application. The first study on LFM in India discussed the application of fine fraction as compost (Manfred and Bhattacharyya 1997). Further, Kurian et al. (2007) suggested using fine fraction for growing inedible plants or use as landfill cover material. The study for compost application for fine fraction had a similar observation that contaminants like heavy metals are present in high concentration in the fine fraction, making its use as compost a challenge (Datta et al. 2021b; Singh and Chandel 2021).

The fine fraction use as earth fill material has been discussed in various literature (Rawat and Mohanty 2019; Somani et al. 2019, 2020; Saluja et al. 2020; Datta et al. 2021a, b). The studies have shown that prior treatment would be required to reduce the contamination levels in the fine fraction. Rawat and Mohanty (2021) carried out experimental investigations on strength characteristics of MSW fines and suggested that the low specific gravity of fine fraction makes them a suitable candidate for filling material. Further, Shubham et al. (2022) studied the seismic response of fine fraction using the DEEPSOIL platform to understand the impact of waste age. Havanagi et al. (2017) studied the application of fine fraction as road embankment material. The study showed that 65–75% segregated MSW can be utilized for building embankment. However, the study recommended carrying out pilot-level tests for 2 years before field-scale trials.

Further studies by Cheela et al. (2021) and Singh and Chandel (2020a) suggested the use of combustible fractions from legacy waste as refuse-derived fuel (RDF). However, a suitable market needs to be located for the same as they contain lots of impurities. Ranjan et al. (2014) studied emissions from Ghazipur landfill and suggested that methane can be trapped and utilized as biofuel while the tapped combustible resources can be used for making RDF. Most of the studies carried out for valorization are preliminary in nature, suggesting application based on physico-chemical traits of legacy waste. However, lab-scale and pilot-scale tests need to be carried out to study the real-life impact of the legacy waste on proposed routes. Furthermore, the contamination of waste seems to be a major hurdle in waste valorization, hence alternative routes also need to be studied which can take care of such contamination.

### 3.2.2 *Major Field Initiatives in Landfill Mining*

Besides the research studies, there has been an increase in LFM (also referred to as biomining in India) projects in India. Private organizations like Zigma Global Environ Solutions Private Limited, Prakash Constrowell Limited, Biomining India Private Limited, Antony Lara Enviro Solutions Private Limited, and many others are getting involved with the government for possible mining and remediation of the dumpsites. Remediation projects on dumpsites like Kumbakonam, Sembakkam, Poonamallee, Pammal, Gorai, Indore and Nashik have been carried out by tender to private parties or by the civic body itself. However, complete details of many projects are not available, making it challenging to identify the reclamation route. LFM projects for which data was available have been discussed to understand the general approach towards LFM.

The Kumbakonam biomining project is the very first mining project initiated in India and was carried out by Zigma Global Environ Solutions Private Limited. Kumbakonam is situated in Thanjavoor district of Tamilnadu. The dumpsite was estimated to be 30 years old and spread over an area of 32.5 acres. The mining project started in July 2016 using a mobile processing plant and was completed in August 2018. The process included stabilizing waste in windrows by using bio culture to enhance waste decomposition for approximately 30 days. Stones, glass and other large metal objects were manually removed from the stabilized waste. The stabilized waste was then further screened into different size fractions using trommels. The segregated waste (secondary resource) from the landfill was sent to appropriate markets, e.g., cement companies, recyclers, steel-making companies, farmers, pyrolysis plants etc. The reclaimed land was used to establish a 70 TPD processing plant and a 5 TPD biogas plant to handle the incoming waste. The Ramapuram remediation project was awarded to Zigma Global Environ Solutions Private Limited at Rs 932 per metric tonnes for 2 lakhs metric tonnes of waste. The project involved biomining and reclamation of the existing MSW dumpsite in a scientific manner. By January 2020, ~106,750 tonnes of waste were processed and segregated into different fractions (Fig. 3.2).

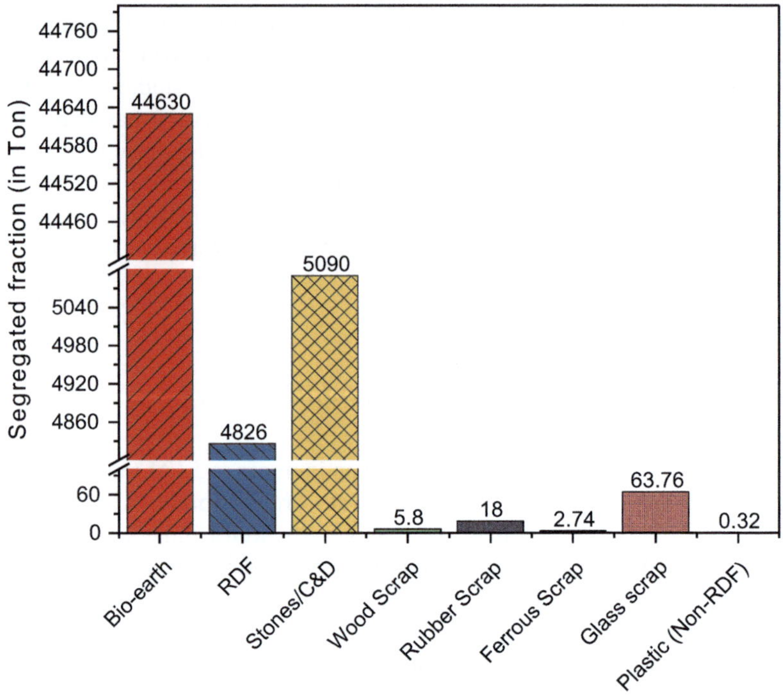


**Fig. 3.2** Different fractions obtained from April 2019 to January 2020 in the excavated waste from Ramapuram remediation project

### *3.2.3 Landfill Mining Potential for India*

According to the data available with the Ministry of Housing and Urban Affairs (MoHUA), the total legacy waste present in dumpsites of 472 cities with a population greater than 1 lakh is 142 million tonnes (CPHEEO 2020). Major resources which can be extracted from these dumpsites include combustible fraction (plastic, rubber, wood and textiles) and inert fraction (soil and C&D waste) along with metal and glass in a very low quantity. However, the secondary resource potential of the dumpsite will change from location to location. The compositional analysis carried out in Sect. 3.2.1.1 indicates that fine fraction constitutes nearly half of the excavated waste, followed by stone/C&D waste and plastic waste. These fractions have the potential of recovering energy or being reused as a construction material; hence managing these fractions of excavated waste effectively is the key to the success of any LFM project. The potential applications of the excavated waste are shown in Fig. 3.3.

Plastics from the landfill can be potentially reprocessed as a waste-derived fuel to recover energy or cracked chemically to produce fuels (Bhatnagar et al. 2017). Zhou et al. (2014) also suggested the incineration and refuse-derived fuel (RDF) preparation for recovering energy is the most feasible way for processing the excavated plastic waste. Another way the plastic fraction can be used is as construction materials (filler and binder). Tiles, paver blocks, bricks, etc., can be manufactured

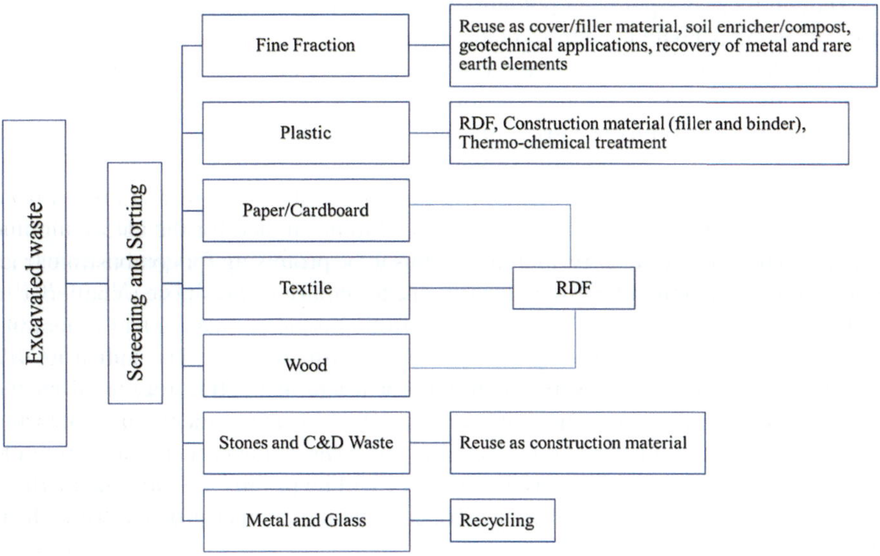


**Fig. 3.3** Potential applications of excavated waste from dumpsites and landfills

by mixing plastic and fine fraction in definite proportion followed by melting and casting. The other combustibles (wood, paper, textiles) obtained can be converted into RDF and utilized for energy recovery or co-combustion. The fine fraction has a huge potential for reuse as reuse as cover/filler material (Burlakovs et al. 2017), reuse in geotechnical applications (Somani et al. 2018) and reuse as soil enricher/compost (Singh and Chandel 2020a). Additional value can be made by recovering metals and rare earth elements from fine fraction (Bhatnagar et al. 2017). The stone/C&D fraction can be reused as a construction material. In the case of Pirana dumpsite biomining project Ahmedabad, the C&D fraction obtained is sorted, crushed and cleaned in a separate C&D waste plant and later utilized for producing paver blocks, precast walls, manhole covers, etc. (CPHEEO 2020).

## 3.3 Benefits and Challenges in Landfill Mining

The primary benefit of LFM is on environmental and social level. LFM can provide extended landfill capacity by void space recovery (VSR), which avoids the time and cost for locating, designing and constructing a new landfill. An additional advantage is provided by reducing carbon footprint due to avoided landfill emission and prevention of further environmental contamination. The cleared landfill can also be used for the development of recreational areas with increased land value (Burlakovs et al. 2017). The other benefits of LFM include revenue generation by valorizing excavated waste, job creation, and avoided costs of closure, after closure care and landfill monitoring.

Though there is significant potential for recovering resources by adopting LFM, it involves many challenges from planning, execution, and project completion. The primary challenge arises in recovered waste valorization. As most of the landfilled waste in India is fine fraction, its valorization is very essential for the success of LFM. Most of the projects carried out for LFM are dumping this material in low-lying areas or back to the landfill. This approach can lead to contamination of unpolluted low-lying areas as fine fraction is polluted with heavy metals and organic contaminant. Efficient reclamation routes need to be studied for the valorizing fine fraction. The waste-to-energy approach seems to be promising for combustible fraction, but the variation in feedstock quality due to impurities may pose a challenge to sell the combustible fraction. This leads to a further issue of locating and generating a suitable market for valorizing the recovered waste fraction. The industries are reluctant to accept the recovered materials for usage, e.g., RDF recovered by the combustible fractions is sometimes not accepted by the cement manufacturing industries as fuel for co-firing due to quality concerns. If the proper market is made available for the recovered materials, the LFM will be profitable to a certain extent. A significant amount of rejects are generated, creating another problem for its handling and disposal.

There may be opposition from the local public due to LFM activities. Odour and dust are released while performing LFM operations imposing health risks to workers and nearby residents; hence it needs to be controlled by water sprinklers for dust settlement and waste stabilization by spraying bio culture solution regularly. Space constraints on the existing landfills for the setting up of the plant may pose another challenge. The majority of LFM projects have more costs involved compared to the benefits. LFM requires a vast amount of equipment that needs to be customized according to the landfill. However, such customizable machinery is not readily available in India. Proper safety precautions should be taken while excavating the legacy waste as it releases gases like methane, sulfur dioxide and other gases which can cause explosion and fire. Excavated waste leads to leachate generation, which needs to be collected and appropriately treated. Also, excavating an area can weaken the adjacent cell structure, which can collapse into the excavated area. Hence proper planning needs to be carried out for developing an excavation plan. LFM operations might lead to the spread of unwanted materials like heavy metals if the technology fails to remove them (Krook et al. 2007). Furthermore, the climatic conditions can cause a delay in the project's completion.

## 3.4 Future Directions for Landfill Mining

Based on the current literature review, it could be observed that the research on the economics and environmental aspect of LFM is lacking in the Indian context. The impact of LFM activities on the environment needs to be comprehended before applying any large-scale project, as it can cause irreplaceable damage. Further, valorization and proper disposal of the mined waste needs to be carried out to remediate dumpsites successfully. Research in this area needs to be picked up before making any long-term decision on LFM in India. Furthermore, the feasibility of a LFM project depends on the revenue generated from recovered waste. Hence proper market needs to be identified by developing partnerships/collaborations with different stakeholders and companies such as material recyclers, metal extractors, energy producers to take recovered secondary resources.

More research in this field is required to demonstrate the capacity, efficiency and suitability of these treatment technologies. In a country like India, where there are thousands of landfills with varying sizes, ages and composition; effective frameworks need to be developed for evaluating the effect of critical factors on the economic and environmental performance of LFM.

## 3.5 Conclusions

Landfill mining seems to be a promising concept for managing old landfills and dumpsites in India. However, the feasibility of landfill mining depends on the dumped waste and will change from case to case. Further, the literature available for landfill mining is limited and mainly deals with waste characterization and valorization studies. Most of the valorization studies suggest using the fine fraction as filler material. However, the long-term impact of such applications needs to be understood as they are contaminated with heavy metals and other pollutants. New avenues for fine fraction valorization need to be explored to reduce the risk of contaminant spread or pretreatment methods need to be studied for treating fine fraction before application. The use of combustible fraction as refuse-derived fuel seems to be the most promising avenue. However, the recovered combustible fraction needs to meet government standards. There is a push from the government to remediate the old landfills and dumpsites. Combining the remediation with reclamation, an integrated approach can be developed to handle the legacy waste.

## References

Bhatnagar A, Kaczala F, Burlakovs J, Kriipsalu M, Hogland M, Hogland W (2017) Hunting for valuables from landfills and assessing their market opportunities: a case study with Kudjape landfill in Estonia. Waste Manag Res 35:627–635. https://doi.org/10.1177/0734242X17697816

Burlakovs J, Kriipsalu M, Klavins M, Bhatnagar A, Vincevica-Gaile Z, Stenis J, Jani Y, Mykhaylenko V, Denafas G, Turkadze T, Hogland M (2017) Paradigms on landfill mining: from dump site scavenging to ecosystem services revitalization. Resour Conserv Recycl 123:73–84. https://doi.org/10.1016/j.resconrec.2016.07.007

Cheela VRS, John M, Dubey B (2021) Quantitative determination of energy potential of refuse derived fuel from the waste recovered from Indian landfill. Sustain Environ Res 31:24. https://doi.org/10.1186/s42834-021-00097-5

Choudhury AR, Rao BK, Kumar DL, Singh N, Kumar MVP, Leelavathi A, Boyina LP, Begum SUA (2021) Techno-commercial assessment of concurrent municipal brown reclamation procedures: a pivotal case study of Jawahar Nagar Dump Site Field. J Toxicol Environ Sci 1(1):23–33

CPHEEO (2020) Advisory on landfill reclamation. Ministry of Housing and Urban Affairs, Government of India, New Delhi

CSO (2021) EnviStats-India 2021. Central Statistics Office, Ministry of Statistics & Programme Implementation, Government of India, New Delhi

Datta M, Gupta G, Somani M (2021a) Geoenvironmental considerations for bulk reuse of MSW residues from old dumps and WTE plants in geotechnical applications. Indian Geotech J 51:63–83. https://doi.org/10.1007/s40098-020-00491-7

Datta M, Somani M, Ramana GV, Sreekrishnan TR (2021b) Feasibility of re-using soil-like material obtained from mining of old MSW dumps as an earth-fill and as compost. Process Saf Environ Prot 147:477–487. https://doi.org/10.1016/j.psep.2020.09.051

Dubey A, Chakrabarti M, Pandit D (2016) Landfill mining as a remediation technique for open dumpsites in India. Procedia Environ Sci 35:319–327. https://doi.org/10.1016/j.proenv.2016.07.012

Esakku S, Selvam A, Joseph K, Palanivelu, K (2005) Assessment of heavy metal species in decomposed municipal solid waste. Chemical Speciation & Bioavailability 17(3):95–102. https://doi.org/10.3184/095422905782774883

Gaikwad RS, Patil DS (2020) The economics of landfill mining. Int J Res Eng Technol Sci XIII:1–7

Guleria A, Chakma S (2021) Probabilistic human health risk assessment of groundwater contamination due to metal leaching: a case study of Indian dumping sites. Hum Ecol Risk Assess 27:101–133. https://doi.org/10.1080/10807039.2019.1695193

Havanagi VG, Sinha A, Parvathi GS, Satish C (2017) Municipal solid waste in road embankment construction – a case study article. J Indian Roads Congr 2:79–90

Joseph K, Nagendran R (2007) Management of Open Dumps in Asia Challenges and Opportunities. In: Proceedings from Kalmar ECO-TECH'07: Technologies for Waste and Wastewater Treatment, Energy from Waste, Remediation of Contaminated Sited, Emissions Related to Climate 305–314. https://doi.org/10.15626/Eco-Tech.2007.030

Joseph K, Esakku S, Nagendran R (2007) Mining of compost from dumpsites and bioreactor landfills. Int. J Environ Technol Manag 7:317. https://doi.org/10.1504/IJETM.2007.015149

Kanmani S, Gandhimathi R (2013) Assessment of heavy metal contamination in soil due to leachate migration from an open dumping site. Appl Water Sci 3:193–205. https://doi.org/10.1007/s13201-012-0072-z

Karthikeyan OP, Swati M, Nagendran R, Joseph K (2007) Performance of bioreactor landfill with waste mined from a dumpsite. Environ Monit Assess 135:141–151. https://doi.org/10.1007/s10661-007-9709-z

Krook J, Mårtensson A, Eklund M (2007) Evaluating waste management strategies—A case of metal contaminated waste wood. Resour Conserv Recycl 52:103–118. https://doi.org/10.1016/J.RESCONREC.2007.03.002

Kurian J, Esakku S, Nagendran R, Chettiyappan V (2005) A decision making tool for dumpsite for dumpsite rehabilitation in developing countries. In: Tenth international waste management and landfill symposium. CISA, Environmental Sanitary Engineering Centre, Italy

Kurian J, Esakku S, Palanivelu K, Selvam A (2003) Studies on landfill mining at solid waste dumpsites in India. In: Proceedings Sardinia Vol. 3, pp 248–255

Kurian J, Esakku S, Nagendran R (2007) Mining of compost from dumpsites and bioreactor landfills. Int J Environ Technol Manag 7:317. https://doi.org/10.1504/IJETM.2007.015149

Mandpe A, Pratap V, Lakshmikanthan P (2021) Chapter 9 – Landfill mining. In: Kumar S, Kumar R, Pandey A (eds) Current developments in biotechnology and bioengineering. Elsevier, pp 197–229. https://doi.org/10.1016/B978-0-12-821009-3.00010-5

Mandpe A, Lakshmikanthan P, Kumar S, Hettiarachchi H (2019) Mining for recovery as an option for dumpsite rehabilitation: a case study from Nagpur, India. J Environ Eng Sci 15:1–9. https://doi.org/10.1680/jenes.19.00021

Manfred JS, Bhattacharyya (1997) Reuse of decomposed waste. In: Coad A (ed) Lessons from India in solid waste management. Water, Engineering and Development Centre, England, pp 16–17

Mohan S, Joseph CP (2020) Biomining: an innovative and practical solution for reclamation of open dumpsite BT. In: Kalamdhad AS (ed) Recent developments in waste management. Springer, Singapore, pp 167–178

Mor S, Ravindra K, Dahiya RP, Chandra A (2006) Leachate characterization and assessment of groundwater pollution near municipal solid waste landfill site. Environ Monit Assess 118:435–456. https://doi.org/10.1007/s10661-006-1505-7

Naveen BP, Apoorva G (2018) Appropriate model of determination of hazard rating system for landfill rehabilitation. J Environ Eng Sci 15:1–50. https://doi.org/10.1680/jenes.18.00041

Peter AE, Shiva Nagendra SM, Nambi IM (2019) Environmental burden by an open dumpsite in urban India. Waste Manag 85:151–163. https://doi.org/10.1016/j.wasman.2018.12.022

Pujari PR, Deshpande V (2005) Source apportionment of groundwater pollution around landfill site in Nagpur, India. Environ Monit Assess 111:43–54. https://doi.org/10.1007/s10661-005-8037-4

Ranjan MR, Ramanathan AL, Tripathi A, Jha PK (2014) Landfill mining: a case study from Ghazipur landfill area of Delhi. Int J Environ Sci 4:919–925. https://doi.org/10.6088/ijes.2014040404530

Rawat P, Mohanty S (2019) Utilization of municipal solid waste as backfill material. In: Proceedings of the Indian geotechnical conference 2019: IGC-2019 volume II. Springer, p 49

Rawat P, Mohanty S (2021) Experimental investigation on MSW fine mixed with fibers: fiber reinforced waste. J Hazard Toxic Radioact Waste 25:04021009. https://doi.org/10.1061/(asce)hz.2153-5515.0000609

Saluja S, Gaur A, Ahmad K (2020) Physico-chemical characterization of stabilized MSW of an Okhla landfill. Mater Today Proc 44:4287–4292. https://doi.org/10.1016/j.matpr.2020.10.546

Shubham K, Prashad D, Metya S (2022) Seismic response of soil-like material in MSW landfill using equivalent linear approach. In: Choudhary AK, Mondal S, Metya S, Babu GLS (eds). Springer, Singapore, pp 293–301

Singh A, Chandel MK (2020a) Effect of ageing on waste characteristics excavated from an Indian dumpsite and its potential valorisation. Process Saf Environ Prot 134:24–35. https://doi.org/10.1016/j.psep.2019.11.025

Singh A, Chandel MK (2020b) Landfill mining potential of legacy waste and its associated challenges. In: Conference on 'Geoenvironment & Sustainability' at geoenvironment-2020. New Delhi

Singh A, Chandel MK (2021) Physicochemical and FTIR spectroscopic analysis of fine fraction from a municipal solid waste dumpsite for potential reclamation of materials. Waste Manag Res J Sustain Circ Econ 39:374–385. https://doi.org/10.1177/0734242X20962844

Singh R, Rao B, Maillacheruvu K, Asolekar SR (2019) Reusing fractions of legacy municipal solid wastes in India: a circular economy approach. Productivity 60(3):227–238. https://doi.org/10.32381/PROD.2019.60.03.1

Somani M, Datta M, Ramana GV, Sreekrishnan TR (2018) Investigations on fine fraction of aged municipal solid waste recovered through landfill mining: case study of three dumpsites from India. Waste Manag Res J Sustain Circ Econ 36:744–755. https://doi.org/10.1177/0734242X18782393

Somani M, Datta M, Ramana GV, Sreekrishnan TR (2019) Leachate characteristics of aged soil-like material from MSW dumps: sustainability of landfill mining. J Hazard Toxic Radioact Waste 23:04019014. https://doi.org/10.1061/(asce)hz.2153-5515.0000452

Somani M, Datta M, Ramana GV, Sreekrishnan TR (2020) Contaminants in soil-like material recovered by landfill mining from five old dumps in India. Process Saf Environ Prot 137:82–92. https://doi.org/10.1016/j.psep.2020.02.010

Sughosh P, Anusree N, Prathima B, Sivakumar Babu GL (2021) Sustainable remediation of a dumpsite In: Patel S, Solanki CH, Reddy KR, Shukla SK. (eds). Springer, Singapore, pp 65–76. https://doi.org/10.1007/978-981-33-6590-2_6

Zhou C, Fang W, Xu W, Cao A, Wang R (2014) Characteristics and the recovery potential of plastic wastes obtained from landfill mining. J Clean Prod 80:80–86. https://doi.org/10.1016/j.jclepro.2014.05.083

# Chapter 4
# Environmental Issues Due to Open Dumping and Landfilling

**Mohammad I. Al-Wabel, Munir Ahmad, Hina Rasheed, Muhammad Imran Rafique, Jahangir Ahmad, and Adel R. A. Usman**

**Abstract** Open dumping and landfilling are the most widely used practices for municipal solid waste (MSW) management due to low costs and less treatment efforts. MSW comprises of metals/metalloids, organic compounds, micropollutants, inorganic macro-components, xenobiotic compounds, and dioxins. The vertical and lateral transmission of toxin-laden leachates, emission of greenhouse gases (GHG), and generation of odor and dust from open dumping and landfilling result in the contamination of water and soil resources globally. To be in line with the circular economy, the impacts of open dumping and landfilling of MSW on the environment need to be explored in detail. Therefore, the composition of MSW and its leachates, as well as the hazardous impacts of MSW landfills and open dumps on water, soil, air, and human beings are discussed in this chapter. The concentration of potentially hazardous pollutants in soil and water resources after releasing from MSW landfills and open dumps is presented. Further, the extent of GHG emissions from various waste types is presented. Mining landfills could be used to reduce the hazardous environmental impacts of open dumping and landfilling of MSW. However, the majority of landfill mining practices require higher costs and energy which are emitting ample amounts of $CO_2$ into the environment. The development of modern closed-loop systems such as enhanced landfill mining could be used as an option to transform waste into energy and useful materials for a sustainable environment and circular economy.

M. I. Al-Wabel (✉) · M. Ahmad · M. I. Rafique · J. Ahmad
Soil Sciences Department, College of Food & Agricultural Sciences, King Saud University, Riyadh, Saudi Arabia
e-mail: malwabel@ksu.edu.sa

H. Rasheed
Department of Soil Science, Faculty of Agriculture and Environment, The Islamia University of Bahawalpur 63100, Bahawalpur, Pakistan

A. R. A. Usman
Department of Soils and Water, Faculty of Agriculture, Assiut University, Assiut, Egypt



P. Pathak, S. G. Palani (eds.), *Circular Economy in Municipal Solid Waste Landfilling: Biomining & Leachate Treatment*, Radionuclides and Heavy Metals in the Environment, https://doi.org/10.1007/978-3-031-07785-2_4

**Keywords** Waste management · Landfill mining · Environmental pollution · greenhouse gases · human health

## Abbreviations

| | |
|---|---|
| BOD | Biological oxygen demand |
| $CH_4$ | Methane |
| CO | Carbon monoxide |
| $CO_2$ | Carbon dioxide |
| GHG | Greenhouse gases |
| GW | Ground water |
| $H_2S$ | Hydrogen sulfide |
| LFM | Landfill mining |
| MSW | Municipal solid waste |
| $N_2O$ | Nitrous oxide |
| $NH_3$ | Ammonia |
| PAHs | Polyaromatic hydrocarbons |
| $SO_2$ | Sulfur dioxide |
| SW | Surface water |
| TDS | Total dissolved salts |

## 4.1 Introduction

Rapid increase of industrial and trade production has generated larger volumes of industrial and municipal waste in the last few decades (Alobaid et al. 2018; Vaverková 2019). International solid waste association and United environment program have reported that global urban waste production is approximately 10 billion tons annually, and waste generation per person has been increased significantly over the last five decades (Wilson and Velis 2015). The annual volume of waste produced in developed countries has reached to 300–800 kg per person in the second half of the 1990s, whilst this volume was recorded at about 200 kg per capita in the developing countries (Ercolano et al. 2018). The world is focusing on reducing the production of waste to ensure a sustainable environment. According to the World Bank's prediction, the municipal solid waste (MSW) generation would reach $3.4 \times 10^9$ tons in the year 2050 (Kaza et al. 2018). The situation is alarming in the urban as well as in the rural areas equally due to the lack of scientifically available waste management techniques, specifically in developing and underdeveloped countries. As an example, 0.3 to 0.4 million tons of solid waste is being produced in the rural areas of India (NIRD&PR 2016). Solid waste generation contains a trivial global footprint and huge management costs. On a global level, solid waste management costs about $205.4 per ton, which is estimated to attain a five-fold rise by 2025–2030 (Hoornweg and Bhada-Tata 2012). Thus, appropriate management of waste is of significant importance for a sustainable ecosystem and environment.

Open dump landfilling is the main practice for the disposal of MSW around the world. For instance, 52.6% of MSW is discarded through landfilling in the USA, 94% in Malaysia, 79% in China, 59.1% in Brazil, and 85% in the Kingdom of Saudi Arabia (Costa et al. 2019; Havukainen et al. 2017; Ouda et al. 2016; Sun et al. 2019; Tan et al. 2014; Wichai-utcha and Chavalparit 2019). However, very few nations are practicing the proper management systems for the disposal of waste due to the variability in the environmental, social, and economic features along with technical aspects. Therefore, the selection of the waste disposal methods should consider the technical aspects and costs of implementation as well as the environmental factors that could be affected by the treatment and disposal processes. Moreover, the opinion of the local communities should also be considered while designing a waste disposal method.

## 4.2 Global Waste Generation

Rapidly increasing population, urbanization, and improved living standards of the communities are the main factors contributing to the swift generation of MSW (Ouda 2013; Sharholy and Ahmad 2008). An increase in population, higher economic development, and urbanization ultimately lead to the production as well as change in the composition and treatment technologies of MSW (World Energy Council 2013). At present, 1.3 billion tons global MSW is being generated worldwide every year, which is expected to increase to 2.2 billion tons annually by the year 2025 (Chan 2016; Hoornweg and Bhada-Tata 2012). This is because of the increase in population as well as per capita increase in waste generation, which has been increased from 1.2 to 1.42 kg per person per day (Hoornweg and Bhada-Tata 2012). The production, quality, and quantity of the waste generation vary with and between geographic locations, due to the economic status of the country, living style of the community, age factor, and literacy rate (Kolekar et al. 2017). It has been reported that developed countries contribute 40% of the global waste production, whilst, the developing and underdeveloped countries account for 37% and 23%, in global waste generation, respectively. Per capita, high GDP income in the developed countries is directly related to the higher MSW production. For instance, MSW generation has reached 0.64 MT per day in the USA, followed by Germany, Mexico, and Japan with MSW production rates of 0.14, 0.13, and 0.10 MT per day, respectively. The situation is getting worse in developing countries. China is among the largest MSW generating countries with an annual MSW production of 2 billion tons (Zhu et al. 2021). Similarly, India is producing about 0.5 MT of waste on daily basis. Alike, about 15.3 million tons of MSW is being generated in the Kingdom of Saudi Arabia annually (Mallick 2021). Likewise, Qatar is producing more than 2.5 MT MSW annually, which is among the highest per capita waste generation per day in the world (Rehrah et al. 2016). Bangladesh, Nigeria, Ethiopia, and Sudan are among those underdeveloped countries which are contributing a large part of waste in the global MSW generation. MSW production per capita ranges from 0.09 to 5.70 kg per day for Sub-Saharan Africa and the Middle East/North Africa, respectively. In Sub-Saharan Africa, the MSW generation per capita is usually low but still reaches an average of 1.54 kg per capita per day (Hoornweg and Bhada-Tata 2012). The differences in waste generation and management among the countries are dependent upon the economic status of countries; however, waste management in developing countries is an issue due to the lack of facilities and infrastructure.

## 4.3 Waste Management

In recent times, waste management has been the main concern of environmentalists globally. Anthropogenic activities and changing lifestyles are increasing the amount of solid waste production worldwide. By keeping in view the current situation of MSW generation, many countries have planned to manage the waste by increasing resource efficiency and minimizing the worst effect on the environment and citizens (Romano and Molinos-Senante 2020). Waste production monitoring is the first step that is used in the waste management strategies, irrespective of the region and country. For instance, radio frequency identification, geographic information systems, ultrasonic sensors, and international systems for mobile and general radio packet service have recently been developed for the collection of waste from trucks and bins (Hassan et al. 2016). Nevertheless, these techniques cannot be used in poor countries because of their higher cost. Furthermore, waste prevention, waste minimization, recycling and reuse, biological treatment, and incineration are the main strategies that are used commonly for waste management, whilst landfilling and open dumping are the final methods of waste management. In Europe, waste prevention has been added to the waste management policies in 2008 (waste framework directive 2008/98/CE) and it was stated that waste prevention is the first step towards managing the waste (Gentil et al. 2011). Likewise, in the UK, the government has funded a vast research program for the prevention of waste (Cox et al. 2010). Alike, the US environment protection agency has commenced a program in which waste management organizations and local authorities are supported for waste prevention; however, it doesn't include the environmental impact categories (USEPA 2006). Taiwan, South Australia, New Zealand, and some other countries have aimed towards "zero waste" as a form of strategic waste prevention (Gentil et al. 2011).

Another aspect of minimizing the waste is through recovering the useful materials from the waste. Non-reusable waste can be used for the production of energy in the form of heat and electricity in waste to energy transformation plants, whereas, metals and organic compounds can also be extracted from the waste. In this way, the adverse environmental impacts of waste can be reduced on the one hand, while the financial benefits can be availed by recovering the useful materials on the other hand. Furthermore, awareness and education about increasing waste is also an important aspect from a global perspective of resource management. Inappropriate handling of the waste could result in adverse impacts on human beings and contamination of the soil, water, and air. Waste is mostly dumped or burned in the open air, which causes the pollution of the atmosphere. Therefore, public education is needed to avoid environmental pollution and human health risks.

## 4.4 Waste Management via Landfilling and Open Dumping

Landfilling is the compression, embankment, and disposal of waste at appropriate sites, whereas, open dumping is dumping the waste in an open environment. Landfilling and open dumping for the moment are easy and less expensive as compared to the other waste disposal methods. Landfilling and open dumping are the most famous practices for waste disposal in the majority of the world due to their cost-effectiveness and less technical requirements (Gonzalez-Valencia et al. 2016; Jovanov et al. 2018). Despite widely accepted methods for the disposal of MSW, some environmental concerns are associated with landfilling and open dumping. Therefore, to protect the environment, "sanitary landfills" were introduced in 1959 by the American Society of Civil Engineers, which involves the disposal of MSW in defined layers where each layer of the waste is compacted and covered with the soil before making the other layers of waste (Vallero and Blight 2019). The basic difference between an open dumping and a sanitary landfilling is that in an open dump, the waste is directly dumped in soil and there is no effort made to separate the waste from reaching the underground water levels, while, in the sanitary landfilling system, there are defined layers of waste with proper collection and removal of leachates (Vallero and Blight 2019).

Leachate and biogas are the major emissions that are influenced by the biological processes in the landfills. The disposal of MSW with any pre-treatment results in emissions of various gases and volatile compounds during the dumping process and even after the landfills have been closed (Laner et al. 2012; Vaverková 2019; Ziyang et al. 2015). These emissions are produced on an average circa? 150 $m^3$ (in the range of 70–300) of biogas per 103 kg of waste and approximately 5 $m^3$ $ha^{-1}$ $d^{-1}$ of severely contaminated leachates, by keeping in view the composition of waste and climatic conditions (Skibniewska 2011). These values are based on the highly compacted landfills in central Europe with annual precipitation of 550–750 mm (Vaverková 2019). Additionally, there are some major environmental concerns from the MSW related to the transport of contaminated leachate and landfill gas, which is why the environmental impacts of the existing landfills cannot be flouted.

## 4.5 Composition of MSW and Its Leachate

The MSW consists of heterogeneous by-products that are generally produced as the result of different anthropogenic activities. This waste is usually composed of two components i.e., biodegradable and non-biodegradable. The dry matter including glass, plastic, metal, paper, leather, textile, ash, paper packing material, and household wastes are non-biodegradable waste, whereas, food material, vegetables, and fruits are biodegradable waste. Generally, the chemical composition of MSW consists of organic matter, chlorinated aliphatic compounds, sulfate, pesticides, volatile fatty acids, phenols, plasticizers, aromatic hydrocarbons, and heavy metals (Kjeldsen et al. 2002). The physical and chemical composition of the MSW is shown in Tables 4.1 and 4.2, respectively. The composition of the waste varies according to the topography, season, food habits, and commercial activities of the area. Thus, due to the heterogeneous nature of the waste, it becomes very problematic to handle and dispose of the waste.

Landfill leachate originates from MSW treatment plants, anaerobic digesters, landfills, and composting piles (Tałałaj et al. 2019). Similar to the MSW, the composition and properties of leachate depend upon the waste contents and the climatic situation of that place as well (Aziz et al. 2010). Generally, leachate contains higher amounts of organic matter (dissolved solids and carboxylic acids), inorganic salts, heavy metals, toxic chemicals, ammonia, xenobiotic organic compounds, and minerals, and therefore, it poses serious environmental risks (Wiszniowski et al. 2006). Further, various organic compounds such as chlorinated aliphatic compounds, aromatic compounds, phenols, and pesticides are also present in leachate (Kulikowska and Klimiuk 2008). It has been reported that the non-biodegradable compounds (humic substances) are dominant over the organic fraction in the landfill leachate (Kumar et al. 2011). These pollutants originate from pharmaceuticals, household chemicals, and personal care products and are found abundantly in the leachate. The composition and quantity of leachate are influenced by several other factors as well, such as the quality of solid waste, degree of waste compaction in the landfills, temperature, humidity level, pH level, and biological processes (Hunce et al. 2012). Landfill leachate can permeate into the groundwater and may cause acute and chronic toxicity which leads to biomagnification (Mishra et al. 2019).

## 4.6 Environmental Impacts of MSW Landfills and Open Dumps

Urbanization, rapid population growth, and changing life patterns have resulted in a substantial increase in the production and complexity of solid waste including municipal, domestic and industrial waste worldwide. The lack of an efficient and stable system is a key factor for encouraging communities to illegal dumping of solid waste in open field areas, roadside, and freshwater resources. It has been

**Table 4.1** Physical and chemical compositions of the waste generated in different countries around the world

| Country | Year | Location | Organic material (%) | Paper and paperboard (%) | Plastics (%) | Glass & ceramics (%) | Metals (%) | Textiles, ash & others (%) | References |
|---|---|---|---|---|---|---|---|---|---|
| Mauritius | 2009 | Nationwide | 62.2 | 11.3 | 11 | 5.6 | 2.9 | 6.8 | Shekdar (2009) |
| United Arab Emirates | 1995 | Abu Dhabi | 49 | 6 | 12 | 9 | 8 | 16 | Qdais et al. (1997) |
| India | 2019 | Kolkata | 29 | 4 | 25 | 1 | 35 | 24 | Ali and Ahmad (2019) |
| Malaysia | 2005 | Nationwide | 44.8 | 15 | 15 | 3 | 2 | 10.4 | Fazeli et al. (2016) |
| Tanzania | 2017 | Dar es Salaam | 57.21 | 6.12 | 13.08 | 2.32 | 1.02 | 20.25 | Kazuva and Zhang (2019) |
| Nigeria | 2006 | Uyocity | 73.7 | 9.6 | 22 | 10 | 4.3 | 0.6 | John et al. (2006) |
| Brazil | 2006 | Sao Carlos, SP | 77 | 7.4 | 10.5 | 1.7 | 1.3 | 20 | Fresca (2007) |
| Nepal | 2007 | Khatmandu | 67.8 | 7 | 0.3 | 1.3 | 4.9 | 19.2 | Dangi et al. (2011) |
| Canada | 2015 | Saskatchewan | 65 | 26 | 8 | – | 35 | 2 | Wang et al. (2016) |
| Bangladesh | 2017 | Khulna city | 78.9 | 9.5 | 3.1 | 0.5 | 1.1 | 1.14 | Ashik et al. (2017) |
| United Kingdom | 2011 | – | 64 | 29 | 7 | 10 | 8 | 3 | Blight (2011) |
| Turkey | 2016 | Songnlik Yeni | 59.13 | 8.54 | 11.45 | 6.90 | 2.11 | – | Ozcan et al. (2016) |
| Ghana | 2015 | Nationwide | 61 | 5 | 14 | 3 | 3 | 14 | Miezah et al. (2015) |

**Table 4.2** Chemical compositions of the waste generated in different countries around the world

| Country | Moisture content (%) | Volatile matter (%) | Organic carbon (%) | Organic nitrogen (%) | C (%) | H (%) | O (%) | S (%) | Ash (%) | References |
|---|---|---|---|---|---|---|---|---|---|---|
| India | 30.76 | – | 30.46 | 1.04 | – | – | – | – | 46.7 | Katiyar et al. (2013) |
| Ireland | 16.36 | – | 38.7 | 1.2 | – | 7.2 | 48.1 | 0.2 | 4.6 | Yusuff et al. (2014) |
| South Africa | 63.47 | – | 45.03 | 1.98 | – | – | 41.2 | 0 | 5.56 | Ayeleru et al. (2016) |
| Korea | 2.30 | 87 | 40.3 | 0.2 | 40.3 | 5.6 | 53 | – | – | Kim et al. (2012) |
| Pakistan | 3.3 | 79.9 | – | – | 63.6 | 8.19 | 27 | 0.1 | 9.1 | Azam et al. (2020) |
| Mexico | 65 | – | 50.4–75.5 | – | 61.2 | 15.4 | 7.45 | 0.02 | 13 | Moreno et al. (2013) |
| Australia | 28.7–46 | 65.5–71.1 | – | – | 43.9–47.5 | 5.2–5.4 | 35.31–38 | 0.12–0.14 | 7.9–14.2 | Hla and Roberts (2015) |
| Kazakhstan | 2.6 | 72 | – | – | 47.9 | 5.9 | 43.6 | 0.2 | 11.2 | Abylkhani et al. (2019) |
| Zimbabwe | 30.9–36 | 79.2–84 | – | – | 44.6–47.9 | 4.8–5.1 | 30.8–37.3 | 0.8–0.8 | 8.3–12.5 | Makarichi et al. (2019) |
| Bangladesh | 0–74.86 | – | 24.3–55 | – | – | 0.1–22.8 | 0.4–44 | 0.2–90.5 | 0.46–98.6 | Islam (2016) |
| Malaysia | 14.6 | – | – | – | – | – | – | – | 7.05 | Khamis et al. (2019) |

estimated that around 70% of world waste ends up in dumps and landfills, which is a huge environmental concern (Kaza et al. 2018). Statistical data shows that about 7–9 billion tons of solid waste are being produced annually around the world (Wilson and Velis 2015). With economic development, an increase in solid waste production was also observed such as 635 Mt solid waste was produced in 1965 which elevated to 1999 Mt in 2005 and is likely to reach 3539 Mt by 2050 (Chen et al. 2020).

The MSW, which is specifically categorized as waste stemming from households and industries (partially) accounted for about 2 billion tons of waste produced in 2016 (Wilson and Velis 2015; Chen et al. 2020). Uncontrolled and open dumping and poorly engineered sanitary landfills of MSW lead to numerous health-associated problems in human beings, deterioration of freshwater resources, GHG emission, heavy metal contamination, and subsequently environmental pollution. Children, senior citizens, and those with chronic health conditions are more susceptible to health issues derived by illegal waste dumping and sanitary landfill sites since some malaria, diarrhea and severe respiratory issues have been reported by long term exposure to MSW (Mataloni et al. 2016). Moreover, odor from landfills builds social disturbance and unrest among communities and residents surrounding the dumping sites. Untreated leachate from MSW heaps at open dumping sites, subsequently polluting the soil and water bodies in the surroundings. Specifically, in urban areas, such waste heaps result in clogging of water pipes, enhanced insects breeding, and flooding during the rainy season. Additionally, mismanagement of MSW badly impacts social, environmental, and economic development, as well as confines ecological growth and sustainable development.

The MSW comprises various types of waste materials which impact the environment in variable ways depending on their composition, source, disposal, and dumping method. For instance, plastic waste, which is usually classified for its rigid and persistent nature, is ingested by living organisms and causes potential health threats via the food chain (Wagner and Lambert 2018). Likewise, the waste discharge from wastewater treatment plants causes 3–4% GHG emissions (Monni et al. 2006), and nitrogen (N) pollution poses long-term environmental impacts causing a nutrient imbalance in water bodies (Kjeldsen et al. 2002). Moreover, the open burning of MSW emits a significant amount of toxic gases and hazardous air pollutants which strongly impact human health (Wiedinmyer et al. 2014). Waste management systems include the application of several practices to minimize the negative impacts of dumped waste on human health and the environment. Some previously reported potentially possible solution practices include employing organic waste retrieving programs with compost and biogas production, implementation of technologies to use waste and biomass for energy production, and making briquettes from glass waste (Hettiarachchi et al. 2018; Ouda et al. 2016; Sawadogo et al. 2018). The impacts of MSW landfills and open dumps have been discussed in subsequent sections.

### 4.6.1 *Water Contamination*

Generally, MSW is dumped in lowland regions deprived of any certain precautions, which results in surface and groundwater contamination. Furthermore, rainfall severs the pollution scenario by percolating contaminated leachate to the soil strata and into the aquifer with it. Water contamination by MSW is divided into two categories as groundwater contamination and surface water contamination.

#### 4.6.1.1 Groundwater Contamination

Irrational dumping of MSW gives rise to groundwater (GW) contamination when coming in contact with precipitation and moisture. Usually, leachate from open dumped MSW sites is complex wastewater in nature and contains a higher concentration of toxic heavy metals, organic carbons, soluble organic matter, inorganic ions, ammonia ($NH_3$), N, and volatiles (Torretta et al. 2017; Mojiri et al. 2016). Moreover, GW contamination becomes inevitable in landfill areas owing to the underflow of waste and infiltration of leachate in soil, and consuming such polluted GW as potable water source proves dangerous for human health. Such kind of GW contamination by seepage of concentrated leachate from unlined landfills is generally higher in underdeveloped and developing countries (Singh and Mittal 2011). Many factors influence the characteristics and composition of leachate including type and composition of deposited waste, pretreatment of solid waste, climate, phase of the landfill, and other site-related conditions (Singh and Mittal 2011).

Risk assessment of MSW dumped landfill areas in GW resources can be interpreted by monitoring physiochemical characteristics and water quality with time. GW in landfill areas generally contains acidic pH of groundwater, an extremely higher concentration of total dissolved salts (TDS), and higher sodium ($Na^+$). Additionally, high values of heavy metals were found in leachate while in GW only iron ($Fe^{+2}$) was found to exceed WHO tolerance level in 75% of analyzed samples (Aderemi et al. 2011; WHO 1997, 2003). In a review study, Han et al. (2016) showed a large number of GW pollutants including metal ions, heavy metals, inorganic pollutants, xenobiotic organic compounds, inorganic salts, and bacteriological pollutants as described in Table 4.3. In reference to uncontaminated GW, Abiriga et al. (2020) found that the pH was in a range of weak acid to weak alkaline (4.9–8.8), with low biological oxygen demand (BOD), higher bicarbonate contents while ammonium ($NH_4^+$), sulfate ($SO_4^{-2}$), nitrate ($NO_3^-$), total N, manganese ($Mn^{+2}$), potassium ($K^+$), calcium ($Ca^{+2}$), magnesium ($Mg^{+2}$), chloride ($Cl^-$), $Fe^{+2}$ and $Na^+$ were found in moderately polluted range (Table 4.3). In a nutshell, to avoid negative impacts on the environment, landfill areas should be lined and managed systematically. Moreover, the prevalence of potentially toxic pollutants including inorganic acids, heavy metals, and metal ions should be monitor regularly.

**Table 4.3** Various contaminants and their concentrations in ground and surface water resources after releasing from open dumping and landfilling

| Categories | Pollutants | Concentration (mg $L^{-1}$) | | Standard Value ($mgL^{-1}$) | WHO Standard value (mg $L^{-1}$) | References |
|---|---|---|---|---|---|---|
| | | Ground water | Surface Water | Ground water | Surface Water | |
| Organic matters | COD | 59.780 | 18.411 | 20 | 5 | Han et al. (2016) |
| | BOD | – | 21.501 | – | <5 | |
| | Permanganate index | 2.073 | – | 3 | – | Mekonnen et al. (2020) |
| Inorganic salts | Sulfate | 140.997 | 44.541 | 250 | 200 | |
| | Fluoride | 0.669 | 0.8414 | 1 | 1.5 | |
| | Chloride | 116.609 | 6.251 | 250 | 250 | Kola-Olusanya (2012) |
| | Iodide | 0.104 | – | 0.2 | – | |
| | Nitrite | 0.077 | 1.611 | 0.02 | – | Han et al. (2016) |
| | Nitrate | 14.746 | 10.552 | 20 | 30 | |
| | Potassium | 31.014 | 10.953 | – | 12 | |
| | Ammonium | 2.694 | – | 0.2 | – | Mekonnen et al. (2020) |
| | Bicarbonate | 213.002 | – | – | – | |
| | Total phosphorus | 0.444 | 0.453 | 0.2 | – | |
| | Total Nitrogen | 14.2 | – | – | – | Abiriga et al. (2020) |
| | Total hardness | 449.428 | 178.611 | 450 | – | Han et al. (2016) |
| | Total dissolved solids | 1123.952 | 526.814 | 1000 | 500 | |
| Heavy metals | Iron | 0.715 | 0.065 | 0.3 | – | Mekonnen et al. (2020) |
| | Manganese | 0.462 | ND | 0.1 | – | Amano et al. (2021) |
| | Cadmium | $0.002 \times 10^{-3}$ | 0.059 | 0.1 | – | Han et al. (2016) |
| | Mercury | $3.154 \times 10^{-3}$ | ND | 0.001 | – | |

(continued)

**Table 4.3** (continued)

| Categories | Pollutants | Concentration (mg $L^{-1}$) | | Standard Value ($mgL^{-1}$) | WHO Standard value (mg $L^{-1}$) | References |
|---|---|---|---|---|---|---|
| | | Ground water | Surface Water | Ground water | Surface Water | |
| | Hexavalent chromium | 0.014 | ND | 0.05 | – | Mekonnen et al. (2020) |
| | Lead | 0.017 | ND | 0.05 | – | Kola-Olusanya (2012) |
| | Nickel | ND | – | 20 | – | |
| | Copper | ND | ND | – | – | AQSIQ (1993) |
| Bacteriological pollutants | Total coliform | 299.373 | – | 100 | – | |
| | Bacterial count | 3640.378 | – | 3 | – | EPA and AQSIQ (2002) |
| Xenobiotic organic compounds | Volatile phenol | 0.018 | – | 0.002 | – | WHO (2004) |
| | Trichlorobenzene | $10.819 \times 10^{-3}$ | – | 20 | – | |

#### 4.6.1.2 Surface Water Contamination

Water from the dumped and landfill areas usually mixes up with a variety of organic and inorganic colloids by percolation generally referred to as leachate. Lateral and downward movement of this leachate results in the contamination of the surface and groundwater resources in the vicinity of landfill areas. In addition to surface water (SW) contamination, unpleasant odor is also a serious concern for residents in the surroundings. The SW in the neighborhood of landfill areas was found with a slightly acidic pH value which might be due to organic matter decomposition in MSW and the addition of carbonic acids by carbon dioxide ($CO_2$) reaction with water released from waste decomposition (Bhalla et al. 2014; Amano et al. 2021). Leachate with acidic pH indicates that landfill area is fresh while leachate from aged landfills has a pH of about 7.5, suggesting the release of methane ($CH_4$) (Vathsalan et al. 2017). Higher contents of heavy metals were found in SW in surroundings of young landfills which might be due to the higher solubility of heavy metals at low pH (Amano et al. 2021). The other chemical parameters were found below the maximum permissible limits of WHO, while water hardness was recorded as 200 mg $L^{-1}$, which was exceeding the maximum permissible limits of WHO, indicating the formation and deposition of soap and scum, which subsequently at high temperature causes deposition of calcium carbonate scales (WHO 2011). Additionally, erosion and steep slopes together with soil poor texture and permeability around landfills further enhance SW contamination during precipitation and permeation. Erosion and steep slopes might also be the reason for severe contamination of SW away ($\geq$1.35 km) from landfill sites. The accumulation of clay, silt, algae and other microscopic microorganisms in SW by permeation of leachate from dumped waste heaps increases SW turbidity and make it unsuitable for drinking purposes (Sharma et al. 2019). Therefore, proper governance and regular monitoring of MSW dumping are required to mitigate landfill-derived water contamination and improve drinking water quality.

### *4.6.2 Soil Contamination*

Despite developing appropriate waste management techniques, open disposal of MSW in landfills is a common practice globally, which subsequently pollutes the terrestrial and aquatic environment and generates airborne pollutants. An open dumping site proves particularly hazardous for the environment and causes natural resource destruction, soil contamination, and ultimately land degradation (Shayler et al. 2009). Generally, a higher concentration of toxic heavy metals, inorganic solid compounds, and various salts are found in soil at MSW landfill areas. In comparison to virgin soil, the soils in the surroundings of open dumped areas are usually found with weak strength, poor texture, and less fertile. Cumulatively, a regular practice of dumping MSW in landfills results in insufficient land use, addition, and accumulation of lots of organic and inorganic pollutants, which deteriorate soil health, and ultimately decrease soil productivity.

#### 4.6.2.1 Heavy Metals and Organic Compounds Pollution

Generally, MSW contains a higher concentration of heavy metals and various organic compounds, which upon their release in soil are consumed by soil microorganisms, create competition for plant nutrient uptake and induce soil toxicity. Some byproducts are generated from MSW by anaerobic conditions such as $NH_3$, organic sulfur compounds, hydrogen sulfide, and volatile organic compounds, which consequently affects the health inhabitant adversely. Various studies have reported higher concentrations of heavy metals on surface soil, while with increasing the soil depth, heavy metals concentrations tend to decrease indicating the influence of dumping activities (Ferronato and Torretta 2019). Ferronato and Torretta (2019) reported that soil in the vicinity of landfills contains about 5–11 times more contents of Pb and Cd than the average concentration in normal soils. On contrarily, Kanmani and Gandhimathi (2013) reported a higher concentration of heavy metals (Pb, Mn, Cu, and Cd) in sub-soil as compared to surface soil, which showed the mobility of heavy metals along with leachate migration released from MSW. Disposal of used batteries and electronic devices as solid waste enormously increase heavy metals (Hg, Ag, Cu, Au, Pd, Li, and Ni) concentration in landfills soil (Zand and Abduli 2008; Olafisoye et al. 2013). Some previously reported concentrations of heavy metals in soils as rereleased from open dumping and landfills are shown in Table 4.4.

Additionally, the open burning of MSW heaps releases some organic volatiles, airborne pollutants, and harmful gases such as $N_2O$, $SO_2$, etc. (Shakya et al. 2008). In general, organic compounds released by solid waste are degraded by soil microorganisms and gases ($CH_4$) are released into the atmosphere. Latterly these gases can be collected separately and used as a power source, while on the other hand, organic waste compounds migrate to sub-soil along with leachate permeated from MSW heaps.

#### 4.6.2.2 Vegetation Decline

An anticipated decline in vegetation growth has been recorded in the surroundings of landfills, which was mainly due to the larger number of contaminants discharged from openly dumped solid waste. These contaminants commonly contain a large number of xenobiotic organic compounds, inorganic salts, and heavy metals which prove toxic for plant growth and often compete for nutrient uptake in plants. Hazardous gases release as a result of the burning of waste heaps declines oxygen percentage in air, which latterly subside plants growth and abate animal population as well (Iravanian and Ravari 2020). Additionally, landfills' operational phase movement of heavy vehicles creates soil compaction which poses a destructive impact on vegetation growth. A study conducted by Ali et al. (2014) about the

**Table 4.4** Heavy metals (mg kg$^{-1}$) concentrations in soil after releasing from open dumping and landfilling

| City | Country | Contaminants | Concentration (mg kg$^{-1}$) | Standard limit (mg kg$^{-1}$) | References |
|---|---|---|---|---|---|
| Karachi | Pakistan | Pb | 51.3 | 50 | Jilani and Rashid (2020) |
| | | Cr | 65.7 | 75 | |
| | | Cu | 45.1 | 100 | |
| | | Zn | 168.9 | 300 | |
| | | Ni | 40.4 | 100 | |
| | | Cd | 03.3 | 03 | |
| Islamabad | Pakistan | Pb | 67.06(ug g$^{-1}$) | 70 (ug g$^{-1}$) | Ali et al. (2014) |
| | | Cd | 11.92(ug g$^{-1}$) | 1.4 (ug g$^{-1}$) | |
| | | Cu | 09.46(ug g$^{-1}$) | 63 (ug g$^{-1}$) | |
| | | Ni | 27.15(ug g$^{-1}$) | 50 (ug g$^{-1}$) | |
| | | Cr | 00.85(ug g$^{-1}$) | 64 (ug g$^{-1}$) | |
| | | Zn | 196.99(ug g$^{-1}$) | 200 (ug g$^{-1}$) | |
| Chennai | India | Zn | 0.48 | 50 | Parameswari et al. (2015) |
| | | Cu | 0.59 | 100 | |
| Tiruchirappalli | India | Mn | 171 | 500 | Kanmani and Gandhimathi (2013) |
| | | Pb | 291.3 | 50 | |
| | | Cd | 47.7 | 04 | |
| Havana | Cuba | Co | 08.4 | 20 | Díaz et al. (2013) |
| | | Ni | 50 | 30 | |
| | | Cu | 252 | 100 | |
| | | Zn | 489 | 50 | |
| | | Pb | 276 | 50 | |
| Uyo | Nigeria | Pb | 11.8 | 50 | Ihedioha et al. (2017) |
| | | Zn | 146 | 50 | |
| | | Ni | 12.6 | 30 | |
| | | Cr | 04.1 | 01 | |
| | | Cd | 12.2 | 04 | |
| | | Mn | 94 | 500 | |
| Omuooke | Nigeria | Cu | 12 | 38.9 | Afolagboye et al. (2020) and Kabata-Pendias (2011) |
| | | Ni | 0.03 | 29 | |
| | | Cr | 8450 | 59.5 | |
| | | Zn | 120.5 | 70 | |
| | | Pb | 65.5 | 27 | |
| | | Mn | 1221 | 418 | |
| | | Fe | 48,700 | 58,100 | |
| Pilsen | Czech Republic | Hg | 00.03 | – | Adamcová et al. (2016) |
| | | Cd | 00.07 | – | |
| | | Pb | 06.56 | – | |
| | | Cu | 34.07 | – | |
| | | Zn | 25.67 | – | |
| | | Co | 04.89 | – | |
| | | Ni | 39.13 | – | |
| | | Cr | 69.99 | – | |
| | | Mn | 689.55 | – | |

hazardous impact of open dumped solid waste on vegetation growth in Islamabad (Pakistan) noted a decline in vegetation diversity and several species from 44 to 32 species due to the destructive impact of contaminants released from dumpsites. The dumped waste pollutants function as the external force which deteriorates the physicochemical properties of soil and makes it less supportive for vegetation growth. On the other hand, consecutive accumulation of such pollutants in plants negatively affects their vigor and nutritional value.

#### 4.6.2.3 Slope Instability

Slope instability due to MSW has become a common environmental calamity causing some ecological and health-related issues. Various factors are involved in the failure of slope in landfills, including overloading and loose filling of waste dumps, and ultimate strength posed by massive loads of waste heaps in landfills, emission of gases from landfills, leachate outcome from MSW as affected by rainfall, and poor drainage systems in landfills sites (Raviteja and MunwarBasha 2017; Shu et al. 2021). Jafari et al. (2013) reported that about 330 people were killed in Quezon City, Philippines due to slope instability in Payats landfills.

#### 4.6.2.4 Biological Vectors

Landfills and MSW dumped sites encourage the growth of certain disease-causing macro and microorganisms by providing them with breeding sites, habitat, and food (Krystosik et al. 2020). Generally, fresh landfills escalate the risk of spreading infectious and vector-borne disease to the immunologically immature population in surroundings. One of the examples of such vector-borne disease is dengue, which is threatening about half a billion people globally. Dengue is transmitted by a mosquito (*Aedes aegypti* L), which commonly breeds on man-made recycled plastic, trash, and old tires dumped containers (Ramos et al. 2008). In another study conducted by Abbasi et al. (2019), 33 species of disease-causing arthropods breeding in MSW landfills in Iran were identified. Hence, open dumps and landfills of MSW can be a potential threat to disease due to the presence of various disease-carrying vectors.

### *4.6.3 Air Contamination*

Many countries have developed strict rules and regulations regarding landfills; however, the open dumping of MSW waste is still an issue in developing countries. Open burning and auto ignition of wastes in such landfills not only create an unpleasant smell but also release hazardous gases and airborne pollutants enormously, which latterly cause severe air pollution and pose health-related threats to residents. Emission of dust and several flammable gases including methane ($CH_4$), hydrogen sulfide ($H_2S$), carbon dioxide ($CO_2$), and nitrous oxide ($N_2O$) as a result of burning solid waste and decomposition of waste by soil microorganisms contaminate air quality and disturb atmosphere gas balance (Khaustov et al. 2019; Speight 2019). Gases released from landfills have a range of different components which largely depend on the type and composition of MSW. Moreover, gases emitted from landfills can replace $O_2$ in the surroundings and make it less available for residents and living organisms raising suffocation and respiratory problems. Numerous atmospheric problems are associated with open dumps and landfills of MSW such as fire, explosions, GHG, and odors.

#### 4.6.3.1 Fire and Explosion

Autoignition and open fire are common issues of MSW landfills which are frequently reported by waste pickers. In general, aerobic decomposition of waste occurs in presence of $O_2$, while the aerobic decomposition occurs in absence of $O_2$ and generates $CH_4$. Due to the microbial activities and decomposition of organic components, the concentration $O_2$ increases within the landfill. The enrichment of $O_2$ with methane gas results in generation of heat, which subsequently results in the development of fire (Fig. 4.1). The presence of a fraction of combustible waste encourages ignition and sometimes leads to an explosion. In Nepal, the common practice of open-air burning of waste tires is a serious issue, which subsequently results in strong heat and release of many harmful gases including $N_2O$, carbon monoxide (CO), sulfur dioxide ($SO_2$) along with polyaromatic hydrocarbons (PAHs) and organic volatiles (Shakya et al. 2008). According to a research study conducted by Cooper (1980), the open wood fireplace can release several toxic pollutants such as $NO_x$-$NO_2$ (1.8 g $kg^{-1}$), $SO_x$-$SO_2$, and CO (11–40 g $kg^{-1}$) into the atmosphere. Another study conducted in Spain to investigate open burning of MSW impact on local population documented 21–92 times higher concentration of ZnO and CO reaching 933 μg $m^{-2}$ against 13 μg $m^{-2}$ in areas far from burning sites (Artíñano et al. 2017). Additionally, higher $SO_2$ contents (1371 μg $m^{-2}$) were also detected, which were 11 times higher than those in the control zone (Artíñano et al. 2017).

#### 4.6.3.2 Odor or Smell

Generally, gas discharged from landfills and dumped waste areas moves freely in the air and creates immense odor, which is one of the environmental and social issues of waste landfills. Among these produced gases, $H_2S$ is commonly pronounced with an extreme odor (Aderemi and Falade 2012). Odor causes discomfort among communities and sometimes creates health issues such as headaches, vomiting, and respiratory problems in residents. Another odor-related drawback is the depreciation of real estate prices in the vicinity of landfill areas (Kumar et al. 2018).

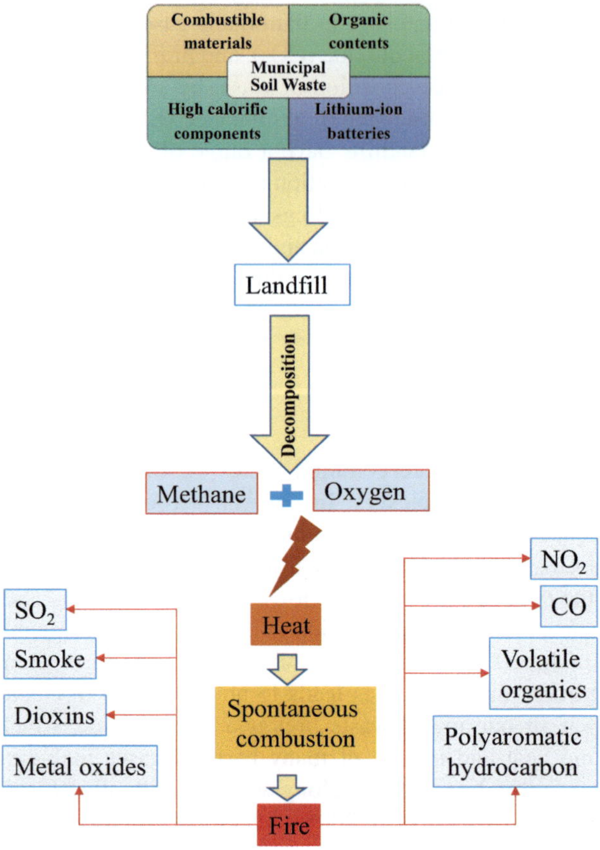


**Fig. 4.1** Schematic presentation of fire generation and subsequent emission of gases and toxic components from municipal solid waste landfills

#### 4.6.3.3 Greenhouse Gases

In some developed countries, gases released from landfills are collected by some facilities and are used as an energy source; however, if these gases remain uncontrolled and freely discharged into the atmosphere, they create a GHG effect and cause global warming. Experimental studies indicated that $CO_2$, $N_2O$, and $CH_4$ are the major gases directly emitted to the atmosphere from landfills (Themelis and Ulloa 2007). Ferronato and Torretta (2019) reported that $CH_4$ is the most important GHG being emitted from MSW open dumps and landfills. They reported that about 200 $Nm^3$ of $CH_4$ is released per ton of waste biomass. Hence, landfills and uncontrolled waste dumped sites are the main source GHGs emissions. As aforementioned, the emission of GHGs from landfill and waste dumped areas mainly depends

**Table 4.5** Emissions of greenhouse gases into the atmosphere from landfills and open dumps

| | Greenhouse gases emission (Gg/y) | | | |
|---|---|---|---|---|
| Type of waste | $CO_2$ (Gg year$^{-1}$) | $N_2O$ (Gg year$^{-1}$) | Total (Gg $CO_2$-eq year$^{-1}$) | Reference |
| Paper | 27.10 | 0.09 | 53.92 | Chen (2018) |
| Textile | 12.61 | 0.00 | 12.61 | |
| Food waste | 0.00 | 0.03 | 8.94 | |
| Plastic | 868.70 | 0.02 | 874.66 | |
| Metal | 0.00 | 0.04 | 11.92 | |
| Glass | 0.00 | 0.01 | 2.98 | |

on the type of waste such as about 1000 kg $CO_2$-eq.t$^{-1}$ are released from an open dumpsite followed by 300 kg $CO_2$-eq t$^{-1}$ from conventional landfill and 70 kg $CO_2$-eq t$^{-1}$ from landfill with low organic waste (Manfredi et al. 2009; Ferronato and Torretta 2019). Chen (2018) reported that various types of waste can release varying volumes of GHGs into the atmosphere (Table 4.5). In Beijing City, annually around 55,000 tons of $CH_4$ is being emitted into the atmosphere from MSW mainly coming from sanitary landfills since 60% of overall waste is dumped in sanitary landfills (Zhao et al. 2011). In addition to creating bad smell, suffocation, and global warming, acid rain is also caused by GHGs released from landfills, which can badly impact the environment causing acidification of soil, loss of stomatal control, enzyme inhibition, reduction in photosynthesis, and modification of synthetic pathways in plants (Danthurebandara et al. 2012). Therefore, it is important to formulate policies and create facilities to mitigate GHGs emissions by recovering and transformation of the organic components of waste into energy.

### *4.6.4 Impacts on Human Health*

Open dumping and landfilling of MSW is of serious concern for human health owing to the introduction of potentially toxic substances into the environment (Talang and Sirivithayapakorn 2021). MSW-based contaminants can enter the human body through ingestion, inhalation, and dermal contact. The composition of waste and leachate, the nature of decomposed products, as well as processing, transportation, and disposal determine the nature and severity of adverse effects on human beings (Cointreau 2006). The potent hazardous impacts of open dumping and landfilling of MSW on human health are presented in Fig. 4.2.

The consumption of landfill leachate contaminated groundwater is considered as the major pathway for pollutants transmission from landfills to humans. Moreover, the metals and other potent contaminants may also accumulate in surrounding soils, and subsequently, be taken up by plants. Consuming these plants by human beings

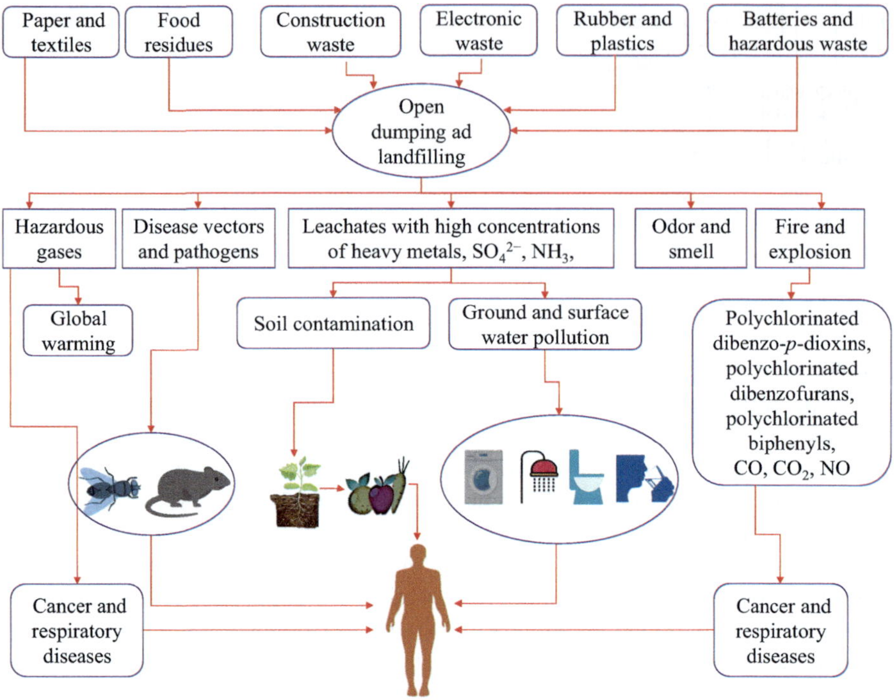


**Fig. 4.2** Schematic presentation for the generation of various hazardous gases, diseases, and odors from municipal solid waste landfills and their introduction to humans

can lead to serious health issues. For instance, Anikwe and Nwobodo (2002) found 214%–2040% higher levels of Pb, Cu, Fe, and Zn in dumpsite soil as compared to the normal soil. Odai et al. (2008) observed that growing vegetables on dumpsite soils resulted in the accumulation of very higher amounts of heavy metals. In developing countries, the waste-picker residents around solid waste dumps are comprised of children and women. Thus, ingestion and inhalation of toxic compounds may result in impaired nervous, reproductive, and immune systems (Cointreau 2006). Due to direct and indirect contact with the waste, the waste-pickers are more likely to develop tuberculosis, pneumonia, bronchitis, chronic cough, eye soreness or redness, diarrhea, jaundice, dysentery, and asthma. Furthermore, eye burning, diminished vision, parasitic disorders, and malnutrition could also be associated with ill-management of MSW (Cointreau 2006). Direct contact with MSW may result in birth defects and infant mortality due to the occurrence of pathogens and other hazardous compounds (Schantz and McAuley 1991). Another study demonstrated that the generation and release of bio-aerosols from MSW dumpsites can develop allergic and asthmatic pulmonary diseases in human beings under its exposure (Poulsen et al. 1995). Likewise, higher levels of Pb can also be seen in blood samples of people residing closer to the MSW dumpsites. The release of volatile organics, certain inorganic gases, and heavy metals from MSW can pose a potential risk for cancer development in human beings (Cucu et al. 1996). Therefore, appropriate management of MSW is of prime importance due to its potential hazardous impacts on human health.

## 4.7 Landfill Mining: A Way to Reduce Environmental and Financial Impacts

The concept of zero waste manufacturing has recently gained significant attraction in the world to ensure sustainable ecosystem health, reduce environmental pollution, and mitigate climate change. Though, development of a zero-waste society is hard to achieve in practice; however, the concept leads to a circular economy system through the development of innovative technologies and systems to reduce waste to its minimum level by reusing and recycling. Therefore, it has been estimated that a circular economy can provide an opportunity of one trillion USD along with other social and environmental advantages (Veleva et al. 2017). In this context, landfill mining (LFM) has been adopted by various countries of the world to recycle larger volumes of waste. LFM refers to the extract, recovery, and process/treatment of dormant materials from landfill deposits. It has been recognized as a mitigating strategy for the adverse environmental impacts of landfill pollution. LFM is an older technology and is being practiced in various countries of the world; however, the main focus of it has not always been the recovery of resources. LFM can potentially bring the materials into a productive cycle, which subsequently results in various economic, environmental, and social advantages. It does not only recover useful materials but improves environmental health and enhances the life span of the landfill (Krook et al. 2012). The extraction of metals and rare-earth metals significantly contributes to economic growth, while extraction of organic compounds results in a reduction of greenhouse gases from dumpsites, which consequently helps mitigate climate change (Burlakovs et al. 2018; Weng et al. 2015). Overall, LFM reduces the potential environmental pollution conserves the landfill space, reduces the area of the landfill, recovers energy and other useful materials, and reduces the cost of waste management (Hogland et al. 2004).

Despite the advantages, LFM involves various environmental and economic risks as well. Extraction, recycling, treatment, and transportation of waste could result in collapses in the excavated areas. Moreover, the release of methane and other gases, odor, particulate matter, as well as the dispersal of organic and inorganic contaminants could adversely affect human health and the environment (Krook et al. 2012; Ford et al. 2013). Ratcliffe et al. (2012) reported that LFM is practically uneconomical for the recovery of useful materials, especially owing to the lower prices of the extracted end-products. Therefore, the economic benefits, potential recoverable materials, capacity of waste-to-material/energy, as well as social and environmental aspects of LFM must be evaluated in order to get the maximum economic gains with minimal damage to the environment. Though, a lot of studies have evaluated the social benefits of LFM, a common framework for landfill evaluation and comprehensive environmental analysis are lacking. Thus, scientific insights are needed to develop a technologically innovative and sustainable system for efficient LFM with minimal environmental burden. Further, regulatory, socio-political, and economic aspects should be studied for the appropriate implementation of LFM technologies.

Integrating thermal valorization, gasification, and residue upcycling techniques could serve as an efficient approach to increase the economic, social, ecological, and environmental benefits of LFM. In surge to develop sustainable LFM technology, the concept of enhanced LFM has been evolved recently, which involves the recycling of materials both into waste-to-materials and waste-to-energy through a combined valorization. It is an advanced technique than that of the traditional LFM where processing, excavation, and treatment of deposited materials is performed. Therefore, the complete valorization in enhanced LFM provides a range of environmental and economic benefits. During the enhanced LFM process, various fractions such as glass and metals are separated from the waste directly, while the intermediate products are further treated to enhance their quality (Jones et al. 2013). The solid material in the waste can be treated and applied as fertilizers, while the materials with higher calorific values can be subjected to incineration, pyrolysis, gasification, and plasma technologies for energy production (Krook et al. 2012). With the help of enhanced LFM, the new landfills can serve as future mines, from where the materials can be extracted more effectively. The enhanced LFM has numerous advantages including climate change mitigation, recovering and utilization of materials and energy, creation of jobs, and reclamation of land. Nevertheless, due to higher heterogeneity in the composition and characteristics of the excavated waste, selecting a suitable valorization technology is of critical importance. Danthurebandara et al. (2015a) reported that plasma gasification was much more efficient than tradition incineration in terms of human toxicity, climate change, fossil depletion, and particulate matter formation. In another study, Danthurebandara et al. (2015b) stated that thermal treatment of excavated waste could provide more environmental and economic benefits in enhanced LFM processes. However, coupling various techniques such as plasma gasification and pyrolysis, could yield high-quality syngas along with other high quality end products. A general process flow diagram for the enhanced LFM is shown in Fig. 4.3. Detailed studies are needed to explore the critical environmental and economic factors to evaluate the enhanced LFM performance.

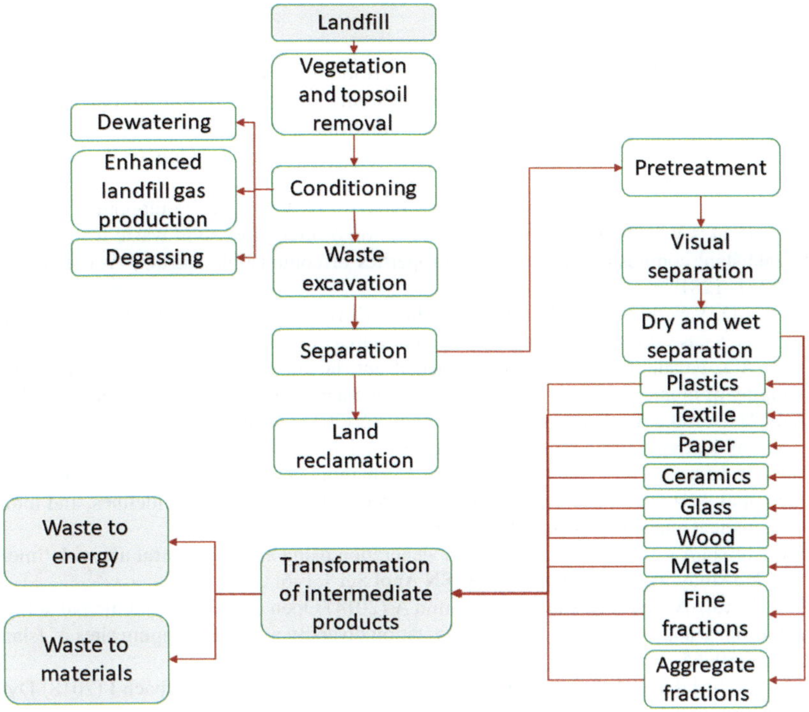


**Fig. 4.3** General process flow diagram for the enhanced landfill mining

## 4.8 Conclusions

Despite the severe hazardous impacts, open dumping and landfilling of municipal solid waste (MSW) is the most common practice for waste management in many parts of the world. Landfills and open dumps are threatening the health of the ecosystem and environment due to the vertical and lateral transmission of pollutant-laden leachate, emission of gases, generation of odor and dust, and presence of disease-carrying vectors. Moreover, the emission of greenhouse gases from the MSW heaps is significantly contributing to global warming. Therefore, efforts should be done to minimize the generation of MSW and recycling the waste to transform it into energy and useful materials by implementing the recent concepts of enhanced landfill mining technologies. Knowledge gaps should be identified to formulate new strategies for sustainable waste management technologies considering the circular economy. Modern closed-loop innovative systems could push the community towards zero waste to ensure a sustainable environment for a better tomorrow.

## References

Abbasi E, Rafinejad J, Hosseinpoor S, Gholami-Borujeni F, Gholizadeh S (2019) Diversity of arthropods in municipal solid waste landfill of Urmia, Iran. J Medical Entomol 56(1):268–270

Abiriga D, Vestgarden LS, Klempe H (2020) Groundwater contamination from a municipal landfill: effect of age, landfill closure, and season on groundwater chemistry. Sci Total Environ 737:140307

Abylkhani B, Aiymbetov B, Yagofarova A, Tokmurzin D, Venetis C, Poulopoulos S, Sarbassov Y, Inglezakis VJ (2019) Seasonal characterization of municipal solid waste from Astana city, Kazakhstan: composition and thermal properties of combustible fraction. Waste Manag Res 37:1271–1281

Adamcová D, Vaverková MD, Bartoň S, Havlíček Z, Břoušková E (2016) Soil contamination in landfills: a case study of a landfill in Czech Republic. Solid Earth 7:239–247

Aderemi AO, Falade TC (2012) Environmental and health concerns associated with the open dumping of municipal solid waste: a Lagos, Nigeria experience. Am J Environ Eng 2:160–165

Aderemi AO, Oriaku AV, Adewumi GA, Otitoloju AA (2011) Assessment of groundwater contamination by leachate near a municipal solid waste landfill. Afr J Environ Sci Technol 5:933–940

Afolagboye LO, Ojo AA, Talabi AO (2020) Evaluation of soil contamination status around a municipal waste dumpsite using contamination indices, soil-quality guidelines, and multivariate statistical analysis. SN Appl Sci 2:1–6

Ali SA, Ahmad A (2019) Forecasting MSW generation using artificial neural network time series model: a study from metropolitan city. SN Appl Sci 1:1–6

Ali SM, Pervaiz A, Afzal B, Hamid N, Yasmin A (2014) Open dumping of municipal solid waste and its hazardous impacts on soil and vegetation diversity at waste dumping sites of Islamabad city. 26:59–65

Alobaid F, Al-Maliki WA, Lanz T, Haaf M, Brachthäuser A, Epple B, Zorbach I (2018) Dynamic simulation of a municipal solid waste incinerator. Energy 149:230–249

Amano KO, Danso-Boateng E, Adom E, Kwame Nkansah D, Amoamah ES, Appiah-Danquah E (2021) Effect of waste landfill site on surface and ground water drinking quality. Water Environ J 35:715–729

Anikwe MA, Nwobodo KC (2002) Long term effect of municipal waste disposal on soil properties and productivity of sites used for urban agriculture in Abakaliki, Nigeria. Bioresour Technol 83:241–250

Artíñano B, Gómez-Moreno FJ, Díaz E, Amato F, Pandolfi M, Alonso-Blanco E, Coz E, Garcia-Alonso S, Becerril-Valle M, Querol X, Alastuey A (2017) Outdoor and indoor particle characterization from a large and uncontrolled combustion of a tire landfill. Sci Total Environ 593:543–551

Ashik MA, Nazmul MH, Rafizul IM (2017) Prediction of solid waste generation rate and determination of future waste characteristics at south-western region of Bangladesh using artificial neural network. WasteSafe:1–9

Ayeleru OO, Ntuli F, Mbohwa C (2016) Utilization of organic fraction of municipal solid waste (OFMSW) as compost: a case study of Florida. South Africa. In Proceedings of the World Congress on Engineering and Computer Science 2

Azam M, Jahromy SS, Raza W, Raza N, Lee SS, Kim KH, Winter F (2020) Status, characterization, and potential utilization of municipal solid waste as renewable energy source: Lahore case study in Pakistan. Environ Int 134:105291

Aziz SQ, Aziz HA, Yusoff MS, Bashir MJ, Umar M (2010) Leachate characterization in semi-aerobic and anaerobic sanitary landfills: a comparative study. J Environ Manag 91:2608–2614

Bhalla B, Saini MS, Jha MK (2014) Leachate contamination potential of unlined municipal solid waste landfill sites by leachate pollution index. Int J Sci Environ Technol 3:444–457

Blight G (2011) Landfills–yesterday, today and tomorrow. In: Waste. Academic Press, pp 469–485

Burlakovs J, Jani Y, Kriipsalu M, Vincevica-Gaile Z, Kaczala F, Celma G, Ozola R, Rozina L, Rudovica V, Hogland M, Viksna A (2018) On the way to 'zero waste' management: recovery

potential of elements, including rare earth elements, from fine fraction of waste. J Clean Prod 186:81–90
Chan JK (2016) The ethics of working with wicked urban waste problems: the case of Singapore's Semakau Landfill. Landsc Urban Plan 154:123–131
Chen YC (2018) Evaluating greenhouse gas emissions and energy recovery from municipal and industrial solid waste using waste-to-energy technology. J Clean Prod 192:262–269
Chen DM, Bodirsky BL, Krueger T, Mishra A, Popp A (2020) The world's growing municipal solid waste: trends and impacts. Environ Res Lett 15:074021
Cointreau S (2006) Occupational and environmental health issues of solid waste management: special emphasis on middle-and lower-income countries. Urban Papers 2
Cooper JA (1980) Environmental impact of residential wood combustion emissions and its implications. J Air Pollut Control Assoc 30:855–861
Costa AM, Alfaia RG, Campos JC (2019) Landfill leachate treatment in Brazil–An overview. J Environ Manag 232:110–116
Cox J, Giorgi S, Sharp V, Strange K, Wilson DC, Blakey N (2010) Household waste prevention—a review of evidence. Waste Manag Res 28:193–219
Cucu M, Iacob I (1996) National Integrated Programme on Environmental and Health in Romania, 1993–1996: final Report on the Study of Acute Environmental Health Problems from Waste Management in Romania. Institute of Hygiene. Public Health, Health Services and Management–Bucharest. Sponsored by the World Health Organization, Regional Office for Europe, Nancy Project Office, France, 1996
Dangi MB, Pretz CR, Urynowicz MA, Gerow KG, Reddy JM (2011) Municipal solid waste generation in Kathmandu. Nepal J Environ Manage 92:240–249
Danthurebandara M, Van Passel S, Nelen D, Tielemans Y, Van Acker K (2012) Environmental and socio-economic impacts of landfills. Linnaeus Eco-Tech:40–52
Danthurebandara M, Van Passel S, Machiels L, Van Acker K (2015a) Valorization of thermal treatment residues in enhanced landfill mining: environmental and economic evaluation. J Clean Prod 99:275–285
Danthurebandara M, Van Passel S, Vanderreydt I, Van Acker K (2015b) Assessment of environmental and economic feasibility of Enhanced Landfill Mining. Waste Manag 45:434–447
Díaz Rizo O, Fonticiella Morell D, Arado López JO, Borrell Munoz JL, D'Alessandro Rodríguez K, López Pino N (2013) Spatial distribution and contamination assessment of heavy metals in urban topsoils from Las Tunas City, Cuba. Bull Environ Contam Toxicol 91(1):29–35
Ercolano S, Gaeta GL, Ghinoi S, Silvestri F (2018) Kuznets curve in municipal solid waste production: an empirical analysis based on municipal-level panel data from the Lombardy region (Italy). Ecol Indic 93:397–403
Fazeli A, Bakhtvar F, Jahanshaloo L, Sidik NA, Bayat AE (2016) Malaysia′ s stand on municipal solid waste conversion to energy: a review. Renew Sust Energ Rev 58:1007–1016
Ferronato N, Torretta V (2019) Waste mismanagement in developing countries: a review of global issues. J Environ Res Public Health 16(1060)
Ford S, Warren K, Lorton C, Smithers R, Read A, Hudgins M (2013) Feasibility and viability of landfill mining and reclamation in Scotland. Scoping Study, Final Report.
Frésca FRC (2007) Analysis of Municipal Solid Waste Generation in São Carlos, SP, Considering Its Physical Characterization Master Dissertation. University of São Paulo, São Carlos
General Administration of Quality Supervision, Inspection and Quarantine of the People's Republic of China (AQSIQ) (1993) Quality standard for ground water (GB/T14848- 93)
Gentil EC, Gallo D, Christensen TH (2011) Environmental evaluation of municipal waste prevention. Waste Manag 31:2371–2379
Gonzalez-Valencia R, Magana-Rodriguez F, Cristóbal J, Thalasso F (2016) Hotspot detection and spatial distribution of methane emissions from landfills by a surface probe method. Waste Manag 55:299–305
Han Z, Ma H, Shi G, He L, Wei L, Shi Q (2016) A review of groundwater contamination near municipal solid waste landfill sites in China. Sci Total Environ 569:1255–1264

Hassan SA, Jameel NG, Şekeroğlu B (2016) Smart solid waste monitoring and collection system. Int J 6(10)

Havukainen J, Zhan M, Dong J, Liikanen M, Deviatkin I, Li X, Horttanainen M (2017) Environmental impact assessment of municipal solid waste management incorporating mechanical treatment of waste and incineration in Hangzhou, China. J Clean Prod 141:453–461

Hettiarachchi H, Meegoda JN, Ryu S (2018) Organic waste buyback as a viable method to enhance sustainable municipal solid waste management in developing countries. Int J Environ Res Public Health 15:2483

Hogland W, Marques M, Nimmermark S (2004) Landfill mining and waste characterization: a strategy for remediation of contaminated areas. J Mater Cycles Waste Manag 6:119–124

Hoornweg D, Bhada-Tata P (2012) What a waste: a global review of solid waste management. World Bank. Accessed on August 15, 2021 at: https://openknowledge.worldbank.org/handle/10986/17388

Hunce SY, Akgul D, Demir G, Mertoglu B (2012) Solidification/stabilization of landfill leachate concentrate using different aggregate materials. Waste Manag 32:1394–1400

Ihedioha JN, Ukoha PO, Ekere NR (2017) Ecological and human health risk assessment of heavy metal contamination in soil of a municipal solid waste dump in Uyo, Nigeria. Environ Geochem Health 39:497–515

Iravanian A, Ravari SO (2020) Types of contamination in landfills and effects on the environment: a review study. InIOP Conf Ser Environ Earth Sci 614:012083

Islam KM (2016) Municipal solid waste to energy generation in Bangladesh: possible scenarios to generate renewable electricity in Dhaka and Chittagong city. J Renew Energy

Jafari NH, Stark TD, Merry S (2013) The July 10 2000 Payatas landfill slope failure. ISSMGE Int J Geoeng Case Hist 2:208–228

Jilani S, Rashid R (2020) Municipal solid waste dumping and its impact on soil quality in Karachi. EQA- J Environ Qual 36:9–14

John NM, Edem SO, Ndaeyo NU, Ndon BA (2006) Physical composition of municipal solid waste and nutrient contents of its organic component in Uyo municipality, Nigeria. J Plant Nutr 29:189–194

Jones PT, Geysen D, Tielemans Y, Van Passel S, Pontikes Y, Blanpain B, Quaghebeur M, Hoekstra N (2013) Enhanced Landfill Mining in view of multiple resource recovery: a critical review. J Clean Prod 55:45–55

Jovanov D, Vujić B, Vujić G (2018) Optimization of the monitoring of landfill gas and leachate in closed methanogenic landfills. J Environ Manag 216:32–40

Kabata-Pendias A (2011) Trace elements in soils and plants, 4th edn. CRC Press, Boca Raton

Kanmani S, Gandhimathi R (2013) Assessment of heavy metal contamination in soil due to leachate migration from an open dumping site. Appl Water Sci 3(1):193–205

Katiyar RB, Suresh S, Sharma AK (2013) Characterisation of municipal solid waste generated by the city of Bhopal, India. Int J Chem tech Res 5:623–628

Kaza S, Yao L, Bhada-Tata P, Van Woerden F (2018) What a waste 2.0: a global snapshot of solid waste management to 2050. World Bank Publications

Kazuva E, Zhang J (2019) Analyzing municipal solid waste treatment scenarios in rapidly urbanizing cities in developing countries: the case of Dar es Salaam, Tanzania. Int J Environ Res Public Health 16:2035

Khamis SS, Purwanto H, Rozhan AN, Rahman MA, Salleh HM (2019) Characterization of municipal solid waste in Malaysia for energy recovery. IOP Conference Series: Environ Earth Sci 264:012003. IOP Publishing

Khaustov VV, Bredikhin VV, Khaustova TV (2019) Treatment technology of liquid phase at industrial waste landfill. InIOP Conference Series: Mater Sci Eng 687(6):066063. IOP Publishing

Kim D, Prawisudha P, Yoshikawa K (2012) Hydrothermal upgrading of Korean MSW for solid fuel production: effect of MSW composition. J Combust

Kjeldsen P, Barlaz MA, Rooker AP, Baun A, Ledin A, Christensen TH (2002) Present and long-term composition of MSW landfill leachate: a review. Crit Rev Environ Sci Technol 32:297–336

Kola-Olusanya A (2012) Impact of municipal solid wastes on underground water sources in Nigeria. Eur Sci J 8(11):1–9
Kolekar KA, Hazra T, Chakrabarty SN (2017) Prediction of municipal solid waste generation for developing countries in temporal scale: a fuzzy inference system approach. Glob Nest J 19:511–520
Krook J, Svensson N, Eklund M (2012) Landfill mining: a critical review of two decades of research. Waste Manag 32:513–520
Krystosik A, Njoroge G, Odhiambo L, Forsyth JE, Mutuku F, LaBeaud AD (2020) Solid wastes provide breeding sites, burrows, and food for biological disease vectors, and urban zoonotic reservoirs: a call to action for solutions-based research. Public Health Front 7:405
Kulikowska D, Klimiuk E (2008) The effect of landfill age on municipal leachate composition. Bioresour Technol 99:5981–5985
Kumar S, Chiemchaisri C, Mudhoo A (2011) Bioreactor landfill technology in municipal solid waste treatment: An overview. Critic Rev Biotech 31(1):77–97.
Kumar A, Datta M, Nema AK, Singh RK, Gurjar BR (2018) Closure of municipal solid waste dumps. Site rating for odor impact. Environ Prot Eng 44(2)
Laner D, Crest M, Scharff H, Morris JWF, Barlaz MA (2012) A review of approaches for the long-term management of municipal solid waste landfills. Waste Manag 32:498–512
Makarichi L, Kan R, Jutidamrongphan W, Techato KA (2019) Suitability of municipal solid waste in African cities for thermochemical waste-to-energy conversion: the case of Harare Metropolitan City. Zimbabwe Waste Manag Res 37(1):83–94
Mallick J (2021) Municipal Solid Waste Landfill Site Selection Based on Fuzzy-AHP and Geoinformation Techniques in Asir Region Saudi Arabia. Sustainability 13:1538
Manfredi S, Tonini D, Christensen TH, Scharff H (2009) Landfilling of waste: accounting of greenhouse gases and global warming contributions. Waste Manag Res 7(8):825–836
Mataloni F, Badaloni C, Golini MN, Bolignano A, Bucci S, Sozzi R, Forastiere F, Davoli M, Ancona C (2016) Morbidity and mortality of people who live close to municipal waste landfills: a multisite cohort study. Int J Epidemiol 45:806–815
Mekonnen B, Haddis A, Zeine W (2020) Assessment of the effect of solid waste dump site on surrounding soil and river water quality in Tepi town, Southwest Ethiopia. J Environ Public Health
Miezah K, Obiri-Danso K, Kádár Z, Fei-Baffoe B, Mensah MY (2015) Municipal solid waste characterization and quantification as a measure towards effective waste management in Ghana. Waste Manag 46:15–27
Ministry of Environmental Protection of the People's Republic of China, General Administration of Quality Supervision (EPA), Inspection and Quarantine of the People's Republic of China (AQSIQ) (2002) Environmental Quality Standards for Surface Water (GB3838-2002)
Mishra S, Tiwary D, Ohri A, Agnihotri AK (2019) Impact of municipal solid waste landfill leachate on groundwater quality in Varanasi, India. Groundw Sustain Dev 9:100230
Mojiri A, Aziz HA, Zaman NQ, Aziz SQ, Zahed MA (2016) Metals removal from municipal landfill leachate and wastewater using adsorbents combined with biological method. Desalin Water Treat 57:2819–2833
Monni S, Pipatti R, Lehtilä A, Savolainen I, Syri S (2006) Global climate change mitigation scenarios for solid waste management
Moreno AD, Rodríguez MG, Velasco AR, Enriquez JC, Lara RG, Gutiérrez AM, Hernández NA (2013) Mexico City's municipal solid waste characteristics and composition analysis. Rev Int Contam Ambient 29(1):39–46
NIRD&PR (2016) Solid waste management in rural areas a step-by-step guide for Gram Panchayats Hyderabad, India
Odai SN, Mensah E, Sipitey D, Ryo S, Awuah E (2008) Heavy metals uptake by vegetables cultivated on urban waste dumpsites: case study of Kumasi, Ghana. Res J Environ Toxicol 2:92–99
Olafisoye OB, Adefioye T, Osibote OA (2013) Heavy metals contamination of water, soil, and plants around an electronic waste dumpsite. Pol J Environ Stud 22(5)

Ouda OK (2013) Assessment of the environmental values of waste-to-energy in the Gaza strip. Curr World Environ 8(3):355
Ouda OK, Raza SA, Nizami AS, Rehan M, Al-Waked R, Korres NE (2016) Waste to energy potential: a case study of Saudi Arabia. Renew Sust Energ Rev 61:328–340
Ozcan HK, Guvenc SY, Guvenc L, Demir G (2016) Municipal solid waste characterization according to different income levels: a case study. Sustainability 8:1044
Parameswari K, Padmini TK, Mudgal BV (2015) Assessment of soil contamination around municipal solid waste dumpsite. Indian J Sci Technol 8:1–5
Poulsen OM, Breum NO, Ebbehøj N, Hansen ÅM, Ivens UI, van Lelieveld D, Malmros P, Matthiasen L, Nielsen BH, Nielsen EM, Schibye B (1995) Collection of domestic waste. Review of occupational health problems and their possible causes. Sci Total Environ 170:1–9
Qdais HA, Hamoda M, Newham J (1997) Analysis of residential solid waste at generation sites. Waste Manag Res 15:395–406
Ramos MM, Mohammed H, Zielinski-Gutierrez E, Hayden MH, Lopez JL, Fournier M, Trujillo AR, Burton R, Brunkard JM, Anaya-Lopez L, Banicki AA (2008) Epidemic dengue and dengue hemorrhagic fever at the Texas–Mexico border: results of a household-based seroepidemiologic survey, December 2005. Am J Trop Med Hyg 78:364–369
Ratcliffe A, Prent OJ, van Vossen W (2012) Feasibility of material recovery from landfills (MFL) in the European Union. Int the ISWA World Waste Congress 2012
Raviteja KV, MunwarBasha B (2017) Probabilistic back analysis of Koshe landfill slope failure. Indian geotechnical conference
Rehrah D, Bansode RR, Hassan O, Ahmedna M (2016) Physico-chemical characterization of biochars from solid municipal waste for use in soil amendment. J Anal Appl Pyrolysis 118:42–53
Roberts D (2015) Characterisation of chemical composition and energy content of green waste and municipal solid waste from Greater Brisbane, Australia. Waste Manag 41:12–19
Romano G, Molinos-Senante M (2020) Factors affecting eco-efficiency of municipal waste services in Tuscan municipalities: an empirical investigation of different management models. Waste Manag 105:384–394
Sawadogo M, Tanoh ST, Sidibé S, Kpai N, Tankoano I (2018) Cleaner production in Burkina Faso: case study of fuel briquettes made from cashew industry waste. J Clean Prod 195:1047–1056
Schantz PM, McAuley J (1991) Current status of food-borne parasitic zoonoses in the United States. Southeast Asian J Trop Med Public Health 22:65–71
Shakya PR, Shrestha P, Tamrakar CS, Bhattarai PK (2008) Studies on potential emission of hazardous gases due to uncontrolled open-air burning of waste vehicle tyres and their possible impacts on the environment. Atmos Environ 42:6555–6559
Sharholy M, Ahmad K (2008) Municipal solid waste management in Indian cities—a review 28:459–467
Sharma A, Ganguly R, Gupta AK (2019) Characterization and energy generation potential of municipal solid waste from nonengineered landfill sites in Himachal Pradesh. India J Hazard Toxic Radioact Waste 23:04019008
Shayler H, McBride M, Harrison E (2009) Sources and impacts of contaminants in soils
Shekdar AV (2009) Sustainable solid waste management: an integrated approach for Asian countries. Waste Manag 29(4):1438–1448
Shu S, Li Y, Sun Z, Shi J (2021) Effect of gas pressure on municipal solid waste landfill slope stability. Waste Manag Re 24:0734242X211001414
Singh V, Mittal AK (2011) Groundwater pollution by municipal solid waste landfill leachate: a case study of Okhla Landfill Delhi InIWRA World Water Congress Proceedings 2011
Skibniewska K (2011) Some aspects of environmental impact of waste dumps. University of Warmia and Mazury in Olsztyn
Speight JG (2019) Unconventional gas. Natural Gas, edited by: Speight, JG, Gulf Professional Publishing, Boston, pp 59–98
Sun W, Wang X, DeCarolis JF, Barlaz MA (2019) Evaluation of optimal model parameters for prediction of methane generation from selected US landfills. Waste Manag 91:120–127

Talang RP, Sirivithayapakorn S (2021) Environmental and financial assessments of open burning, open dumping and integrated municipal solid waste disposal schemes among different income groups. J Clean Prod 2:127761
Tałałaj IA, Biedka P, Bartkowska I (2019) Treatment of landfill leachates with biological pretreatments and reverse osmosis. Environ Chem Let 17(3):1177–1193
Tan ST, Lee CT, Hashim H, Ho WS, Lim JS (2014) Optimal process network for municipal solid waste management in Iskandar Malaysia. J Clean Prod 71:48–58
Themelis NJ, Ulloa PA (2007) Methane generation in landfills. Renew Energy 32:1243–1257
Torretta V, Ferronato N, Katsoyiannis IA, Tolkou AK, Airoldi M (2017) Novel and conventional technologies for landfill leachates treatment: a review. Sustainability 9:9
USEPA (2006) Solid waste management and greenhouse gases: a lifecycle assessment of emissions and sinks. Washington, DC
Vallero DA, Blight G (2019) Chapter 12 – The municipal landfill. In: Letcher TM, Vallero DA (eds) Waste (Second Edition). Academic, pp 235–258
Vathsalan AB, Indu KS, Ramasany EV (2017) Leachate pollution index – an effective tool in determining the phytotoxicity of municipal solid waste leachate. Waste Manag 68: 329–336
Vaverková MD (2019) Landfill impacts on the environment—review. Geosciences 9:431
Veleva V, Bodkin G, Todorova S (2017) The need for better measurement and employee engagement to advance a circular economy: lessons from Biogen's "zero waste" journey. J Clean Prod 154:517–529
Wagner M, Lambert S (2018) Freshwater microplastics: emerging environmental contaminants? Springer Nature
Wang Y, Ng KT, Asha AZ (2016) Non-hazardous waste generation characteristics and recycling practices in Saskatchewan and Manitoba, Canada. J Mater Cycles Waste Manag 18:715–724
Weng YC, Fujiwara T, Houng HJ, Sun CH, Li WY, Kuo YW (2015) Management of landfill reclamation with regard to biodiversity preservation, global warming mitigation and landfill mining: experiences from the Asia–Pacific region. J Clean Prod 104:364–373
WHO (2004) Guidelines for drinking water quality, World Health Organization, WHO Library Cataloguing, Geneva, Switzerland, 3rd edition, 2004
WHO (2011) Guidelines for drinking-water quality, 4th edn. World Health Organisation, Geneva
Wichai-utcha N, Chavalparit O (2019) 3Rs Policy and plastic waste management in Thailand. J Mater Cycles Waste Manag 21:10–22
Wiedinmyer C, Yokelson RJ, Gullett BK (2014) Global emissions of trace gases, particulate matter, and hazardous air pollutants from open burning of domestic waste. Environ Sci Technol 48:9523–9530
Wilson DC, Velis CA (2015) Waste management–still a global challenge in the 21st century: an evidence-based call for action. Waste Manag Res 33(12):1049–1051
Wiszniowski J, Robert D, Surmacz-Gorska J, Miksch K, Weber JV (2006) Landfill leachate treatment methods: A review. Environ Chem Let 4(1):51–61
World Energy Council (2013) World energy resources: waste to energy, world energy council.
World Health Organization (WHO) (1997) Guideline for drinking water quality, 2nd edition volume 2, Health criteria and other supporting information. World Health Organization, Geneva, p 9
World Health Organization (WHO) (2003) Guideline for drinking water quality. WHO/SDE/WSH 03.04, Geneva
Yusuff AS, John W, Okoro O, Ajibade A (2014) Physico-chemical composition and energy content analysis of solid waste: a case study of Castlereagh District, Northern Ireland. Am J Eng Sci Technol Res 2:1–9
Zand AD, Abduli MA (2008) Current situation of used household batteries in Iran and appropriate management policies. Waste Manag 28:2085–2090
Zhao Y, Christensen TH, Lu W, Wu H, Wang H (2011) Environmental impact assessment of solid waste management in Beijing City, China. Waste Manag 31:793–799
Zhu Y, Zhang Y, Luo D, Chong Z, Li E, Kong X (2021) A review of municipal solid waste in China: characteristics, compositions, influential factors and treatment technologies. Environ Dev Sustain 23:6603–6622
Ziyang L, Luochun W, Nanwen Z, Youcai Z (2015) Martial recycling from renewable landfill and associated risks: a review. Chemosphere 131:91–103

# Chapter 5
# Molecular Tools- A Future Perspective Approach for Monitoring Landfill Leachates and Validating Bioremediation Process

**J. Immanuel Suresh and M. S. Sri Janani**

**Abstract** Bioremediation is an alternative to conventional methods that involve the use of consortia of microbial strains to remove or neutralize the pollutants such as landfill leachate from a contaminated site. To improve the contaminant ecosystem and urban areas, remediation processes are indispensable in contaminated locations. Bioremediation is must for protecting the long-term sustainability of global ecosystems. Microbes and plants have huge prospects for the detoxification of natural, metallic and organic wastes in the polluted ecosystems. High-throughput methodologies are enabling the quantitative (e.g., dot blot and fluorescence in situ hybridization, or real-time PCR), qualitative (e.g., DNA fingerprinting techniques), Metaproteomics, Metagenomics, and Microarrays are identifying new catabolic operons of Xenobiotics in environmental bacteria against recalcitrant contaminants. DNA microarrays may analyze thousands of genes in a soil concurrently. Analysis of genes that are being expressed in different conditions in contaminated soils will expose the metabolic status of microorganisms. It indicates environmental modifications accelerating bioremediation. The enlargement of different OMICs approaches provides competent research avenues at dissimilar levels. These tools are useful to identify the microbial communities, to restore environmental health, and to monitor remediation activities.

**Keywords** Metaproteomics · Metagenomics · Microarray · Xenobiotics · Microbial community · Bioremediation · DNA fingerprinting

---

J. Immanuel Suresh (✉)
Assistant Professor, PG Department of Microbiology, The American College, Madurai, Tamil Nadu, India
e-mail: immanuelsuresh1978@gmail.com

M. S. Sri Janani
PG Department of Microbiology, The American College, Madurai, Tamil Nadu, India



P. Pathak, S. G. Palani (eds.), *Circular Economy in Municipal Solid Waste Landfilling: Biomining & Leachate Treatment*, Radionuclides and Heavy Metals in the Environment, https://doi.org/10.1007/978-3-031-07785-2_5

## Abbreviations

| | |
|---|---|
| DGGE | Denaturing gradient gel electrophoresis |
| DNA | Deoxyribonucleic acid |
| FISH | Fluorescent in situ Hybridization |
| PCR | Polymerase Chain Reaction |
| RFLP | Restriction Fragment Length polymorphism |
| RISA | Ribosomal Intergenic Spacer Analysis |
| rRNA | Ribosomal Ribonucleic acid |
| RSGP | Reverse Sample Genome Probing |
| RT-PCR | Real Time Polymerase Chain Reaction |
| SSCP | Single-Strand Conformation Polymorphism |
| T-RFLP | Terminal Restriction Fragment Length Polymorphism |

## 5.1 Introduction

Bioremediation is one of the most excellent biological methods to decrease the impurities of diligence to target levels. This process involves degrading, breaking down, and eliminating contaminants from water and soil. Bioremediation involves domineering natural methods (Abdulsalam et al. 2013). Several contamination problems can be found in biological processes in nature and these contamination issues were solved only by the Bioremediation method. Bioremediation is a natural process that relies on microorganisms such as fungi, bacteria, and plants to modify impurities or contaminants. As these organisms carry out their regular functions of life. Metabolic functions of these organisms are able to degrade the contaminants. Oil contaminants, Heavy metals, Hydrocarbons, and Pesticides are very dangerous and toxic to the environment. It causes severe consequences on the ecosystem (Adams et al. 2015). These contaminants cause mutagenesis and tumorigenesis predominantly in humans. Several compounds having harmful effects and those toxic effects are established into the ecosystem through human activities which also affects human health. Landfill leachates play an imperative role in the environment. Human actions create an incredible diversity of products. Some are lifeless and undisruptive, but many are noxious and extremely critical to the atmosphere, predominantly the water and soil. Bioremediation processes are used to change environmental contaminants in soil and water with the help of microorganisms to degrade the target contaminants especially landfill leachates (Adams and Omokhagbor 2015). Numerous substances are in contact with water and soil. Conventional methods like incineration and extraction are used to eliminate, diminish or moderate harmful substances in water and soil. These conventional methods are used to treat the contaminant environment having some disadvantages and levels of threats (Adebusoye et al. 2007). Bioremediation method is cost- effective and stable when compared to other conventional methods of remediation. It is a similar method used to treat contaminants in industrial, solid waste, soil waste and waste water (AI-Jawhari et al. 2014). Bioremediation can be efficient in ecological circumstances allow bacterial growth and action. Generally, the compensations are superior to that of disadvantages. Microorganisms are engaged to devastate the deliberation of dangerous wastes on a polluted site such as landfill leachates, soils, sludges, water, and waste streams from a variety of sources due to industrialization, augmented inhabitants or populations, and urbanization (Yeonghee et al. 2005). Commonly, various species are explored from dissimilar sites and they are effectual in control mechanism. The diversity of the bacterial populations in surroundings polluted by landfill leachates, heavy metals, oils. Radionuclide's, petroleum and derivates, hydrocarbons etc., these are indispensable to increase a superior vision of probable degraders and to enlarge suitable bioremediation strategies (Kuppusamy et al. 2020).

Nowadays, High-throughput methodologies are used to monitoring and validating Bioremediation process. Advances in molecular biology methods have opened an innovative era in bioremediation techniques (Monu Jariyal et al. 2020). Molecular biology approaches have shown an exponential mount in the past few decades (Tigini et al. 2010). Molecular Nanotechnology is the newest development will lead to additional vision into bioremediation processes (Shekhar et al. 2020). Bioinformatics-associated tools are very significant for bioremediation of landfill leachates and other dangerous wastes. This review chapter focussed on widely used molecular tools from various research areas such as Genomics, Proteomic, Transcriptomics, Molecular phylogenetics, Microarray informatics, Metagenomics, Interactomics, Metabolomics, Fluxomics, and Metametabolomics in Bioremediation process.

## 5.2 Contaminants in the Environment

Pollution is the serious dilemma in our world. The human population in world at an astounding rate will be reach in surfeit of 19 billion in the year 2050. The exhaustive industrial and agricultural areas required to bear such a huge quantity of population will predictably cause a deposition of air, water and soil pollution (Ahalya et al. 2003). 1.3 billion tonnes of wastes are generated in every year. Mainstream of wastes are stored and dumped into the landfill and in oceans. Microbes might be the best solution to manage these types of pollution. Microbes had capability to break down an enormous assortment of organic substances and suck up inorganic compounds. Presently, microorganisms are used to clean up toxic waste management in processes (Hazen and Palmisano 2003). These method is universally called as Bioremediation.

In this modern era, Soil, water and air environment get heavily polluted due to advanced technologies. Bioremediation uses microbes to diminish contamination and pollution through the natural degradation of contaminants into non-toxic materials. This can engage either anaerobic or aerobic microorganisms that frequently utilize this breakdown as an energy source (Arora et al. 2009). Bioremediation is a promising technology that requires microbial processes. Industrial soils can be contaminated by diversity of sources like deposition of heavy metals, landfill leachates and by toxic chemical spillages from industrial emanations (Arvind Kumar 2020). Moreover, agricultural soils can be polluted by heavy metal enclosed in agricultural commodities and by pesticide usage. Environmental air is also contaminated by an assortment of unpredictable organic substances produced by a variety of industries (Asira 2013). Bio filtration is a technique to clean toxic gases released from chimneys of industries. It is a biological method to remediate airborne contaminants. In this technique, polluted air was passed in a culture medium containing microbes that degrade the pollutant and convert into products such as water, salts and $CO_2$ gas. The World Health Organization (WHO) calculates approximately that every year 9, 42,000 people passed away as an effect of diarrhoeal diseases. It might be prevented only by dirt-free water and accurate cleanliness (Atlas and Bartha 1992). Wastewater is aerated to provide oxygen to microorganisms that humiliate pollutants and organic material in a treatment plant (Babita et al. 2020). Microorganisms can take up the organic pollutants which can be filtered off. Toxic gases such as ammonia is abridged to nitrogen and emitted into the environment.

Landfill leachates, Hydrocarbons, Halogenated organic solvents and compounds, Sewage, Litters, Non-chlorinated Pesticides and Herbicides, Nitrogen compounds, Fossil oil, Radionuclides, Heavy metals are the various types of contaminants present in the environment (Janssen and Stucki 2020). Microbes perform momentous toxin removal equipments in air, soil, and water. Bioremediation has several advantages that are carried out particularly by microbes. Microbes are reinstating the innovative normal environment and preventing from pollutions.

## 5.3 Bioremediation

In biological technology, Bioremediation plays an imperative role which is cost effective and reduce the level of pollution in the environment (Baek et al. 2004). Globally, Bioremediation has been used in various sites (Singh et al. 2011). This technique has been followed in worldwide. Microorganisms are used for remove the pollutants. It is a promising innovative methodology in the field of environmental science for environmental safety. Bioremediation can be defined as the organic wastes and heavy metal pollutants are biologically degraded with the help of microbes (Bakermans and Madsen 2002). Bioremediation process develops a variety of naturally occurring improvement processes such as Biostimulation, Bioaugmentation and Natural attenuation.

Abundant natural and man-made factors particularly the development of human industrial evolution had contaminated abundant natural resources to lead modern life (Shemer et al. 2019). To clean the environment and get back to the original level of safety, Bioremediation plays a significant role and the hazardous materials can be degraded by the help of microorganisms. It removes the impending hazards to human healthiness and atmosphere.

Bioremediation can be defined as the organic wastes and heavy metal pollutants are biologically degraded with the help of microbes (Rajendran et al. 2003). Bioremediation process develops a variety of naturally occurring improvement processes such as Biostimulation, Bioaugmentation and Natural attenuation (Belkin 2003). Abundant natural and man-made factors, particularly the development of human industrial evolution had contaminated abundant natural resources to lead modern life. Bioremediation process is divided into two categories. In situ method is suitable for groundwater and land treatment. Ex-situ method is suitable for Biofiltration and Bioreactors of air (Scow and Hicks 2005).

To clean the environment and get back to the original level of safety, Bioremediation plays a significant role and the hazardous materials can be degraded with the help of microorganisms. It removes the impending hazards to human healthiness and atmosphere. Biostimulation is the method of adding electron acceptors to speed up biodegradation by existing bacterial communities (Adams et al. 2015).

Soil contaminated by oil spills can be cleaned up by the process of biostimulation. This method is widely used to treat landfill leachates and polluted soils (Chen et al. 2019). Microbial growth shows great effect and used as a Biostimulating agent to treat pollutants. The reason for choosing microbes as a agent is that they shows increased rate of degradable activity of pollutants and they shows high proliferation rate while culturing the growth in the laboratory.

Introduction of exogenic microorganisms had ability to detoxifying an exacting pollutant is known as Bioaugmentation (Boopathy 2000). On occasion genetically changed microbes can be used in this method. Natural attenuation is the method of bioremediation which takes place without human involvement. It is also specialized method to treat soil and water pollutants. These three methods can be done by both Ex situ (with excavation) and in situ (without excavation). In situ method involves entire bioremediation work is made at the pollutant site (Mareddy 2017). It requires less physical work but it removes landfill leachates and other contaminants by pumping away from spreading to nearby areas.

Biosparging and Bioventing are the classes of In situ method. Ex situ method eliminates hazardous materials from the environment (Cebron et al. 2008). These are all the strategies involved in Bioremediation process. Bioremediation methods are environment friendly when compared than treatment methods using chemicals. Contaminants are easily degraded without any harm to the environment. Microbial bioremediation process gets affected by two factors. Biological factors such as Microbial interaction, Mutation, horizontal gene transfer, enzyme activity and environmental factors such as Moisture, temperature, soil texture, pH, solubility in water, nutrients, oxygen content and redox potential are the factors affecting bacterial bioremediation.

For Bioremediation process, three contents play a significant role to run the process. They are existence of pollutant, an electron acceptor, and the occurrence of microbes that are able to humiliating the definite pollutant (Chandrika et al. 2019).

Usually, Microbes having the ability of biodegradation so that that pollutant can be easily got degrade in the environment. Microorganisms have confines of tolerance for exacting ecological circumstances, in addition to most advantageous situation for high performance (Chen and Wang 2007). Soil microbes have need of pH and moisture for cell development and function. Bioremediation had been enhanced by biotechnology and molecular tools to controlling ecological pollution (Cho and Tiedje 2002). Since bioremediation shows potential substitute to conventional clean-up methodologies, generous research is rising in this Bioremediation field. *Pseudomonas veronii, Kocuria flava, Penicillium canescens, Aspergillus versicolor, Cladophora fascicularis, Spirogyra spp. and Cladophora spp., Saccharomyces cerevisiae, Modococci, Streptomyces, Bacilli*, are some of the microbes used in Bioremediation process for treating heavy metal and petroleum contaminants present in the environment (Juhanson et al. 2009).

Bioremediation is based on the action of microorganisms. Microbes play a momentous role in bioremediation treatments such as to degrade the petroleum products and heavy metals (Das and Chandran 2011). The bioremediation method is a natural procedure that encourages obliging microorganisms to employ detrimental pollutants as their basis of vigour and food. Some microbes consume lethal chemicals and assimilating them and eradicating into risk-free gases such as carbon dioxide by altering their composition.

The major advantage of using bioremediation procedures was their contribution to the atmosphere. It uses life to fix nature. It is a natural biological process and it has no lethal side effects (Davison 2005). It is cost effective and it has lower responsibility (Gavrilescu 2006). This process was carried out in situ for lots of applications. This process has been highly accepted from regulatory authorities because of natural process and slight interruption (Vidali 2001) Contaminants are completely degraded by this process. When compared to conventional methods like land filling and incineration, Bioremediation consumes lesser energy (Demnerova et al. 2005). Equipments are needed for monitoring and for validation process. It converts contaminated soil and water into fresh resources.

## 5.4 Monitoring and Validating Bioremediation Process

Bioremediation systems operate under both anaerobic and anaerobic conditions. Various methods used to improve the capability of microorganisms to eliminate harmful materials like landfill leachates in the environment (Dokic et al. 2010). Moreover geological, hydrological, and chemical studies of the threat prospective of polluted locations, the evaluation of bacterial activity and the estimation of bacterial populations maintain information on the physiological condition of micro flora in soil or whether microbes do survive in the polluted soil (Tingting et al. 2014). Molecular tools are categorized into two types. They are Quantitative and Qualitative methods. Quantitative methods are primarily composed of Dot plot and RT-PCR. Qualitative methods are containing DNA Fringerprinting techniques (Fig. 5.1).

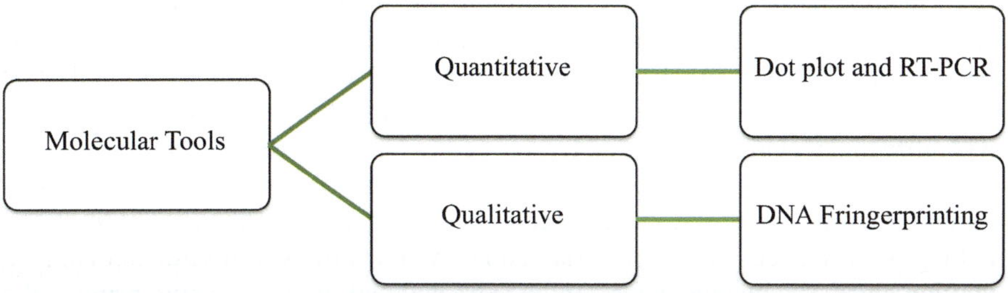


**Fig. 5.1** Molecular tools for validating Bioremediation process

An impediment to the improvement of successful in situ bio remedial systems, It is the fundamental requirement of considerate the desires, limitations and dynamics on the performance of degradative microbes in natural environments like soil and water (Dua et al. 2002). Both original microbes and genetically-modified microbes are used for bio remedial systems.

Even genetically modified organisms may degrade plastics from the ecosystem (Fulekar et al. 2012). Nowadays, Genetic engineering plays a significant role in Bioremediation reporter genes present in genetically modified organisms that are able to monitor pollution levels. (Dubey et al. 2006). The management of polluted soil can be executed consistently by applying some diverse technologies that are presently obtainable. Metaproteomics, Microarrays, Metagenomics are various fields play a role in Bioremediation process.

Recent techniques focus in the direction of molecular methods. It does not necessitate culturing the microbes but supply the methods based on hereditary assortment (Diplock et al. 2009). A variety of approaches that had been developed to monitor and validate environmental microbes against contaminants such as quantitative (e.g., dot blot and fluorescence in situ hybridization, or real-time PCR), qualitative (e.g., DNA fingerprinting techniques), Metaproteomics, Metagenomics, and Microarrays.

## 5.5 Qualitative Method- DNA Fingerprinting

DNA Fingerprinting techniques are tremendously to examine indefinite microbial population. They have restrictions and it is time consuming and labour extensive method (El Fantroussi and Agathos 2005). It is a laboratory based method to monitoring the Bioremediation process. This method used to evaluating the consequence of diverse pollutants present on bacterial communities. Landfill leachates, Heavy metal such as lead, Mercury contaminated soil, Cu pollution in soil, Pesticide contamination by land farming. Poly aromatic hydrocarbons and Oil refinery sludge are the Pollutants monitored by the DNA Fingerprinting.

To analyze ecological samples of unfamiliar microbial population present in landfill leachate, DNA Fingerprinting methods can be classified into two methods. They are as follows: (A) Denaturing gradient gel electrophoresis (DGGE) and Restriction Fragment Length polymorphism (RFLP). Both methods had the ability to identify and quantify fragments. Every species have exclusive position sites in the whole genome. Fingerprinting techniques rely on genome sequence deviations of dissimilar organisms (Fosso-Kankeu et al. 2014). Those deviations effect in diverse melting performances in Double stranded DNA and various restriction enzyme recognition sites. The nucleic acids can be extracted and improved using primers that target precise molecular markers (for example, 16S rRNA). After that the samples can be loaded in the gel and identify the various bands on the gel. Those separated bands can be excised from the gel and examined for sequencing and cloning (Karigar and Rao 2011). Polymerase Chain Reaction (PCR) plays a significant role to amplify the specific genes of taxonomic clusters of organisms. Those amplified genes can be used to validate for the Bioremediation process.

## 5.6 Quantitative Method

Recent techniques focus in the direction of molecular methods, which do not necessitate culturing the microbes based on hereditary functions (Kate and Kristin 2005). The gene sequences can be used to recognize organisms. These techniques assure to afford an enhanced perceptive and superior control of environmental biotechnology methods (Terry Gentry et al. 2004; Fig. 5.2).

Advanced molecular methods for validating Bioremediation such as Nucleic Acid or DNA hybridization, PCR-based methods (Quantitative PCR (qPCR) and RT-PCR), Single-Strand Conformation Polymorphism (SSCP), Terminal Restriction Fragment Length Polymorphism (T-RFLP), DNA microarrays, Ribosomal Intergenic Spacer Analysis (RISA),), Reverse Sample Genome Probing (RSGP), Fluorescent In Situ Hybridization, DNA cloning and sequencing. These methods are cost effective and bio-remediate all noxious landfill leachates and polluted surroundings.

### *5.6.1 DNA Hybridization*

DNA hybridization is an imperative quantitative and qualitative implementation in molecular microbial ecosystem (Shanker et al. 1998). DNA, RNA or in situ can be extracted by this method. It is based on genome to genome sequence evaluations. An additional approach is in situ molecular level hybridization (Kulshreshtha 2013). It measures the hereditary complexity of the bacterial population. Total DNA had been extracted and purified from the environmental samples (Kumar et al. 2011). Further the samples get denatured and authorized to amplification. The speed of

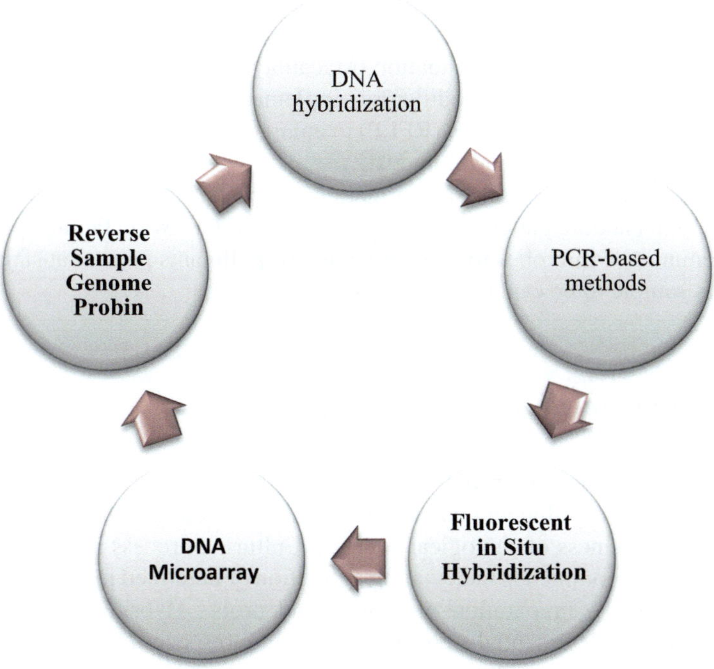


**Fig. 5.2** Advanced molecular methods for validating bioremediation

re-annealing based on the relationship of sequences present. Increase in complexity of DNA sequences then there is a decrease in re-annealing.

### *5.6.2 RT-PCR*

The main principle of Real-Time Polymerase Chain Reaction (PCR) is amplification of DNA. This method is used to perceive, differentiate, and recognize pollutant-degrading bacterial communities present in landfill leachates (Ledin 2000). The foremost apprehension in this technique was the endurance of accumulating microbes and the preservation of their degrading property throughout the Bioremediation process (Lee et al. 2006). The most frequently used methods for the recognition and identification of microbes is the PCR amplification technique of bacterial ribosomal RNA genes such as 23S rRNA, 18S rRNA, and 16S rRNA. The rRNA genes are the foundation for analyses of bacterial phylogeny.

Microorganisms containing rRNA genes are widely used in Bioaugmentation process. Monitoring the microbial communities and the amount of active working genes are accountable for toxic pollutant degradation facilitates the competence of the bioremediation process (Lovley 2003). Bacterial monooxygenases and dioxygenases can be detected by PCR methods. Flourishing performance of bioremediation method involves appreciative activities of bacterial communities accountable for target contaminant degradation (Macaulay 2014). It is essential to estimate the rRNA intensification commodities by supplementary procedures like Restriction Fragment Length Polymorphism (RFLP) to enlarge the accuracy of recognition and identification (Madhavi and Mohini 2012).

RFLP identifies structural transformations in bacterial population but little bit complexity in banding patterns. Quantitative PCR (qPCR) with Taqman probes also used to enumerate microbes to biodegradation of pollutants present at stained sites. It permits sample to be evaluated in real time.

### *5.6.3 Fluorescent In Situ Hybridization*

This is one of the molecular tools for validating the Bioremediation process. It is a rapid method, susceptible and taxonomic approach that allows the straight hallucination and uniqueness of ecological bacteria. Culturing process is not necessary in this method. It recommends undeviating apparition of unrefined microbes (Margaret and Jeremy 2014). Preparation of sample and probe, Hybridization and Signal Visualization are the essential steps involved in Fluorescent in Situ Hybridization.

By the help of fixatives, bacterial cells are pierced, and hybridized with accurate probes. Normally 12- to 25 base pairs of oligonucleotides with fluorescently labelled probes are added. After that, samples are envisioned under confocal laser microscopy. rRNA – targeted probes are hybridized which improves the categorization of unrefined microbes (Talos et al. 2009). It enables the explanation of intricate bacterial populations It assists the quantification of accurate bacterial communities. This method has been recurrently used with DNA fingerprinting approaches. Fluorescent in situ Hybridization (FISH) used for the enumeration and categorization of bacterial communities from polluted resources. The drawback of this method is that a limited assortment of probes may be utilized in a solitary hybridization testing. Isotope- characterized substrates are used in this method for Bioremediation.

Some samples shows complexity in the fluorescence history. A principal impediment of this method is its constrained sensitivity for the reason that microbial cells with less ribosome content that habitually emerge in oligotrophic atmosphere are not adequately tainted for microscopic examinations.

### *5.6.4 DNA Microarray*

DNA microarray is an advanced molecular tool in the field of Biotechnology. It has been confirmed to be exceedingly prevailing tools for the multiplexed relative examination of gene expression (Zabochnicka-Swiatek and Krzywonos 2014). It led to essential insights to gene expression blueprints associated with bacterial population. This method is widely used to recognize the alterations in gene sequences that frequently connected with meticulous microbes. This method also detect the genes which are turned on or off in microorganisms. Sample analysis, Quantification of gene expression, Examination of quality control and Data mining are methodologies involved in DNA Microarray technology.

The application of DNA microarray technology in environmental microbiology particularly in the assessment of bacterial communities occupied in biodegradation (Nester et al. 2001). This method can recognize microbes in beneficial ecological position. DNA microarray is a novel detection method (Stenuit et al. 2009). Particularly for elevated concentration of protein expression profiling arrays, with sufficient sensitivity are requisite to scrutinize the definite attachment of proteins on the microarray. Many ecological samples are ultimately have restrictions in RNA extraction involve that advances in RNA purification and intensification techniques are essential to create Microarray gene expression estimation. This method is probable for a broader assortment of samples.

Microarray experimentations are commonly less costly, and used in larger-scale research areas (Nicolaou et al. 2010). These techniques play a decisive part in Bioremediation with the aid of Molecular Biotechnology. Microarray is the pre-eminent method for microbial detection and validating the bioremediation process (Stenuit et al. 2009). Microarrays identify the simplest prevailing communities in the atmosphere (Niti et al. 2013). Probes are intended to be specific to recognize sequences can hybridize to related or unidentified sequences (Walworth et al. 2005). Furthermore water, soil and deposits are frequently enclosed acids and dissimilar organic compounds that might also inhibit DNA hybridization on microarrays. These are all the molecular tools for validating and monitoring the Bioremediation process. These techniques play an ultimate role in the field of Biotechnology.

### *5.6.5 Reverse Sample Genome Probing*

Reverse Sample Genome Probing (RSGP) is a method of approach in DNA microarray for understanding a bacterial population. This method has four steps (Cho and Tiedje 2001). Isolation of genomic DNA from samples, Hybridization to achieve DNA fragments, Preparation of genome arrays, and labelling the groups of DNA attained from Microarray are the noteworthy steps involved in Reverse Sample Genome Probing (RSGP). This approach has been used to examine bacterial communities in oil polluted areas and in contaminated soils (Phulia et al. 2013). Microarrays can integrate a numerous target gene sequences. This method uses genome microarrays to research the bacterial system composition of the most important isolated species in an atmosphere.

## 5.7 "Omics" Approaches in Bioremediation

Multi- omics approaches used in various fields of biological research. A variety of 'omics' strategies have opened the system for ecological solution at the molecular level. They have produced a novel hypothesis in design and control for bioremediation (Pankaj Kumar Jain and Vivek Bajpai 2012). The revolution of the 'omics' techniques have forced an altered concept for researchers. Ecogenomics is a novel innovative strategy in 'omics' which means the efficacy of genomic level (DNA, RNA and Protein) to environmental field with a ability of interact between organisms in the environment comparative to evolutionary techniques. 'Omics' methodologies are powerful tools and had the prospective in research areas. 'omics' approaches having ingenious procedures to achieve in various fields.

Nowadays, for successful bioremediation, we need the 'Omics' methodologies. The result of this 'omics' methodologies is solely systematic in biotechnological industrial research areas (Perfumo et al. 2007). 'omics' plays an imperative role in the bioremediation process to clean the landfill leachates and other contaminated atmosphere. Developments in molecular biotechnology methods have opened an innovative era in bioremediation progressions. The improvement of assorted 'omics' approaches provides a proficient opportunities in research areas (Hidayat and Tachibana 2012). 'Omics' plays a fundamental crucial role in estimating biological methods at various levels. Genomics, Metagenomics, Proteomics, Interactomics, Transcriptomics, Metabolomics, Fluxomics, Metametabolomics are the "Omics" approaches plays a vital role in monitoring bioremediation (Fig. 5.3).

A dominant computer-based technology is Genomics. It is widely used to recognize the construction and characteristics of all genes in creatures. Genomics plays

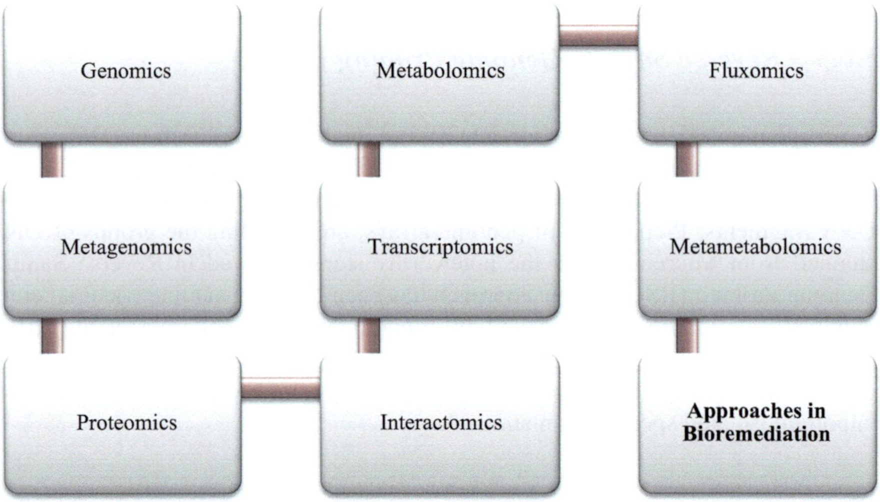


**Fig. 5.3** OMICS approaches monitoring Bioremediation

an imperative role in Bioremediation. The assessment of 16S rRNA genes is a significant relevance to the study of landfill leachates and polluted surroundings, which offered a suggestion of the microbes that logically colonized this atmosphere (Okrent 1999). The atmosphere was influenced to speed up the bioremediation process. Genomics software is used for bioremediation (Priyanka et al. 2020). Next genome sequencing approaches will be decisive in sophisticated the genomic characters of microbes pertinent to bioremediation process. Inclusive genome sequences of some microbes are important in bioremediation.

Examination of mRNA levels for genes are significant in bioremediation can offer the metabolic action of microbes in polluted atmosphere (Slonczewski 2009). Likewise, microorganisms with bioremediation potential have been monitored for entire genome sequencing. Whole genome sequences are enhanced to design whole – genome microarrays to evaluate the expression of genes under different atmospheric conditions. Sequencing occupies an imperative role in genomics (Quintella et al. 2019). Bioremediation processes are researched by isolating microbes from a landfill leachates and contaminated environment.

The segregation and categorization of microbes has been interpreted for molecular analyses in bacterial ecology. The efficacy of molecular techniques is utilized by the bioremediation processes. Microbes are strongly correlated and are significant in bioremediation (Norris and Robert 1993). Genome sequencing of these organisms appraising their genetic approaches can lead to function in landfill leachates. The genetic prospective of microbes in the environment are used in remediation processes (Safiyanu et al. 2015). Applications o genetics shows an assurance to the harmless environment. The domains in 'Omics' targets specific molecules. Metagenomics and genomics targets the DNA molecules. Transcriptomics targets mRNA molecules. Proteomics targets proteins. Metabolomics and Fluxomics targets Metabolites (Satpute et al. 2010). Advances in the high-throughput techniques have significantly enhanced the efficiency purpose and performance of bacterial bioremediation approaches.

In transcriptomics, the separation of genes gets transcribed in any organism which makes link with the proteomes and phenotypes of the microbial cell. The control in gene expression system is must for the existence. This controlling system leads to any alteration those results in environmental circumstance (Kauppi et al. 2011). In transcriptomics, DNA microarrays allowed to identification of RNA expression of every gene of an organism. OMIC approaches are an incredible mission that connects bacterial genes and their products with bioremediation prospective.

The study of entire proteins in an organism is called as Proteomics. Proteins provide signal transduction and catalytic enzymes for metabolic and regulatory pathways of cells. Due to the toxic contaminant present in the environment such as soil and water, there is a variation in cellular expression in an organism (Sandeep Kumar et al. 2013). Then it triggers the physiological response of organisms. Proteomics involved in bioremediation process to identify the abundance of proteins and made it feasible to world-wide alterations in the environment. It down regulate in response to contaminated sites.

Proteomics is particularly helpful for investigating microorganisms that acquires an extensive range of metabolic pathways (Sani et al. 2015). The advantages of this strategy to an ecological model are to clean the environment without any harmful effect (Sayler and Ripp 2000). Environmental proteomics can supply information on the bacterial populations under unambiguous ecological circumstances.

Metabolomics is a recent and exciting field of scientific development. At present, Metabolomic assessment has become very fashionable and new procedures for obtaining and statistics that are helpful for diverse natural biological studies. It is an emerging field. Bioremediation is one of the promising fields and it gets more advances from the metabolomics field. An assortment of bioremediation research can be developed through metabolomics. Bioremediation can be analysed in a more inclusive approach using metabolomics (Quintella et al. 2019). The other OMICS-related approaches offer precious complementary information. The upcoming prospect of metabolic engineering is to assimilate the dissimilar set of statistics to expand most favourable contaminants for use in bioremediation.

Metagenomics is a deliberate approach of analyzing bacterial populations at a genomic level. (Stenuit et al. 2009). It is one of the most excellent scientific advancement to bioremediation (Sukumar and Nirmala 2016). The pollutants are wide spread in world-wide and occur in all resources of life including soil, air and water which are measured as the most essential for human life. Detection and selection of metagenome from contaminated atmosphere are very decisive in a metagenomic study (Samakshi Verma and Arindam Kuila 2019). In bioremediation process, Metagenomics assist in comprehending the features of microbial populations in various types of landfill leachates and polluted surroundings. RT-qPCR methods plays a fundamental role in Metagenomic studies.

Bioreporter technology is one of the monitoring methods for Bioremediation process (Ripp 2019). Genetically engineered microbes generate a quantifiable signal transcriptionally persuaded in response to a precise physical or chemical agent in their atmosphere. This technology is based on the signals produced by an organism (Saranya Kuppusamy et al. 2016). Promoter sequence, Regulatory gene and Reporter gene are the Essential genetic elements in Bioreporter technology. Preliminary use of this technology is for the recognition of noxious materials.

The bioremediation strategy requires using improved and bioengineered microbes. The bioremediation expert has to supply the microorganism to make sure containment and scrutinizing prospective recombinant gene transfer proceedings in the original bacterial communities. This optical tagging can frequently be offered through bioreporter technology (Philp et al. 2014). The handling of landfill leachates can be performed habitually by concerning numerous dissimilar methodologies that are presently obtainable (Menn et al. 2001). For biodegradation, there must be bacterial communities with the metabolic capability to humiliate the contaminant, a surroundings with the accurate rising circumstances for the microorganisms, and the accurate quantity of nutrients and contaminants (Fig. 5.4).

The biological processes used by these microorganisms are extremely unambiguous. Therefore, numerous ecological factors should be taken into description and synchronized as well (Juwarkar et al. 2010). Consequently, bioremediation processes should be specially prepared in agreement to the circumstances at the

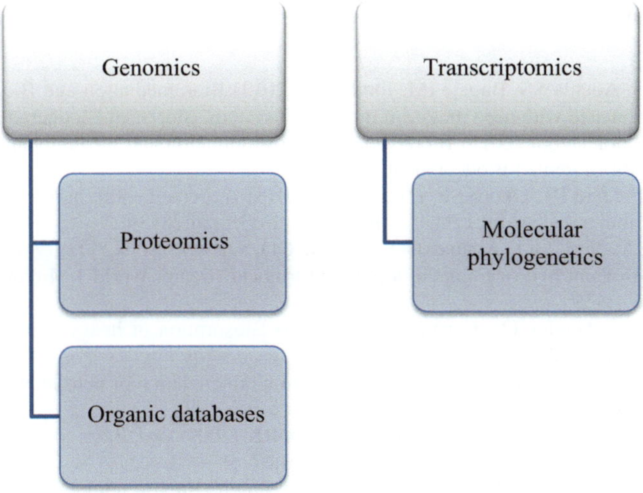


**Fig. 5.4** Branches in Bioinformatics

polluted locations (Steven Ripp et al. 2010). Moreover, numerous factors are small-scale tests, mutually dependent, are frequently executed at the polluted areas. Nevertheless, it can be complicated to extrapolate the grades from the small-scale test methods into big field procedures.

Bioinformatics is a significant element of contemporary biotechnology that merges information technology and biology (Singh et al. 2014a, b). Bioinformatics look for information and maintain all the genetic information, serving to examine and make decision for associations between organic compounds (Paranthaman and Karthikeyan 2015). Bioinformatics has also led to ruling probable cures for detoxification of the surroundings. (Mulligan et al., 2001). Bioinformatics field is an important branch in life science and very crucial to understand (Simarro et al. 2013). Scientist's proficient in evaluating biological information can use the computational tools and softwares to crack the struggles of bioremediation.

## 5.8 Conclusion

In conclusion, the bioremediation of polluted soils, landfill leachates and water is measured an ecologically friendly, flexible, and inexpensive methodology. In addition to restoring the smashed environment at inexpensive, it decreases the danger to healthiness and preserves the biodiversity. Bioremediation shows potential substitute to conventional clean-up methods, generous research is rising in this field hastily. Scientists are annoying to enlarge new bioremediation methods by developing databases for explicit microbes, scheming artificial pathways, and interdisciplinary metabolic pathways that are related to bioremediation process. The OMICs diplomacies play a significant role in monitoring and validating Bioremediation process.

## References

Abdulsalam S, Adefila SS, Bugaje IM, Ibrahim S (2013) Bioremediation and Biodegradation of soil contaminated with used motor oil in a closed system. Bioremed Biodegrad 3:100–172
Adams and Omokhagbor (2015) Bioremediation, biostimulation and bioaugmentation: a review. Int J Environ Bioremed Biodegrad 3(1):28–39
Adams GO, Fufeyin PT, Okoro SE, Ehinomen I (2015) Bioremediation, biostimulation and bioaugmentation: a review. Int J Environ Bioremed Biodegrad:28–39
Adebusoye SA, Ilori MO, Amund OO, Teniola OD, Olatope SO (2007) Microbial degradation of petroleum hydrocarbons in a polluted tropical stream. World J Microbiol Biotechnol 23(8):1149–1159
Ahalya N, Ramachandra TV, Kanamadi RD (2003) Biosorption of heavy metals. Res J Chem Environ:4544–4552
Al-Jawhari IFH (2014) Ability of some soil fungi in biodegradation of petroleum hydrocarbon. J Appl Environ Microbiol 2:46–52
Arora PK, Kumar M, Chauhan A, Raghava GP, Jain RK (2009) OxDBase: a database of oxygenases involved in biodegradation. BMC Res Notes:67
Arvind Kumar (2020) In: Ajay Kumar, Rishikesh Singh, Raghwendra Singh, Shilpi Pandey, Archana Rai, Vipin Kumar Singh, Bhadouria Rahul (eds) Genetically engineered bacteria for the degradation of dye and other organic compounds, pp 331–350
Asira EE (2013) Factors that determine bioremediation of organic compounds in the soil, academic. J Interdiscip Stud:125–128
Atlas RM, Bartha R (1992) Microbial ecology: fundamentals and applications. Benjamin/Cumming Publishing Company, Menlo Park
Babita Kumari, Kriti, Gayatri Singh, Geetgovind Sinam, Singh DP (2020) Microbial remediation of crude oil-contaminated sites, pp 333–351
Baek KH, Lee YK, Lee IS, Oh HM, Yoon BD, Kim HS (2004) Detection of Nocardia sp. H17-1 by PCR during bioremediation of crude oil-contaminated soil. Kor J Microbial Biotechnol:91–95
Bakermans C, Madsen E (2002) Detection in coal tar waste-contaminated groundwater of mRNA transcripts related to naphthalene dioxygenase by fluorescent in situ hybridization with tyramide signal amplification. J Microbiol Methods 50:75–84
Belkin S (2003) Microbial whole-cell sensing systems of environmental pollutants. Curr Opin Microbiol 6:206–212
Boopathy R (2000) Factors limiting bioremediation technologies. Bioresour Technol 74:63–67
Cebron A, Beguiristain T, Leyval C, Norini M (2008) Real-time PCR quantification of PAH-ring hydroxylating dioxygenase (PAH-RHD alpha) genes from gram-positive and gram-negative bacteria in soil and sediment samples. J Microbiol Methods 73:148–159
Chandrika Malkanthi Nanayakkara, Ayoma Witharana (2019) Bioremediation of oil contaminated soil and water
Chen C, Wang JL (2007) Characteristics of Zn2+ biosorption by *Saccharomyces cerevisiae*. Biomed Enviro Sci 20:478–482
Chen B, Ye X, Zhang B, Jing L, Lee K (2019) Chapter 22: Marine oil spills—preparedness and countermeasures. In: Sheppard C (ed) World seas: an environmental evaluation, 2nd edn. Academic, pp 407–426
Cho JC, Tiedje JM (2001) Bacterial species determination from DNA–DNA hybridization by using genome fragments and DNA microarrays. Appl Environ Microbiol 67:3677–3682
Cho JC, Tiedje JM (2002) Quantitative detection of microbial genes by using DNA microarrays. Appl Environ Microbiol 68:1425–1430
Das N, Chandran P (2011) Microbial degradation of petroleum hydrocarbon contaminants: an overview- biotechnology, pp 1–13
Davison J (2005) Risk mitigation of genetically modified bacteria and plants designed for bioremediation. J Ind Microbiol Biotechnol:639–650

Demnerova K, Mackova M, Spevakova V, Beranova K, Kochankova L (2005) Two approaches to biological decontamination of groundwater and soil polluted by aromatics characterization of microbial populations. Int Microbiol 8:205–211

Diplock EE, Alhadrami HA, Paton GI (2009) Application of microbial bioreporters. Environ Microbiol Bioremed:189–209

Dokic L, Savic M, Narancic T, Vasiljevic B (2010) Metagenomic analysis of soil microbial communities. Arch Biol Sci Belgrade 62(3):559–564

Dua M, Singh A, Sethunathan N (2002) Biotechnology and bioremediation: successes and limitations. Appl Microbial Biotechnol 59:143–152

Dubey SK, Tripathi AK, Upadhyay SN (2006) Exploration of soil bacterial communities for their potential as bioresource. Bioresour Technol 97:2217–2224

El Fantroussi S, Agathos SN (2005) Is bioaugmentation a feasible strategy for pollutant removal and site remediation? Curr Opin Microbiol 8:268–275

Fosso-Kankeu E, Mulaba-Bafubiandi AF (2014) Implication of plants and microbial metalloproteins in the bioremediation of polluted waters: a review. Phys Chem Earth:242–252

Fulekar MH, Sharma J, Tendulkar A (2012) Bioremediation of heavy metals using biostimulation in laboratory bioreactor. Environ Monit Assess 184(12):7299–7307

Gavrilescu M (2006) Overview of in situ remediation technologies for sites and groundwater. Environ Eng Manag J 5(1):79–114

Hazen T, Palmisano A (2003) Bioremediation of metals and radionuclides: what it is and how it works, 2nd edn. Lawrence Berkeley National Laboratory, pp 23–25

Hidayat A, Tachibana S (2012) Biodegradation of aliphatic hydrocarbon in three types of crude oil by *Fusarium* sp, F092 under stress with artificial sea water. J Environ Sci Technol 5:64–73

Jain PK, Bajpai V (2012) Biotechnology of bioremediation- a review. Int J Environ Sci 3:535–549

Janssen DB, Stucki G (2020) Perspectives of genetically engineered microbes for groundwater bioremediation. Environ Sci Proc Impact:487–499

Jariyal M, Yadav M, Singh NK, Yadav S, Sharma I, Dahiya S, Thanki A (eds) (2020) Microbial remediation progress and future prospects, pp 187–214

Juhanson J, Truu J, Heinaru E, Heinaru A (2009) Survival and catabolic performance of introduced pseudomonas strains during phytoremediation and bioaugmentation field experiment. FEMS Microbiol Ecol:446–455

Juwarkar AA, Singh SK, Mudhoo A (2010) A comprehensive overview of elements in bioremediation. Rev Environ Sci Biotechnol 9(3):215–288

Karigar CS, Rao SS (2011) Role of microbial enzymes in the bioremediation of pollutants: a review. Enzyme Res:1–11

Kate MS, Kristin AH (2005) Natural attenuation and enhanced bioremediation of organic contaminants in groundwater. Curr Opin Microbiol 16(3):246–253

Kauppi S, Sinkkonen A, Romantschuk M (2011) Enhancing bioremediation of diesel-fuel-contaminated soil in a boreal climate: comparison of biostimulation and bioaugmentation. Int Biodeterior Biodegrad 65(2):359–368

Kulshreshtha S (2013) Genetically engineered microorganisms: a problem solving approach for bioremediation. J Bioremed Biodegr 4:1–2

Kumar A, Bisht BS, Joshi VD, Dhewa T (2011) Review on bioremediation of polluted environment: a management tool. Int J Environ Sci 1:1079–1093

Kuppusamy S, Maddela NR, Megharaj M, Venkateswarlu K, Kuppusamy S, Maddela NR, Megharaj M, Venkateswarlu K (2020) An overview of total petroleum hydrocarbons. In: Total petroleum hydrocarbons. Springer, Berlin/Heidelberg, pp 1–27

Ledin M (2000) Accumulation of metals by microorganisms—processes and importance for soil systems. Earth Sci Rev:1–31

Lee TH, Byun IG, Kim YO, Hwang IS, Park TJ (2006) Monitoring biodegradation of diesel fuel in bioventing processes using in situ respiration rate. Water Sci Technol 53:263–272

Lovley DR (2003) Cleaning up with genomics: applying molecular biology to bioremediation. Nat Rev Microbiol:35–44

Macaulay BM (2014) Understanding the behaviour of oil-degrading micro-organisms to enhance the microbial remediation of spilled petroleum. Appl Ecol Environ Res 13:247–262

Madhavi GN, Mohini DD (2012) Review paper on – parameters affecting bioremediation. Int J Life Sci Pharma Res:77–80

Mareddy AR (2017) Environmental impact assessment. Theory and practice; technology in EIA. Butterworth- Heinemann, Oxford. Chapter 13; ISBN 9780128111390

Margaret W, Jeremy B (2014) Water bioremediation. Elsevier, pp 13–34

Menn F, Easter JP, Sayler GS (2001) Genetically engineered microorganisms and bioremediation. Biotechnol Set:441–463

Mulligan CN, Yong R, Gibbs BF (2001) Remediation technologies for metal contaminated soils and groundwater: an evaluation. Eng Geol:193–207

Nester EW, Denise G, Anderson C, Roberts E Jr, Pearsall NN, Nester MT (2001) Microbiology: a human perspective, 3rd edn. McGraw-Hill, New York, pp 23–35

Nicolaou SA, Gaida SM, Papoutsakis ETA (2010) Comparative view of metabolite and substrate stress and tolerance in microbial bioprocessing: from biofuels and chemicals, to biocatalysis and bioremediation. Metab Eng:307–331

Niti C, Sunita S, Kamlesh K (2013) Bioremediation: an Emerging technology for remediation of pesticides. Res J Chem Environ 17:88–105

Norris, Robert (1993) Handbook of bioremediation, vol 45. CRC Press

Okrent D (1999) On intergenerational equity and its clash with intragenerational equity and on the need for policies to guide the regulation of disposal of wastes and other activities posing very long-term risks. Risk Analysis, Elsevier article, pp 877–901

Paranthaman SR, Karthikeyan B (2015) Bioremediation of heavy metal in paper mill effluent using *Pseudomonas* spp. Int J Microbiol 1:1–5

Perfumo A, Banat IM, Marchant R, Vezzulli L (2007) Thermally enhanced approaches for bioremediation of hydrocarbon-contaminated soils. Chemosphere 66:179–184

Philp JC, Whiteley AS, Ciric L, Bailey MJ (2014) Monitoring bioremediation, pp 237–268

Phulia V, Jamwal A, Saxena N, Chadha NK, Muralidhar (2013) Technologies in aquatic bioremediation. In: Freshwater ecosystem and xenobiotics chapter: technologies in aquatic bioremediation. Discovery Publishing House, New Delhi, pp 65–91

Priyanka, Kumar C, Chatterjee A, Wenjing W, Yadav D, Singh PK (2020) Cyanobacteria:193–202

Quintella CM, Mata AMT, Lima LCP (2019) Overview of bioremediation with technology assessment and emphasis on fungal bioremediation of oil contaminated soils. Environmental Management. Elsevier, pp 156–166

Rajendran P, Muthukrishnan J, Gunasekaran P (2003) Microbes in heavy metal remediation. Indian J Exp Biol:935–944

Ripp S (2019) Bioreporter technology for monitoring soil bioremediation, pp 233–250. https://doi.org/10.1007/3-540-28904-6_11

Ripp S, Di Claudio ML, Sayler GS (2010) Biosensors as environmental monitors, pp 213–233. https://doi.org/10.1002/9780470495117.ch9

Safiyanu I, Isah AA, Abubakar US, Rita Singh M (2015) Review on comparative study on bioremediation for oil spills using microbes. Res J Pharm Biol Chem Sci 6:783–790

Samakshi Verma, Arindam Kuila (2019) Bioremediation of heavy metals by microbial process. Environ Technol Innov. https://doi.org/10.1016/j.eti.2019.100369

Sandeep Kumar, Vikas Kumar Dagar, Yogender Pal Khasa, Ramesh Chander Kuhad (2013) Genetically modified microorganisms (GMOs) for bioremediation, pp 191–218

Sani I, Safiyanu I, Rita SM (2015) Review on bioremediation of oil spills using microbial approach. IJESR 3:41–45

Saranya Kuppusamy, Thavamani Palanisami, Mallavarapu Megharaj, Kadiyala Venkateswarlu, Ravi Naidu (2016) In-situ remediation approaches for the management of contaminated sites: a comprehensive overview, pp 1–115

Satpute SK, Bhuyan SS, Pardesi KR, Mujumdar SS, Dhakephalkar PK, Shete AM, Chopade BA (2010) Molecular genetics of biosurfactant synthesis in microorganisms, pp 14–41. https://doi.org/10.1007/978-1-4419-5979-9_2

Sayler GS, Ripp S (2000) Field applications of genetically engineered microorganisms for bioremediation processes. Curr Opin Biotechnol:286–289

Scow KM, Hicks KA (2005) Natural attenuation and enhanced bioremediation of organic contaminants in groundwater. Curr Opin Biotechnol 16(3):246–253

Shanker R, Purohit HJ, Khanna P (1998) Bioremediation for Hazardous waste management: the Indian scenario. Bioremed Technol Principles Pract:81–96. ISBN 978-1-56676-561-9

Shekhar SK, Godheja J, Modi DR (2020) Molecular technologies for assessment of bioremediation and characterization of microbial communities at pollutant-contaminated sites, Chapter-18. Springer, pp 438–464

Shemer B, Belkin S (2019) Microbial biosensors for the detection of organic pollutants. Springer, Cham, pp 1–24

Simarro R, Gonzalez N, Bautista LF, Molina MC (2013) Assessment of the efficiency of in situ bioremediation techniques in a creosote polluted soil: change in bacterial community. J Hazard Mater 262:158–167

Singh JS, Abhilash PC, Singh HB, Singh RP, Singh DP (2011). Genetically engineered bacteria: an emerging tool for environmental remediation and future research perspectives. Gene 480(1–2):1–9

Singh JS, Abhilash PC, Singh HB, Singh RP, Singh DP (2014a) Genetically engineered bacteria: an emerging tool for environmental remediation and future research perspectives. Gene 480(1-2):1–9

Singh R, Singh P, Sharma R (2014b) Microorganism as a tool of bioremediation technology for cleaning environment: a review. Proc Int Acad Ecol Environ Sci:1–6

Slonczewski JL (2009) Stress responses: pH. In: Schaechter, moselio, encyclopedia of microbiology, 3rd edn. Elsevier, pp 477–484. ISBN 978-0-12-373944-5

Stenuit B, Eyers L, Schuler L, George I, Agathos SN (2009) Molecular tools for monitoring and validating bioremediation:339–353. https://doi.org/10.1007/978-3-540-89621-0_18

Sukumar S, Nirmala P (2016) Screening of diesel oil degrading bacteria from petroleum hydrocarbon contaminated soil. Elsevier Article:18–22

Talos K, Pager C, Tonk S, Majdik C, Kocsis B (2009) Cadmium biosorption on native *Saccharomyces cerevisiae* cells in aqueous suspension. Agric Environ:20–30

Terry Gentry, Christopher Rensing, Ian Pepper (2004) New approaches for bioaugmentation as a remediation technology. Critical Rev Environ Sci Technol:447–494. https://doi.org/10.1080/10643380490452362

Tigini V, Prigione V, Giansanti P, Mangiavillano A, Pannocchia A (2010) Fungal biosorption, an innovative treatment for the decolourisation and detoxification of textile effluents. Water:550–565

Tingting X, Perry N, Chuahan A, Sayler G, Ripp S (2014) Microbial indicators for monitoring pollution and bioremediation, pp 115–136. https://doi.org/10.1016/B978-0-12-800021-2.00005-4

Vidali M (2001) "Bioremediation. An overview" (PDF). Pure Appl Chem 73(7):1163–1172

Walworth J, Pond A, Snape I, Rayner J, Ferguson S, Harvey P (2005) Fine tuning soil nitrogen to maximize petroleum bioremediation, pp 251–257

Yeonghee A, Jung H, Tatavarty R, Choi H, Yang J-w, Kim S (2005) Monitoring of petroleum hydrocarbon degradative potential of indigenous microorganisms in ozonated soil. Biodegradation 16(1):45–56. https://doi.org/10.1007/s10531-004-0428-2

Zabochnicka-Swiatek M, Krzywonos M (2014) Potentials of biosorption and bioaccumulation processes for heavy metal removal. Pol J Environ Stud:551–561

# Chapter 6
# An Overview of Physicochemical and Biological Treatment of Landfill Leachate

**MD Tanvir Hasnine, N. Anand, Ali Zoungrana, Sankar Ganesh Palani, and Qiuyan Yuan**

**Abstract** Landfilling is the most widely used method for solid waste disposal, with a heavy burden of generating high-strength wastewater known as leachate. Leachate consists of various mixtures of organic and inorganic pollutants, heavy metal contaminants, dissolved and suspended materials. Therefore, landfill leachate should be collected and adequately treated to prevent its unfortunate fate associated with groundwater and surface water contaminations. Biological and physical/chemical are two major treatment technologies for leachate. Conventional treatment techniques have been widely used in landfill leachate treatment. Even though these techniques might have failed to remove all pollutants effectively, they can be good pretreatment techniques, while new and evolving processes are promising technologies for the effective treatment of leachate. The first part of the chapter sums up the landfill leachate treatability with physicochemical processes. Conventional physicochemical techniques, including coagulation-flocculation, chemical precipitation, and new technologies such as ion exchange, adsorption by activated carbon, membrane filtration, chemical oxidation, electrochemical processes are illustrated and their application to landfill leachate treatment discussed. The second part of this chapter highlights various aerobic and anaerobic biological treatment technologies that employ microorganisms and fungi that are operated either alone or in combination with the physicochemical treatment methods.

---

M. T. Hasnine · Q. Yuan (✉)
Environmental Engineering Program, Department of Civil Engineering, University of Manitoba, Winnipeg, MB, Canada
e-mail: qiuyan.yuan@umanitoba.ca

N. Anand · S. G. Palani
Environmental Biotechnology Laboratory, Department of Biological Sciences, BITS Pilani, Hyderabad Campus, Hyderabad, Telangana, India

A. Zoungrana
Department of Environmental Engineering, Yildiz Technical University, Istanbul, Turkey



P. Pathak, S. G. Palani (eds.), *Circular Economy in Municipal Solid Waste Landfilling: Biomining & Leachate Treatment*, Radionuclides and Heavy Metals in the Environment, https://doi.org/10.1007/978-3-031-07785-2_6

**Keywords** Landfill leachate · Pretreatment · Physicochemical treatment · Biological treatment · Combined treatment

## 6.1 Introduction

Solid waste production is intrinsically connected to the world population growth and the improvement of living standards. It causes diverse impacts on our daily life, causing malodors and unaesthetic view, source of bacteria and dangerous chemicals, attracting rats and harmful insects such as mosquitoes, responsible for infectious diseases, therefore being hurtful to the environment and living things. Collecting, managing, and proper disposal of solid waste are of great importance in mitigating and preventing its unhealthy effects.

Dumping solid waste into dumpsites is an old waste management method that has proven dangerous and unsuitable for any environment, especially in today's context of industrialization, where solid waste varies a lot in its content. Landfilling is the most commonly used technique for the management and disposal of solid wastes around the globe (Kurniawan et al. 2006b; Renou et al. 2008; Visvanathan et al. 2007). Landfilling is preferred in many countries for solid waste management because it is relatively less expensive, able to receive various types of waste, and a specific location for disposal can be chosen and monitored. In addition, organic and recyclable materials in landfills can be composted, used to produce natural gas or other reasons. Landfills can also be reclaimed, built on, or repurposed as parks or agricultural land. However, landfilling is not without its negatives; it produces future wastes that may contain enormous amounts of toxic chemicals that are difficult to treat., and one of the most common problems attracting environmental scientists (EPA 2011; Kjeldsen et al. 2002; Kurniawan et al. 2006a; Li et al. 2009; Salem et al. 2008; Saquing et al. 2014; Trebouet et al. 2001; Xu et al. 2008). Wastewater generated by a landfill is termed landfill leachate, a liquid that drains or 'leaches' from the landfill in decomposition through which rainwater infiltrates. Its composition varies and depends on the landfill's age and the sort of waste it contains, and it frequently carries both dissolved and suspended materials. Thus, landfill waste management plants require close environmental engineering monitoring in their design and operation because of the leachate they generate, which readily could contaminate nearby groundwater and surface water.

Landfill leachate varies widely in composition but mainly contains inorganic macro components, heavy metals, dissolved organic matter, and organic compounds that are xenobiotic in nature (Kjeldsen et al. 2002; Kurniawan et al. 2006b; Peng 2017). These pollutants are difficult to be treated and generally require complex multiple-stage treatment processes, sometimes combining biological and physicochemical techniques. The various techniques used for leachates treatment can be summarized in two main groups as conventional treatments, including leachate transfer, biodegradation, physicochemical methods, and new treatment by use of membrane processes, adsorption, ion exchange processes, or advanced oxidation processes (Kurniawan et al. 2006b; Renou et al. 2008; Salem et al. 2008; Visvanathan et al. 2007; Xu et al. 2008). The biological processes are used for the treatment of young leachates with high biodegradable matters and are effective for nitrogen (N) removal through the nitrification process. However, the presence of inhibitors such as heavy metals, sulfide commonly found in leachate can reduce the efficiency or block the nitrification process (Kabdasli et al. 2000). Physiochemical processes are competent in treating recalcitrant, inorganic matters, and heavy metals.

The present book chapter intends to renew our understanding of municipal solid waste landfill leachate in our environment and understand conventional and advanced physicochemical and biological treatment techniques used to treat them. The operating mechanisms of the technologies, their role in the treatment, the target pollutants for each process, and their performance will be discussed. The application of conventional processes such as coagulation-flocculation, chemical precipitation, and air stripping and new physicochemical treatment technologies including membrane technology, ion exchange processes, adsorption, and advanced oxidation processes (AOP), and lastly, biological processes, mainly anaerobic, membrane bioreactor, and hybrid processes will be discussed in their application for landfill leachate treatment.

## 6.2 Municipal Solid Waste Landfill and Leachate

### *6.2.1 Landfills*

Solid waste is the undesirable (unusable) solid materials produced through residential, industrial, and commercial activities. It can be classified based on its source (household, commercial, industrial, construction, or institutional); contents (organic material, metal, glass, etc.); or hazard potential (toxic, radioactive, combustible, infectious, etc.). A substantial quantity of garbage is generated each day around the globe and exponentially increases with population growth, industrialization, and changes in consumption patterns. To face the impact of increasing waste amount, many countries are investigating environmentally friendly disposal strategies through Integrated Solid Waste Management (ISWM). Landfilling is among the most well-known solid waste management techniques employed worldwide (Kurniawan et al. 2006b; Renou et al. 2008; Visvanathan et al. 2007). It is a monitored solid waste dumping ground. Landfills are broadly classified as (i) Sanitary Landfills, (ii) Secure Landfills. However, according to the type of waste to dispose of, landfills can be categorized as municipal solid waste (MSW), hazardous, industrial, coal combustion residuals waste landfills, and building and demolition waste landfills (*Landfills | US EPA*n.d.). It is a widely used waste management method due to its low cost, ability to receive various types of waste, and ease of choosing a specific location for waste disposal and monitoring.

Up to 95% of all MSW collected around the world is disposed of in landfills. Over 150,000 landfills, with various capacities, were built globally (Kurniawan et al. 2006a). The Apex Regional Landfill in Las Vegas, United States, is 250 years old, one of the world's largest plants, covering about 2200 acres of land, which holds about 50 million tons of waste (*Statista*n.d.). However, it is worth mentioning that many municipalities, especially in low-income countries, do not have essential services for waste collection and disposal. The generated waste is disposed of improperly on open dumpsites. This has environmental impacts and severe health issues such as diarrheal diseases, low birth weights, some cancers (*Statista*n.d.).

Even though landfilling is a vital waste management technique, it is becoming less attractive in recent years in the profit of increasing recycling. MSW recycling has increased from less than 10% of MSW produced in 1980 to about 34% in 2010, while garbage disposal in landfills has dropped from 89% of MSW generated in 1980 to around 54% in 2010 (EPA 2011; Saquing et al. 2014). Landfilling responded to some of the limits associated with dump yards by mitigating the potential risks connected with the environment and human health. In an engineered landfill, the waste is stored by layers on a prepared site (ground), compacted with heavy equipment, and the garbage collection is covered with land (Pujotomo and Azzahra 2019), while adequate arrangements are set for the leachate management to avoid its infiltration into the ground. After being landfilled, residues decompose over time through a series of combined physicochemical and biological processes and result in leachate formation (Kurniawan et al. 2006a).

### 6.2.2 Landfill Leachate

The major drawback associated with landfilling is the production of leachate, high strength wastewater, which is hard to manage, and a source of environmental threats (Benoit et al. 2016; Bilgili et al. 2006; Li et al. 2009; Salem et al. 2008; Silva et al. 2004). When soluble elements leach or dissolve out of solid material in a landfill site (Salem et al. 2008) by percolating water (generally due to excess rainwater), leachate is formed (Kurniawan et al. 2006b; Renou et al. 2008). Pollutants are transferred from the waste materials to the leaching water due to a combination of physicochemical and microbiological processes in the waste (Kjeldsen et al. 2002). Besides causing malodors and aerosols, untreated leachates that seep from a dumpsite by permeating groundwater or mixing with surface water can result in soil, groundwater, and surface water pollution (Deng and Englehardt 2006; Sheng Li et al. 2009). Due to its composition, this multi-complex wastewater exhibits acute and chronic toxicity, and it is harmful to living things, human life, and the surrounding environment.

Complex series of chemical and biological processes occurring during leachate formation has been studied, and four main phases: the aerobic, the anaerobic acid, the initial methanogenic, and the stable methanogenic phase are proven to occur during leachate decomposition (Kjeldsen et al. 2002; Kurniawan et al. 2006b). The aerobic phase lasts only a few days, during which the oxygen is rapidly consumed in the waste and $CO_2$ is produced. Once all oxygen is consumed, the anaerobic phase starts in the waste, such as fermentation reactions. Cellulose and hemicellulose (representing 45–60% MSW dry weight) decompose to methane ($CH_4$) and $CO_2$ during this phase. When a certain amount of $CH_4$ is produced in the anaerobic phase, the initial methanogenic phase starts. Methanogenic bacteria convert acids formed in the acidic phase to $CH_4$. $CH_4$ production rate reaches its maximal in the stable methanogenic phase. Environmental conditions in the landfill (primarily moisture) significantly impact the rate of refuse decomposition (Kjeldsen et al. 2002).

### *6.2.3 Landfill Leachate Characteristics*

Landfill leachates vary widely in composition depending on the deposited waste type, hydrogeological factors, and mainly landfill age. It usually contains both dissolved and suspended materials. Organic and inorganic materials, including humic-type compounds, ammonia nitrogen ($NH_3$-N), heavy metals, and chlorinated organic and inorganic salts, can all be found in considerable proportions in leachate (Deng and Englehardt 2006; Kurniawan et al. 2006a; Li et al. 2009; Silva et al. 2004; Trebouet et al. 2001). Landfills may receive a mixture of different wastes from various origins, and each of the receiving sources affects the final composition; however, four groups of pollutants are roughly determined in landfill leachate:

- **Dissolved organic matter**: Total Organic Carbon (TOC), refractory compounds such as humic-like and fulvic-like compounds, Chemical Oxygen Demand (COD), and volatile fatty acids (VFA).
- **Inorganic macro components:** common cations and anions including calcium ($Ca^{2+}$), magnesium ($Mg^{2+}$), chloride ($Cl^-$), ammonium ($NH_4^+$), hydrogen carbonate ($HCO_3^-$), and sulfate ($SO_4^{2-}$).
- **Heavy metals:** chromium ($Cr^{3+}$), lead ($Pb^{2+}$), copper ($Cu^{2+}$), cadmium ($Cd^{2+}$), zinc ($Zn^{2+}$), and nickel ($Ni^{2+}$).
- **Xenobiotic organic compounds:** Halogenated organics such as phenols, plasticizers, aromatic hydrocarbons, and insecticides (Kjeldsen et al. 2002; Kurniawan et al. 2006b; Peng 2017).

When the acidic phase is dominant, it is young leachate, rich in BOD and COD with a high BOD/COD ratio (>0.7). Young leachates have a low pH due to the high concentration of VFA in such leachates (Renou et al. 2008). Leachates older than 10 years, termed old leachates, are generally in the methanogenic phase. Old leachates are not suitable for biological treatment due to the low BOD/COD ratio, which is generally below 0.1 (Bohdziewicz and Kwarciak 2008). The classification of leachates based on their age is depicted in Table 6.1.

**Table 6.1** Age-based characteristics of landfill leachate (Bohdziewicz and Kwarciak 2008)

| Characteristics | Young | Middle age | Old |
|---|---|---|---|
| Age (year) | <5 | 5–10 | >10 |
| pH | 6.5 | 6.5–7.5 | >7.5 |
| COD (mg/L) | >10.000 | 4.000–10.000 | <4.000 |
| BOD/COD | >0.3 | 0.1–0.3 | <0.1 |
| Heavy metals | Low-middle | – | Low |
| Biological treatability | High | Middle | Low |

### 6.2.4 *Landfill Leachate Effect on the Environment and Local Ecology*

Leachate can be very toxic to humans and the local ecology if the original waste contains harmful compounds, such as some industrial waste. Its potential dangers have been proven, and treatment is a must before disposing of it in the environment (Li et al. 2009). The pollution of ground and surface waters is the most severe potential environmental risk of landfill leachate, with a great danger of groundwater contamination (EPA 2011). Toxic substances found in stabilized leachate (adsorptive organic halogen compounds, humic acids, $NH_3$–N) are considered essential toxicants to living organisms (Kurniawan et al. 2006b). High $NH_3$ content in leachate can be hazardous to biological leachate treatment. Heavy metals in leachate are a source of concern because of their negative impact on the environment and living organisms, as reported by Haque et al. (2013) by emphasizing the possible entrance of heavy metals from landfill leachate into the food chain through soil contamination (Silva et al. 2004). In addition, the microbiological characteristics of leachate pose significant pollution, which can raise the concern of infection (Deng and Englehardt 2006).

### 6.2.5 *Overview of Landfill Leachate Treatments*

Municipal landfill leachate is complex wastewater that requires adequate treatment technologies to make it suitable for discharge. Mainly used technologies for leachate treatment are biological, physicochemical, AOP, natural systems (Aminul Haque et al. 2013; Hasar et al. 2009; Kjeldsen et al. 2002; Silva et al. 2004; Sun et al. 2015). The leachate characteristics, age, and structure control the efficiency of the treatment. As a result, adequate treatment of leachate is complex, and the application of various processes is generally necessary due to the high contents of $NH_4^+$ and high COD/BOD ratio (Khai 2012).

Biological leachate treatment includes aerobic treatments (activated sludge and aerated lagoons) and anaerobic treatment (anaerobic lagoons, reactors) (Raghab et al. 2013). Biological processes are mainly applied for N removal and are adequate to treat immature leachate rich in biodegradable matters and poor in toxic chemicals and heavy metals (Zielińska et al. 2020). Various physicochemical treatment techniques such as air stripping, chemical precipitation, oxidation, and reduction, coagulation (Raghab et al. 2013; Anqi et al. 2020), on the other hand, are used for the removal of organic pollutants and may be suitable even for stabilized leachate and for the removal of refractory substances (Zielińska et al. 2020). Advanced techniques, including carbon adsorption, ion exchange, AOP, and membrane techniques (Raghab et al. 2013) are growing interest with better efficiency. The application of physicochemical methods can be complementary to the biological methods for the effective treatment of leachate due to the limited biodegradability of leachates (Wu et al. 2010).

## 6.3 Physicochemical Treatment of Landfill Leachate

### *6.3.1 Conventional Physicochemical Treatment Techniques of Landfill Leachate*

The conventional treatment method for landfill leachate is a multistage process that can be divided into three categories: leachate transfer, biodegradation, and physical and chemical approaches. Leachate transfer comprises recycling and leachate treatment, whereas aerobic and anaerobic processes are used in biodegradation (Renou et al. 2008). The selection of biological and/or physicochemical treatment of leachate varies on the level of contamination (Postacchini et al. 2018). For instance, biodegradable materials are removed using a biological treatment process, organic components and $NH_3$ are removed using ion-exchange, metals and colloids are eliminated by coagulation/flocculation process, metals and organics are removed using adsorption, and AOPs are used to remove various organic components (Chai Ken et al. 2016). An integrated approach with different techniques is more favorable because a single technology cannot fulfill the required removal of pollutants (Mojiri 2015).

#### 6.3.1.1 Coagulation-Flocculation

Coagulation-flocculation is a process that disrupts tiny suspended particles, causing accumulation and the formation of flocs, which can then be separated via sedimentation, flotation, or filtration processes. As a result, these processes are thought to aid in eliminating non-biodegradable materials and metal ions (Tatsi et al. 2003). Coagulation-flocculation is a standard pretreatment process, and when paired with membrane separation techniques, it can help prevent membrane fouling and flux decline (Alfaia et al. 2019). Furthermore, the alkalinity of the leachate can be controlled by the coagulating agent, such as calcium oxide, which can lower the alkalinity of the leachate during the coagulation-flocculation process and generate non-toxic, readily disposable sludge (de Almeida et al. 2019).

In the coagulation and flocculation process, suitable coagulant selection and pH effects are critical factors for landfill leachate treatment. In research, it was found that iron salts are much more promising when leachate is treated in coagulation and flocculation than aluminum ones. The removal efficiency for COD was found to be up to 50% for iron salts, whereas for aluminum the removal was in the range of 10–40% (Alfaia et al. 2019). As outlined by Amokrane et al. (1997), a combination of coagulants or the addition of flocculants to coagulants may boost the floc settling rate, resulting in improved COD removal efficacy. This research reported a decrease of COD from 3800 to 1700 mg/L and 2200 mg/L using $FeCl_3$ and $Al_2(SO_4)_3$, respectively, as coagulants.

#### 6.3.1.2 Chemical Precipitation

Chemical precipitation is a cost-effective, uncomplicated process used to remove $NH_4^+$ and some refractory organics from leachate (Barnes et al. 2007; Fard et al. 2016; Ates and Argun 2018). It is used as a pretreatment technique and can effectively precipitate $NH_4^+$ from leachate and quickly form settleable insoluble compounds. $NH_4^+$-N precipitation by forming magnesium ammonium phosphate (MAP) precipitate, named the struvite precipitation technique, is a propitious and effective technology for $NH_4^+$-N removal from leachate (Dogan 2018; Li et al. 2012). Struvite precipitation forms partially or entirely insoluble matters easily removable from the aqueous phase. The insoluble form, $MgNH_4PO_4.6H_2O$, is known in analytical chemistry with low solubility of 0.023 g/100 ml-$H_2O$ at 0 °C and is mineralogically known as struvite (Li et al. 1999). Struvite crystals are white orthorhombic crystalline structures consisting of equal molar concentrations of $Mg^{2+}$, $NH_4^+$, and phosphate ($PO_4^{3-}$). Below is the reaction of struvite precipitation (Dogan 2018; Li et al. 1999, 2012):

$$Mg^{2+} + PO_4^{3-} + NH_4^{+} + 6H_2O <-------- > MgNH_4PO_4 \cdot 6H_2O \downarrow \text{pKs} = 12.6(25°C) \tag{6.1}$$

Struvite precipitation is considered a cost-effective technique for eliminating nutrients (N and $PO_4^{3-}$) from wastewater. Its performance and success depend on the ratio of $Mg^{2+}$:$NH_4^+$-N:$PO^{3-}$-P and the pH of the solution, as reported by Münch and Barr (2001). Li and Zhao (2001) achieved a fast reduction of $NH_4^+$-N concentration from 5618 to 112 mg/L by struvite precipitation with an equal molar ratio of $Mg^{2+}$:$NH_4^+$-N:$PO_4^{3-}$P (1:1:1) (Li et al. 2012). Struvite precipitation effectively removes $NH_4^+$-N, and the process cost is almost similar to the biological nitrification-denitrification process, approximately 20% more expensive than air stripping (Li et al. 1999). The resulting MAP is a suitable fire retardant in fabrics (Li et al. 1999) and since it has a similar composition of Mg, P, and N to that of commercial fertilizers used in the soil, therefore has soil fertilizing properties (Li et al. 1999, 2012; Siegrist 1996).

#### 6.3.1.3 Air Stripping

Air stripping is a technique for purifying wastewater leachate containing volatile compounds that require the mass transfer of undesirable compounds from the liquid to the gas phase into an air stream (Yuan et al. 2016). It can also be used to reduce leachate toxicity by removing $CH_4$, $NH_4^+$, and volatile organic compounds (VOCs) (Campos et al. 2013).

The stripping efficiency can be enhanced by monitoring the temperature, pH, and retention time (RT) (Renou et al. 2008). It is suitable as a pretreatment technique for biological processes. In a research, it was found that when the pH is 11 at 20 °C together with 24 h RT, the reduction of $NH_3$ was 89% (Marttinen et al. 2002). Cheung et al. (1997) also found that with a 24-h RT, 93% of $NH_3$ was removed in the air stripping process. In a different study, they had demonstrated that when the starting $NH_3$ concentration was 2020 mg/L and 800 mg/L, they were able to remove 85% and 99.5% of the $NH_3$, respectively (Ozturk et al. 2003; Silva et al. 2019). However, in the air stripping process, $NH_3$ is released into the atmosphere, which is one of the drawbacks of the process, and to overcome this problem, $NH_3$ should be absorbed previously by $H_2SO_4$ to form $(NH_4)_2SO_4$, which is used as fertilizer for plant life in the form of N and sulfur (S). However, significant drawbacks include the development of calcium carbonate precipitation due to the usage of lime to adjust the pH and the requirement for a vast tower to avoid foaming during stripping.

### *6.3.2 New Physicochemical Treatment Technologies of Landfill Leachate*

#### 6.3.2.1 Pressure Driven Membrane Technology

Ultrafiltration (UF), reverse osmosis (RO), nanofiltration (NF), and microfiltration (MF) are pressure-driven membrane techniques used in water treatment, with driving force as the hydrostatic pressure differences (Madsen 2014). The filtration occurs through these membranes because a pressure gradient, called transmembrane pressure (TMP) is created across the membrane. Figure 6.1 depicts a schematic view of pressure-driven membrane processes.

MF membranes have a larger pore size (0.1–1 μm) and work at lower pressure (0.5–1 bar). They can eliminate particles with a size of around 1 μm and larger. In landfill leachate, MF membranes can be used as a pretreatment technique, and effluents can further be treated by smaller pore size membranes and membrane distillation or some chemical methods (Madsen 2014; Trebouet et al. 2001; Zhang et al. 2014).

Compared to MF membranes, UF membranes have a smaller pore size (0.01–0.1 μm) and higher operating pressure (1–10 bar), so a better rejection efficiency. The membrane can remove macromolecules and particles of pores size larger than 0.002 μm. Because it does not completely eliminate polluting substances,

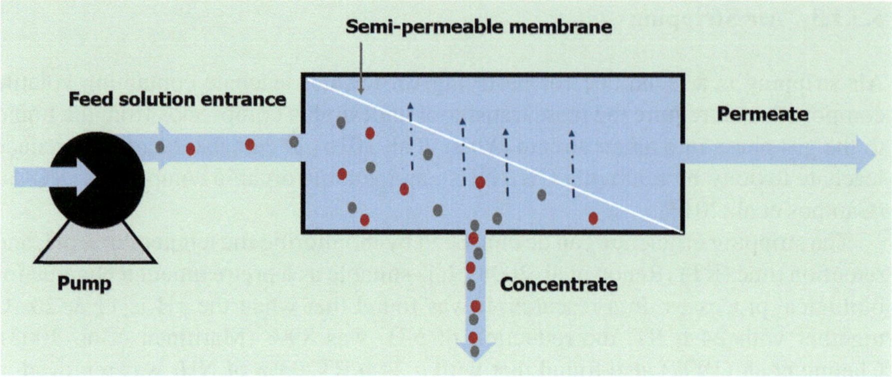


**Fig. 6.1** General simplified schematic view of membrane filtration

UF is suitable to be a pretreatment method further treated by RO and other techniques (Madsen 2014).

NF membranes, as well as RO, are two high-pressure operating membranes. NF membrane pore size is approximately 0.001 μm (between 0.001–0.01 μm) and generally operates at a pressure between 5 and 15 bars, but the pressure can reach 70 bars. Besides particles, colloidal, bacteria, macromolecules, and viruses, NF can remove some divalent ions such as $Mg^{2+}$ and $Ca^{2+}$ (Cakmakci et al. 2009; Madsen 2014; Zulaikha et al. 2014). On the other hand, RO operates at very high pressure between 10 and 100 bars and its pore size varies from 0.1 to 1 nm (Madsen 2014). It is a semipermeable membrane capable of removing monovalent ions and producing pure water. It is the most efficient pressure-driven membrane in terms of rejection and removes nearly all inorganic contaminants from water (Çakmakci et al. 2008). Operated as a pretreatment or final treatment, the membrane application in landfill leachate treatment has grown. They are used as a single and combined treatment such as MF and UF to pretreat water further treated by RO. Except for MF, which membrane pores are not well suited (too large pores) to treat leachate, UF, NF and RO are widely used in leachate treatment, generally as the final treatment process (Anqi et al. 2020). High-pressure NF and RO successfully reject refractory COD found in leachate; however, due to their energy consumption, they appear expensive in most countries (Trebouet et al. 2001). The main limits of pressure-driven membranes in leachate purification are membrane fouling and scaling in a short operating time, caused by the deposition of organic and inorganic particles into the membrane surface and pore, which reduces the membrane viability and its efficiency. Appropriate membranes and treatment conditions and proper membrane cleaning can be used to mitigate membrane fouling.

MF is the cheapest technique, with a low-pressure and high membrane flux due to larger membrane pore size, and it is effective in eliminating colloids and suspended matter, making it suitable as a pretreatment technique. UF, however, is more effective in leachate treatment and can eliminate macromolecules and particles but does not totally remove organic compounds in the leachate (Peng 2017). Moreover, the fundamental issue with MF and UF operation is membrane fouling, which results in a reduction in volumetric permeate flux (Peng 2017; Zielińska et al. 2020). Membrane fouling and scaling are caused by organic and inorganic compounds present in leachate, with a higher impact of inorganic compounds (Kulikowska et al. 2019; Syzdek and Ahlert 1984). During leachate treatment with UF, Syzdek and Ahlert (1984) reported that 99% of the fouling layer was inorganic, which did not prevent organic carbon from passing through the membrane, but it did cause a pressure drop, resulting in a lesser flow. Other studies revealed that mature leachate is well treated by UF compared to the young one, with a two-fold higher final permeability, and the result is attributed to the higher organic content of the young leachate (Collado et al. 2020).

NF and RO are pressure-intensive membrane processes. They are used as end treatment processes to remove recalcitrant, heavy metals, and inorganic pollutants that failed to be removed by biological or other physicochemical methods from landfill leachate. Except for its ineffective removal of monovalent ions, NF is as good as the RO process for leachate treatment, so it is preferred because, in addition to producing high quality permeates, it runs at lower pressure and generates higher flux compared to RO, which to a certain extent results in modest financial investment, operational and maintenance costs (Chaudhari and Murthy 2010; Ince et al. 2010). NF and RO membrane fouling is a serious factor hindering their application in leachate treatment. A wide range of substances, including dissolved organic and inorganic, colloidal, and suspended particles participate in membrane fouling and various pretreatment techniques such as coagulation/flocculation, adsorption, filtration, chemical precipitation, MF, UF (Dolar et al. 2016), and many others are used to prevent NF and RO membrane fouling and scaling (Košutić et al. 2015). Even though their application in leachate treatment has increased in the last decade, these are expensive processes and are mostly preferred for recycling or treatment that results in the beneficial use of the water or a product present in the water. The high treatment efficiency, however, boosts the scientific community in investigating ways to make it cost-effective in wastewater treatment. Mariam and Nghiem (2010) achieved 10–16% total nitrogen (TN) removal, 90% TOC, and 95% turbidity rejection with NF process for leachate treatment. Singh et al. (2012) achieved over 99% dissolved organic matter and 93% salts removal by NF and RO. In another study, Hasar et al. (2009) used RO as a final treatment of leachate and achieved a rejection of 99.7% and 99.2% of conductivity and COD, respectively. RO membranes have an advantage over NF in removing $NH_3$-N (Chaudhari and Murthy 2010; Košutić et al. 2015; Mariam and Nghiem 2010; Dolar et al. 2016), which is a persistent pollutant present in high concentrations in landfill leachate. This is not surprising since RO removes monovalent ions, while NF is limited to divalent and multivalent ions removal.

#### 6.3.2.2 Forward Osmosis

Forward Osmosis (FO) uses the principle of osmosis to induce water transport from the feed solution to the draw solution (DS) through a semipermeable membrane (Mohammadifakhr et al. 2020; Nagy 2019; Woo et al. 2019). FO processes use an asymmetric membrane with a very thin and selective active layer (Nagy 2019). Practically, the solution to be treated (feed solution) is separated from a more concentrated solution (DS) with a semipermeable membrane. An osmotic pressure difference is created between the two solutions due to the high concentration difference of solutes in the solutions. The osmotic pressure difference is the driving force and causes water molecules to migrate through the semipermeable membrane from the feed solution (high water chemical potential – low osmotic pressure) to the draw solution (low water chemical potential – high osmotic pressure). Its principle is totally opposite to RO where high hydrostatic pressure is applied to force the water to move away from its natural direction (Mohammadifakhr et al. 2020; Nagy 2019). While the DS faces the membrane support layer during the FO process, the feed solution to be treated faces the active layer (Nagy 2019). With the main disadvantage of the final product being a diluted draw solution, requiring an additional process to be separated (Mohammadifakhr et al. 2020), the FO process has many advantages over NF or the RO treatment processes (Woo et al. 2019; Iskander et al. 2017; Dong et al. 2014). The attractiveness of the FO process includes the high rejection of pollutants, no need for external hydraulic pressure, minimized membrane fouling (Woo et al. 2019; Dong et al. 2014), and may be more energy-efficient due to the lack of hydraulic pressure (Mohammadifakhr et al. 2020) and with the use of a proper draw solute (Iskander et al. 2017).

FO has proven to be an excellent process for treating landfill leachate for water recovery, inorganics pollutants, and aromatic hydrocarbons removal (Woo et al. 2019). Ibrar et al. (2020) operated the FO for leachate treatment and achieved a rejection of heavy ions and contaminants between 93% and 99%. In a similar study, Dong et al. (2014) achieved approximately 98.6% of COD, 96.6% of total phosphorus, and 76.9% of ammonium rejection efficiency. The goal of FO is to reduce the leachate volume by water molecules transport from the leachate to the DS. The concentrated leachate can further be treated with other methods, generally physicochemical processes (Woo et al. 2019). FO is effective for nutrient rejection, including phosphate, ammonia, and potassium (Mohammadifakhr et al. 2020). The process is generally more effective when operated in a hybrid process with other techniques such as RO, membrane distillation (MD) (Mohammadifakhr et al. 2020). Zhou et al. treated landfill leachate in a hybrid FO-MD system and reported over 96% rejection rate of salt, TOC, and TN rejection rates higher than 98%, while $NH_4^+$-N, Hg, As, and Sb were removed entirely (Zhou et al. 2017).

#### 6.3.2.3 Membrane Distillation

MD is a separation technique in which a hydrophobic microporous membrane separates components at different temperatures present in aqueous solutions (Khayet and Matsuura 2011; Zoungrana et al. 2016). The hot feed solution is circulated tangentially to the hydrophobic membrane surface. Simultaneously, cold water is also circulated tangentially at the permeate side of the membrane. The temperature difference between hot feed and cold water creates a vapor pressure difference. The vapor formed in the feed side across the hydrophobic membrane, once in contact with cold water on the permeate side, condensation occurs, and vapor water is transformed into liquid water (EL-Bourawi et al. 2007; Khayet and Matsuura 2011; Zoungrana et al. 2016). Hydrophobic membranes in MD allow the passage of vapors but not the non-volatile components; consequently, theoretically, they should result in 100% pure effluent (Zoungrana et al. 2017).

Four basic MD configurations exist (Essalhi and Khayet 2015; Sørensen et al. 2014), among which the direct contact membrane distillation (DCMD) is the most popular due to its inherent simplicity (Drioli et al. 2015; Essalhi and Khayet 2015; Sørensen et al. 2014), with the significant disadvantage of heat lost due to direct contact between both heated feed and cold permeate streams with the porous hydrophobic membrane (Bahmanyar et al. 2012; Chung et al. 2014; Jacob et al. 2015). MD systems are widely used in desalination processes, but a game of areas show interest in the process due to its performance and the improvement of the process performance. The barriers to MD application generally caused by the need for appropriate hydrophobic membrane, high energy consumption, and low transmembrane flux are being overcome with the high effluent quality, availability of better membranes at a competitive cost, and mainly the possibility to utilize alternative energy sources which reduced the implantation and operating costs, making the MD system more attractive.

Very few types of research have been conducted on landfill leachate treatment with the MD process. However, theoretically, MD rejects all non-vaporizable compounds and minimizes membrane fouling due to the formation of liquid/vapor layer at the feeding membrane surface, and this may be an advantage for leachate treatment which main treatment problem with pressure driving membranes is membrane fouling (Zoungrana et al. 2016). Jacob et al. (2015) applied an MD system for high strength wastewater treatment and achieved ammonia rejection between 89.6% and 96.3%, and COD rejection between 97.8% and 99.9% and reported that most of the non-VOCs were retained by the membrane. MD is effective in removing all non-volatile compounds from leachate by concentrating it on the concentrate side. Conductivity, COD, sulfate, alkalinity, and hardness were removed at around 85%, 99%, 95%, 90%, and 98%, respectively, in MD raw leachate treatment performed by Zoungrana et al. (2017). The MD principal advantage is the possibility of a single treatment without an additional need for pretreatment. Membrane fouling is also minimized compared to other membrane processes. Another advantage of the MD process in leachate treatment is the possibility of recovering $NH_3$, which can be used as a fertilizer (Zico et al. 2021; Zoungrana et al. 2020). It was reported that

1.5 kg of $NH_3$ can be recovered from each 1 $m^3$ of treated leachate, which is tons of ammonia sulfate from a standard plant (Zico et al. 2021). Two main drawbacks are associated with the MD process: the low membrane flux and high energy consumption, resulting in the high overall cost of the process and treated water cost.

#### 6.3.2.4 Electrodialysis

ED is an electrochemical membrane-based technology, mainly used for desalination, converting salt water into freshwater by transporting ions through ion exchange membranes (IEMs). Cation exchange membranes (CEMs) and anion exchange membranes (AEMs) are alternatively used and separated by spacers in the ED module to induce the movement of charged ions across the membranes, resulting in their separation from the solution (Zoungrana and Çakmakci 2020). Electrical current is used to induce the migration of the ions against the chemical potential. IEMs are regarded as one of the most advanced membranes within the membrane industry (Nagarale et al. 2006; Sata 2004). They have positively or negatively charged groups linked to the membrane backbone and distinguish cations from anions. Two ionic groups are involved in ED, the counter-ions that are different from the one of the fixed charges in the IEM polymer matrix and co-ions similar to the ones of the fixed charges in the polymer matrix. CEMs allow cations (counter-ions) and exclude anions (co-ions) and AEMs allow anions (counter-ions) and exclude cations (co-ions) (Wachinski 2016a; Zoungrana and Çakmakci 2020; Hong et al. 2015). They separate cations and anions from each other in a solution.

The membrane properties are essential in electrodialytic processes such as ED, and a high membrane conductivity and selectivity, low resistance, and cheaply available are adequate IEMs generally desired in an ED process (Nagarale et al. 2006; Sata 2004). Kulhavý et al. (2018) conducted an ED process to reduce the concentration of inorganic salts in municipal landfill leachate to be post-treated by RO they dissolved 18–27% removal of solids content in the leachate. Schoeman et al. (2005) reached similar results with the ED system followed by RO and reported that IEMs fouling is minimized and traditional cleaning methods are available in case cleaning is required. $NH_3$-N is a persistent contaminant in leachate which removal was not a success with many processes. Zhang et al. (2020) successfully investigated the elimination of $NH_4^+$ from landfill leachate with the ED process. They achieved up to 86.17% removal efficiency after 3 h of treatment, but they reported that $Mg^{2+}$ and $Ca^{2+}$ depicted an inhibitory effect on the migration of $NH_4^+$.

#### 6.3.2.5 Ion Exchange Process

Identifying the ion exchange (IE) phenomenon dates back to the middle of the nineteenth century with natural ion exchangers such as clays and aerolites. It was not until the twentieth century that synthetic polymeric IE materials and resins were introduced (Bashir et al. 2010). The IE process is a reversible exchange of ions between a solid phase (ion exchanger or ion exchange resin) and a liquid phase (Wachinski 2016a). The ion exchangers which carry the exchangeable ions are insoluble in the liquid phase and do not take part or undergo significant structural changes during the reaction (Levchuk et al. 2018). They have one major property associated with their structure (polymer matrix), transporting surplus anions or cations, compensated by the counter-ions (Levchuk et al. 2018; Helfferich 1995). In IE processes, ion exchangers are either negatively charged and have cation exchange properties (able to remove positively charged cations from the solution) or negatively charged, which detain anion exchange properties (able to remove negatively charged anions from the solution) (Wachinski 2016a). Because the counter-ions are freely mobile, other ions of the same sign can simply replace them (Levchuk et al. 2018).

IE processes are largely applied to remove organic substances and metals from wastewater and leachates. Natural and synthetic resins developed from high-molecular-weight polyelectrolytes are used in IE processes. The process is effective for the elimination of charged pollutants such as metals. Many studies used IE resins to investigate the elimination of ammonia and metal ions from water and wastewater, including the removal of $NH_3$-N, $NH_4^+$, $Cr^{3+}$, $Cd^{2+}$ and $Ni^{2+}$, $Cu^{2+}$, $Zn^{2+}$ $Ni^{2+}$, $Ca^{2+}$, $Na^+$ from wastewaters and landfill leachate (Jorgensen and Weatherley 2003; Wachinski 2016b; Zamri et al. 2017). Bashir et al. (2010) used synthetic IE resins for landfill leachate treatment and achieved 97%, 88%, and 94% of color, COD, and $NH_3$-N removal rates, respectively. In addition to these pollutants, Zamri et al. (2017) eliminated metal ions from leachate with IE resin. To prevent fouling on the resins that are sensitive to the presence of organic matter, pretreatment is necessary to remove most of the suspended solids and some soluble organics.

#### 6.3.2.6 Adsorption

An adsorption is a form of physicochemical and biological technique that can be effective in landfill leachate treatment. The two most prevalent types of adsorbents that may easily adsorb and remove the organic components in leachates are powdered and granular activated carbon (Bashir et al. 2015). The final polishing phase can also remove heavy metals or organics (Salem et al. 2008). Adsorption by carbon can eliminate approximately 50–70% of COD and $NH_3$-N from leachates (Salem et al. 2008). Other types of adsorbents include kaolinite, municipal waste incinerator, bottom ash, palm oil fuel ash (POFA), and zeolite (Abuabdou et al. 2019). In an adsorption process, net negative charge attracts the positive charge species thus, the high surface area, charges of species, and porosity have a significant contribution to adsorption (Shehzad et al. 2015).

#### 6.3.2.7 Advanced Oxidation Process

Advance Oxidation has been considered as an efficient technology for the mineralization of refractory organics in leachate. They are primarily based on the production of high-oxidizing free radicals, which improve the efficiency of chemical oxidation (Abuabdou et al. 2019). Various non-biodegradable and toxic chemicals can be found in leachate that cannot be treated using physical methods but can be eliminated by chemical oxidation. Electrochemical oxidation technologies, Fenton-based reactions, and ozone-based techniques are the most often used AOPs for leachate treatment.

##### Electrochemical Oxidation

Electrochemical oxidation (direct or indirect oxidation) is among the most promising technologies for landfill leachate treatment. In the direct oxidation process, pollutants electron transfers directly on the anode surface. The indirect oxidation process facilitates the production of oxidants (hydroxyl radical, available chlorine, persulfate, percarbonate, and per phosphate). The anodization of electrolyte ions plays a vital role in the electrochemical oxidation process (Rodrigo et al. 2010). The elimination of COD, $NH_4^+$-N, nitrite-N, and some sort of reducing ions in leachate is aided by indirect oxidation based on available chlorine generated by the oxidation of chloride ions (Li et al. 2010a, b; Deng et al. 2018). The anode material heavily influences the electrochemical oxidation process. In recent years, boron-doped diamond (BDD) electrode and metal oxide anodes have been extensively applied to treat landfill leachate. $CO_2$ and $H_2O$ are generally produced at the end of this process, thus avoiding the issue of pollutants transitioning from one phase to the next (Grimm et al. 1998). Current density and pH are two main factors affecting the electrochemical oxidation process. When a decreased current density is delivered, the anode potential falls below the electrode potential requirements, making oxidation unachievable (Fernandes et al. 2015). In contrast, there is a dramatic drop in the current efficiency at a higher current density than the required limit density because the mass transfer mechanism is primarily responsible for improving reaction efficiency (Ross and Wragg 1965). Chloride ions concentration in landfill leachate is a key component that influences the electrochemical oxidation efficiency. If the $Cl^-$ concentration is low and less amount of available chlorine is produced from the process, then the COD removal efficiency becomes less (Deng and Englehardt 2007). At pH between 6 and 9, chlorine is primarily available in the form of hypochlorous acid and hypochlorite (Deborde and von Gunten 2008), which indicates excellent oxidation. When the pH rises, available chlorine is transformed to chlorate ($ClO_{3-}$), which has weak oxidation characteristics (Eqs. 6.2–6.4) (Lima et al. 2010). Furthermore, when the pH is below 4, available chlorine is mostly in the form of free chlorine ($Cl_2$), which quickly overflows from the liquid phase and affects oxidation (Deng et al. 2019).

$$2Cl_2 + OH^- \rightarrow Cl^- + H^+ + ClO^- \tag{6.2}$$

$$2HClO + ClO^- \rightarrow 2Cl^- + ClO_3^- + 2H^+ \tag{6.3}$$

$$6ClO^- + 3H_2O \rightarrow 4Cl^- + 6H^+ + 3/2O2 + 6e^- + 2ClO_3^- \tag{6.4}$$

However, electrochemical oxidation needs more energy and cost, limiting its usage in full-scale capacities for leachate treatment (Grimm et al. 1998). In addition, some of the toxic by-product produced from the electrooxidation technique requires necessary steps to eliminate (Zolfaghari et al. 2016). Combining electrochemical oxidation with an adequate biological process can largely reduce the overall process cost (Chong et al. 2012). The oxidative processes could be used as a pre-step before biological processes, which will enhance the biodegradability of the solution (Cortez et al. 2011b).

Ozonation

Ozone ($O_3$) is a very selective oxidant, having standard potentials in acidic solutions of 2.07 V (vs NHE) and basic solutions of 1.25 V (vs. NHE) (Pi et al. 2007) and stands out for its high oxidation efficiency. There are two different types of oxidation mechanisms, specifically direct electrophilic attack by molecular $O_3$ and indirect attack through the formation of hydroxyl radicals (Pera-Titus et al. 2004). The effectiveness of the process is controlled by the solution pH, which can alter the reaction's kinetics and routes. Direct ozonation occurs at low pH, while indirect ozonation happens at higher pH (Agustina et al. 2005). Oxidation can be employed as a pre-or post-treatment due to its usefulness. Because in pretreatment stages, less energy is required for the partial mineralization of the recalcitrant compounds. In the following phases, this mechanism decomposes the refractory chemicals into simpler molecules that the microbes may consume (Geenens et al. 2001). One study employed $H_2O_2$ ozonation to increase leachate biodegradability before treating it with activated sludge (Geenens et al. 2001). Though the removal of COD was lower in that study, the leachate biodegradability increased. A similar explanation was given in another study where they used Fenton and Ozonation as pretreatment for leachate and found improved biodegradability up to 93% (Cortez et al. 2011b). However, in a different study, they employed $O_3$ and hydrogen peroxide ($H_2O_2$) to eliminate leachates recalcitrant compounds after stabilization ponds where they found higher removal efficiency for Color, COD, and TOC, reaching up to 99%, 88%, and 66%, respectively (Gomes and Schoenell 2018).

Fenton Processes

The Fenton process is also an AOP used in leachate treatment. For the generation of oxidizing species and subsequent breakdown of refractory organic materials, the Fenton process uses $H_2O_2$ and iron salts. In the Fenton process, $H_2O_2$ reacts with $Fe^{2+}$ to produce strong reactive species such as hydroxyl radicals, while other compounds such as ferryl ions can also be produced. The following reactions are involved in the Fenton process:

$$Fe^{2+} + \mathrm{H_2O_2} \rightarrow Fe^{3+} + OH + OH \tag{6.5}$$

$$Fe^{3+} + \mathrm{H_2O_2} \rightarrow Fe^{2+} + HO_2 + \mathrm{H}^{+} \tag{6.6}$$

OH is produced by electron transfer from Eq. 6.5, but it can be changed because of Fenton reagents, as demonstrated in Eq. 6.6. UV irradiation is combined with the classic Fenton system in the photo- Fenton reaction, with the goal of boosting the UV-induced reduction of dissolved $Fe^{3+}$ to $Fe^{2+}$. Either one or both Fenton reagents may be generated through electrochemical methods in the electro-Fenton reaction.

Though there is some operational complexity involved in Fenton processes such as pH adjustment (low pH required) and post-treatment neutralization, coagulation-precipitation to ensure high COD elimination, but Fenton process is still more feasible compared with the ozonation process (Zhou et al. 2010). The process is technically plain with no mass transfer obstruction (homogeneous nature), and both iron and $H_2O_2$ are cheap and non-toxic (Ghasemzade and Pazoki 2017).

## 6.4 Biological Treatment of Landfill Leachate

Similar to the bioprocessing of other organic wastes, landfill leachate is also treated through biological treatment methods because this approach is simple, reliable and cost-effective, compared to physicochemical treatment methods (Renou et al. 2008). Biological treatment primarily employs microorganisms to remove organic compounds and inorganic substances such as ammoniacal nitrogen, phosphorous, heavy metals, and other refractory organics present in landfill leachate (Luo et al. 2020). In aerobic treatment, organic compounds are mineralized to carbon-di-oxide and water with the production of sludge. In anaerobic treatment, the organics get converted to methane-rich biogas and a much lower volume of sludge. Although they provide reliable and cost-effective solutions, the major challenge in the biological treatment of landfill leachate is its poor BOD/COD ratio (0.1–0.5), caused by the refractory organics and the presence of excess ammoniacal nitrogen (500–1500 mg/L) (Luo et al. 2020; Renou et al. 2008). Table 6.2 shows the efficiency of different biological treatment methods available for landfill leachate treatment.

**Table 6.2** Efficiency of different biological treatment methods for landfill leachate treatment

| Treatment method | Substrate(s) | HRT | Temperature °C | Removal Efficiency (%) | | References |
|---|---|---|---|---|---|---|
| | | | | COD | $NH_3$–N | |
| Activated sludge process | Synthetic young leachate | 8 h | 20 | 83 | 99 | Ren et al. (2017) |
| | Landfill leachate | 24 h | – | 93–96 | 80–88 | Kanyatrakul et al. (2020) |
| Moving bed biofilm reactor | Mature leachate | 20 days | 27–30 | – | 56–82 | Xiong et al. (2018) |
| | Leachate | – | – | 40 | 68–72 | Xu et al. (2018) |
| Lagoons | Leachate | 11.5–254 days | – | 75 | 80 | Mehmood et al. (2009) |
| | Leachate | – | 14.3–22.8 | 40 | 77 | Frascari et al. (2004) |
| Sequencing batch reactor | Compost leachate | 24 h | – | 92 | 74 | Hashemi et al. (2017) |
| | Young leachate | 11.6 days | – | 50 | 90 | Remmas et al. (2018) |
| | Leachate + municipal sewage | 6 days | | 84 | 93 | Ranjan et al. (2016) |
| Trickling filters | Leachate | – | – | 49 | 59 | Aluko and Sridhar (2014) |
| | Sewage and leachate | 24 h | – | 80 | 90 | Ferraz et al. (2014) |
| Constructed wetlands | 1:10 diluted leachate | – | 17–40 | 21–30 | 60–85 | Camaño et al. (2019) |
| | Leachate | – | – | 39–92 | 30–54 | Akinbile et al. (2012) |
| Upflow anaerobic sludge blanket | High strength leachate | 4.5–5.8 days | 35 | 83 | – | Ye et al. (2011) |
| | Leachate | – | 25 | 51 | – | Ismail and Tawfik (2016) |
| | Leachate and domestic septage | 1.5 days | 35 | 42 | 48 | Lin et al. (2000) |
| Membrane bioreactor | Old leachate | – | 21 | 54–78 | 98 | Saleem et al. (2018) |
| | Leachate | 18 h | >17.5 | 63 | 98 | Zolfaghari et al. (2016) |

(continued)

**Table 6.2** (continued)

| Treatment method | Substrate(s) | HRT | Temperature °C | Removal Efficiency (%) | | References |
|---|---|---|---|---|---|---|
| | | | | COD | $NH_3$–N | |
| Anammox | Leachate | 24 h | 30 | 64 | 94 | Wang et al. (2016) |
| | Leachate | – | 35 | – | 95 | Miao et al. (2016) |
| | Leachate | 12–19 h | – | 30 | 93 | Nhat et al. (2014) |
| Phytoremediation | Leachate | – | – | 92 | 93 | Aronsson et al. (2010) |
| Rotating biological contactor | Leachate | 24 h | 20 | 43 | 99 | Kulikowska et al. (2010) |
| | Leachate | 10 h | 28 | Low | 95 | Cortez et al. (2011a) |
| Fungal bioremediation | Old leachate + Glucose/cellulose | 72 h | 20–25 | 54–63 | – | Bardi et al. (2017) |
| | Mature leachate + wood chips/wheat straws | – | 23 | 48 | – | Islam and Yuan (2020) |
| | Mature and intermediate leachate | 13 days | 23 | 70–90 | – | Siracusa et al. (2020) |
| Anaerobic co-digestion | Leachate + acid mine drainage | 20 h | 35 | 83 | – | Zhou et al. (2021) |
| | Mature leachate + sewage sludge | – | 35 | 38 | 40 | Berenjkar et al. (2019) |
| | Industrial landfill leachate + Glycerine | 30 days | 30 | 92 | – | de Castro et al. (2020) |
| Algal bioremediation | Landfill leachate | 13 days | 25 | 60 | 90 | El Ouaer et al. (2017) |
| | Leachate (dilutions) | 14 days | 25 | – | 77/90/66 | Nordin et al. (2017) |
| | Leachate | 14 days | 42 | – | 99.9 | Pan et al. (2021) |

### *6.4.1 Conventional Biological Treatment Techniques of Landfill Leachate*

Many studies have reported that landfill leachate can be successfully treated both aerobically and anaerobically using conventional techniques. Aerobic treatment methods such as activated sludge process, sequential batch reactors, and membrane reactors operate by converting part of the biodegradable components to biomass and completely removing ammoniacal nitrogen by nitrification and denitrification process (Luo et al. 2020). Anaerobic treatment converts biodegradable components to biogas and a small part to biomass production. The biomass production in anaerobic reactors is much lesser than aerobic treatment methods because the growth rate of aerobic microorganisms is much higher than that of anaerobes. The important conventional biological treatment types are described below.

#### **6.4.1.1 Lagoons**

Aerated and anaerobic lagoons are the commonly used onsite landfill leachate treatment methods. Aerobic lagoons are usually constructed with a depth of 2.5–3 meters, allowing diffusion of sunlight and oxygen without external aeration, allowing the biological oxidation of organic and inorganic compounds. The anaerobic lagoons are deep to provide anaerobic conditions at the bottom of the lagoons, which helps organic compounds to get oxidized and release their products. These released compounds are converted to biomass, $CO_2$ and water in the aerobic zone, thereby reducing the odour issues. Lagoons provide easy onsite leachate treatment, but the issue lies with higher hydraulic retention time (HRT) and their inability to control the process temperature (Renou et al. 2008). With higher HRT, only a limited volume of leachate could be treated. The temperature variation will affect the microbial growth and activity, which severely impacts the process's efficiency. Usually, these lagoons are more efficient in treating diluted leachate with an ammoniacal nitrogen concentration of less than 300 mg/L. When undiluted leachate is treated in lagoons, the efficiency decreases, and the disposal standards cannot be met (Mojiri et al. 2021).

A study on treating leachate using aerobic lagoons performed by Mehmood et al. (2009) achieved 75% and 80% COD and nitrogen removal with a loading rate of 1740 mg/L COD and 1241 mg/L ammoniacal nitrogen with an average HRT of 11 days. Frascari et al. (2004) achieved removal efficiencies of 40%, 64%, 77%, and 77% of COD, BOD, ammoniacal nitrogen and TKN, respectively, with a sequential system of anaerobic and facultative lagoons.

#### 6.4.1.2 Activated Sludge Process

The activated sludge process (ASP) consists of a mixed culture of microorganisms grown under aerobic conditions, responsible for decomposing organic matter into biomass and mineralizing these compounds to carbon dioxide, water, and other minerals. ASP is a more robust treatment method than lagoons, as it uses a rapidly growing mixed culture of microorganisms acclimated to the leachate conditions. Well adapted microbial biomass and extensive aeration result in efficient leachate treatment even at lower HRT (Wang et al. 2018). The treated effluent and the suspended biomass are separated in a clarifier, and the sludge is partially recycled, and the excess sludge is disposed of or utilized as land supplements after drying. ASP is primarily used for sewage treatment, but co-treatment of landfill leachate with sewage in ASP is not uncommon (Luo et al. 2020). However, the use of ASP for leachate treatment has decreased in recent years because of the following reasons: (i) process inhibition by high ammoniacal nitrogen in leachate, (ii) high energy requirement for aeration to treat high strength leachate, and (iii) production of high quantities of sludge which requires further treatment.

Ren et al. (2017) studied an activated sludge process that achieved a COD removal efficiency of 83% within the first 73 days from the start of the system. At the same time, ammoniacal nitrogen removal was 99% when the AN concentration was around 300 mg/L. As it increased, the removal efficiency fluctuated between 76% and 99%. The phosphorous removal was 89% during the initial stages, and it decreased to 76% over time. Kanyatrakul et al. (2020) could achieve COD, BOD, TKN, ammoniacal nitrogen and phosphorous removal of 93–96%, 95–98%, 78–83%, 80–88% and 76–84%, respectively. It also shows that this process can remove 81–100% of most micropollutants harmful to many living organisms.

#### 6.4.1.3 Sequencing Batch Reactor (SBR)

SBR can be considered as an upgraded treatment method compared to ASP for the treatment of landfill leachate. The four phases of SBR are feeding, aeration, settling and discharge. Leachate treatment occurs in a single reactor with controlled aeration and settling phases. During the aeration phase, the organic contents are utilized for microbial growth and ammonical nitrogen gets converted to nitrite. As the settling phase starts, the reactor assumes anoxic conditions and converts nitrite to nitrate and then to nitrogen gas. This method also allows the growth of phosphate accumulating microorganisms which can effectively remove the phosphates in the landfill leachate (Luo et al. 2020). While SBR can treat leachate emanating from young, intermediate and mature landfills, it is widely used in treating mature leachate.

Hashemi et al. (2017) studied the performance of SBR in treating leachate from composting plants and achieved COD, BOD, TN and phosphorous removal efficiencies of 92%, 96%, 74% and 67%, respectively, over 11 runs that were performed in 280 days. Remmas et al. (2018) studied landfill leachate treatment in a full-scale SBR and observed that BOD and COD removal efficiencies were 90%

and <50%, indicating that leachate had higher amounts of recalcitrant compounds. However, the nitrogen removal was more than 70%. In a study by Ranjan et al. (2016), more than 84% of COD removal was achieved, with 93% ammoniacal nitrogen and 80% phosphorous removal.

#### 6.4.1.4 Moving Bed Biofilm Reactor (MBBR)

MBBR is a type of attached growth system, where carrier elements made of unique material act as media for the growth of microorganisms and biofilm formation. The carrier elements are kept in circulation throughout the reactor using an aeration system. Biofilms in the carrier elements increase the contact time and interactions between landfill leachate and microorganisms, leading to higher treatment efficiency than ASP or lagoons. The biofilm also provides insulation to the microorganisms against toxic compounds present in the leachate. In contrast to suspended growth processes, MBBR produces meagre sludge with low settling time and high biomass content (Renou et al. 2008; Wang et al. 2018). With the characteristics mentioned above, MBBR outperforms other aerobic landfill treatment systems.

Xiong et al. (2018) studied the efficiency of nitrification and removal of refractory organics from mature landfill leachate treated in an MBBR system. The nitrification efficiency was 56–82% whereas, the removal efficiencies of humic and fulvic acid concentrations were 31% and 19%. Xu et al. (2018) studied simultaneous nitrification and denitrification of landfill leachate in an MBBR system. The nitrogen removal was 68–72%, and COD removal was more than 40%.

#### 6.4.1.5 Trickling Filters

Trickling filters (TF) or biological filters consists of filter media made out of cuttings or slags of blast furnaces or any specially designed plastic media that can allow the attachment and growth of microorganisms. The microorganisms growing over the filter media will enable the decomposition of organic matter and remove ammoniacal nitrogen present in leachate. The aerobic conditions are usually maintained through diffusion from the atmosphere or by external aeration systems. The filter media is porous to increase the filtration efficiency and provide more surface area for the growth of microorganisms. The effluent from the filter is further treated in a settling tank to remove the biomass and other solids that are suspended. High organic content in the influent can lead to fouling of the filters; hence trickling filters are not suitable for treating young or intermediate landfill leachate (Luo et al. 2020; Renou et al. 2008). Additionally, the high maintenance cost of the trickling filter makes it not the right choice for landfill leachate treatment.

In a bench-scale study for landfill leachate treatment using TF and SBR to evaluate their efficiencies, Aluko and Sridhar (2014) achieved significant reductions in COD, BOD, TKN, suspended solids and turbidity (49%, 76%, 59%, 73% and 71%, respectively). Ferraz et al. (2014) studied the treatment efficiency of aerobic biofilters for the co-treatment of sewage and leachate. The treatment efficiencies were 80%, 90%, 98%, and 90% of COD, suspended solids, BOD, and ammoniacal nitrogen.

#### 6.4.1.6 Constructed Wetlands

Constructed wetlands utilize the vegetation soil, microorganisms and the plantation to treat landfill leachate. Constructed wetlands have been successfully employed to treat leachate in many developing countries, whereas their popularity in developing countries is considerably low due to their poor performance during winter. Studies at lab, pilot and full-scale have been done on the treatment of landfill leachate, and it has been observed that they can effectively remove the phenolic compounds and total nitrogen (Mojiri et al. 2021). Constructed wetlands can be easily maintained and operated with high treatment efficiency yet at a low cost.

In a study by Dan et al. (2017), the removal of phenolic compounds such as phenol, bisphenol – A and 4-tert-butyl-phenol was studied using a vertical flow constructed wetland from synthetic landfill leachate. It was found that the removal efficiencies of these compounds were in the order of phenol (>88%), 4-tert-butyl-phenol (18–100%) and bisphenol – A (9–99%). In a pilot-scale study by Camaño et al. (2019), the efficiency of constructed wetland was studied with two different plant species and their ability to treat leachate. Mean COD removal of 21–25% for *T. domingensis* and 24–30% for *C. indica* was observed. Also, the ammonia and nitrogen removal for *T. domingensis* was 64–80% and 16–53%, whereas for *C. indica* it was around 60–85% and 18–48%, respectively. In another study by Akinbile et al. (2012), a sub-surface flow constructed wetland was evaluated for leachate treatment using *C. hyspan*. The ranges of COD, BOD, ammonia, phosphorous and total nitrogen removal were 39–92%, 60–78%, 30–54%, 60–99%, and 34–67%, respectively. In this study, heavy metal removal from landfill leachate was also higher, and their removal efficiencies were Fe – 35–59%, Mn – 51–71%, Mg – 29–75% and Zn – 74–90%.

#### 6.4.1.7 Upflow Anaerobic Sludge Blanket (UASB) Reactor

In UASB reactors, microorganisms form granules that are suspended throughout the reactor, forming a sludge blanket. The UASB reactor can efficiently treat high COD and BOD wastewaters at lower HRTs and mesophilic conditions, ideal for landfill leachate treatment. Harnessing methane-rich biogas as one of the products could reduce the operational expenses of the UASB reactor (Luo et al. 2020; Mojiri et al. 2021; Renou et al. 2008). Removal of COD and ammoniacal nitrogen got enhanced when the UASB reactor was used with other methods such as SBR, membrane filtration, and chemical treatment.

Ye et al. (2011) treated high strength leachate from an incineration plant using a UASB reactor operated at mesophilic conditions and observed around 83% COD removal with an average methane yield of 0.31 $LCH_4$/gCOD removed. Ismail and Tawfik (2016) studied the effects of leachate's pretreatment (advanced oxidation process) using a UASB reactor. The pretreatment increased the COD removal by 51% and the VSS removal by 15%. In another scenario, Lin et al. (2000) studied the co-digestion of leachate with septage using a UASB reactor operated at mesophilic conditions. COD, VS, ammoniacal nitrogen removal efficiencies were 42%, 68% and 48%. It was observed that the addition of septage ameliorated the inhibitory effects of leachate.

#### 6.4.1.8 Membrane Bioreactor (MBR)

MBR usually consists of two stages in the treatment system: (i) Bioreactor: Landfill leachate is treated by microorganisms, leading to the oxidation of the pollutants and (ii) Membrane: Sludge and microorganisms are separated from treated leachate. MBR is highly efficient with 90% and 75% removal of ammoniacal nitrogen and COD, respectively, from leachate compared to other aerobic systems (Luo et al. 2020; Renou et al. 2008). Most of the reported studies on the treatment of leachate by MBR were at a lab scale. Hence detailed research has to be performed to validate MBR's efficiency and affordability at an industrial scale to treat higher volumes of landfill leachate.

In a study by Saleem et al. (2018), landfill leachate treatment using a dynamic membrane bioreactor was evaluated for its ammonium and total nitrogen removal. It was observed that there was more than 98% ammonium removal along with more than 90% total nitrogen removal efficiency. They also observed that at a pore size of 52 μm, the turbidity removal was higher and aided in better solid-liquid separation. In another study by Zolfaghari et al. (2016), membrane bioreactor and electrochemical oxidation methods' efficiencies were evaluated during leachate treatment. After membrane bioreactor treatment, COD, phosphorous, and ammoniacal nitrogen removal efficiencies were 63%, 52%, and 98%, respectively.

#### 6.4.1.9 ANAMMOX Process

The ANAMMOX process is commonly used to remove ammoniacal nitrogen by utilizing ammonium and nitrite as electron donors and acceptors. Either by using the ANAMMOX process solely or combined with other treatment methods, 90% nitrogen and 80% COD removal can be achieved. The ANAMMOX process is better than other treatment methods for treating mature landfill leachate with lower COD and higher ammoniacal nitrogen contents. In a combined nitrification and anammox process to treat mature leachate, Wang et al. (2016) achieved total nitrogen removal of 94% along with COD removal of 64%. Miao et al. (2016) assessed a two-stage lab-scale SBR operated at mesophilic conditions for ammonia removal. With an initial COD and ammoniacal nitrogen concentrations of 3000 mg/L, the system achieved total nitrogen (TN) removal of 95%, with effluent having less than

20 mg/L TN concentration. In a study by Nhat et al. (2014) using combined partial nitrification and ammonia removal process, they achieved a removal efficiency of 93% with 4.2 Kg $m^{-3}$ $d^{-1}$ initial load of TN and the efficiency decreased to 81% when the initial load was increased to 8.3 Kg $m^{-3}$ $d^{-1}$.

#### 6.4.1.10 Phytoremediation

Phytoremediation utilizes phytoplanktons or plants to remove pollutants in landfill leachate by various processes, including evapotranspiration, rhizofiltration or bioaccumulation. The plant system can efficiently remove the soluble or volatile compounds via foliar uptake or by evaporation via root uptake along with nutrients such as ammonium ion, potassium and phosphorous. Higher BOD can lead to the growth of microorganisms present in the rhizosphere, leading to the removal of pollutants and sequestration of some metals. Phytoremediation decreases leachate infiltration and groundwater contamination. The commonly used techniques in phytoremediation are constructed wetlands and hydroponic treatment methods. In constructed wetlands, higher plants such as willows and poplar trees can absorb organic contents, nutrients (K and $PO_3^-$) and remove/sequester metals (Fe and Zn). The nitrogen content of landfill leachate can be further reduced by providing a denitrifying zone near the root system (Jones et al. 2006; Kim and Owens 2010).

Willows and poplar are the primary plants that are used in the remediation of leachate. Aronsson et al. (2010) studied the effects of irrigation of landfill leachate on willows grown on the clay layer and poplar grown on the sand layer for 2 years. It has been found that willows retained a total of 93.8%, 99.8% and 92.5% of nitrogen, phosphorous and total organic carbon supplied each year. Another study by Song et al. (2018) assessed various macrophytes for their ability to remediate leachate. It was observed that *Typha angustifolia* had a very high remediation potential by accumulating Na, Mg, P, Mn, Fe, and Zn, at higher concentrations than any other plant species. Nissim et al. (2021) also studied the ability of willows and poplar in treating leachate. In this study, both the species had similar treatment efficiency with an average of 340–680 $m^3$/ha of leachate in the start-up period (1st year), and the subsequent year the treatment capacity increased to 2470–4950 $m^3$/ha of leachate. Also, they yielded the same aboveground biomass, but poplar can extract Cd, Cu, P and N at a higher rate than willows.

#### 6.4.1.11 Rotating Biological Contactors (RBC)

RBC is considered an alternative aerobic activated sludge treatment method, in which the microorganisms attach to the media present and form a biofilm. This treatment method is contact-based, where the attached biomass is submerged partially or wholly in leachate, leading to the oxidation of the contaminants in leachate. The aerobic condition is maintained by the rotors that rotate the media with attached biomass, leading to oxygen transfer. The excess biomass is shed due to the turbulence created by the rotation of the motors (Luo et al. 2020; Mojiri et al. 2021; Renou et al. 2008). RBC can treat low strength wastewater only, so young or intermediate leachate cannot be treated using RBC.

To study the efficiency of an RBC to remove nitrogen and organic compounds, Dorota Kulikowska et al. (2010) treated leachate with high amounts of ammonia and refractory organics. They observed a 99% nitrogen removal at a low ammonium load of 1.92 g $NH_4 - N/m^2.d$, whereas it decreased to 75% when the ammonium load was increased above 4.79 g $NH_4 - N/m^2.d$. Along with this, COD removal decreased from 43% to 32%, with 20–30% removal in humic substances. Cortez et al. (2011a) treated leachate with high nitrate concentration using RBC and checked its denitrification efficiency. A high denitrification rate was observed with 95% nitrate removal. However, the COD removal efficiency was low and could not be directly discharged into the environment.

## 6.4.2 *New Biological Treatment Techniques of Landfill Leachate*

### 6.4.2.1 Fungal Bioremediation

Fungal bioremediation is a considerably newer treatment method for treating landfill leachate by lignin-degrading fungi. The fungi that are widely used in leachate treatment belong to Ascomycetes and Basidiomycetes sub-kingdoms. These fungi secrete lignin-degrading enzymes such as lignin peroxidase, laccase and manganese peroxide. These enzymes are utilized to degrade refractory organics, including humic and fulvic acids present in landfill leachate, as they resemble lignin molecules (Luo et al. 2020; Renou et al. 2008). In addition to removing refractory organics, fungal bioremediation can remove COD and nitrogen from leachate.

Many fungal species such as *Dichomitus squalens*, *Trametes trogii*, *Phanerochaete chrysosporium*, and *Bjerkandera adusta* can remove more than 60% carbon content, so many of these strains are used for the treatment of mature or old landfill leachate with high concentrations of refractory organics. In a study by Bardi et al. (2017), the effects of glucose and cellulose addition as a co-substrate during leachate treatment were assessed, and it was observed that glucose addition has a higher COD and sCOD removal compared to the addition of cellulose. In another study by Islam and Yuan (2020), wood chips and wheat straws were evaluated as co-substrates during the fungal treatment of leachate. The addition of wood chips increased the COD, sCOD removal, and lignocellulosic enzymes compared with glucose and wheat straw additions. Mycoremediation of old and intermediate leachate using fungal isolate *Lambertella spp.* was evaluated by Siracusa et al. (2020). Old leachate treatment with the isolate yielded 70% removal of sCOD after 13 days of the treatment. The treatment of intermediate leachate in lab-scale reactors with immobilized *Lambertella spp.* had a total organic carbon removal of 90%.

#### 6.4.2.2 Anaerobic Co-digestion of Landfill Leachate

Anaerobic co-digestion is a proven technology where two or more substrates are treated simultaneously to provide stability and increase treatment efficiency. Co-digestion provides the required nutrient balance, maintains optimal C/N ratio, increases OLR, ensures syntrophism between microorganisms and eventually enables better methane production. The co-digestion process needs to be optimized by focusing on the following parameters: chemical properties of the landfill leachate, functional parameters (pH, temperature, C/N ratio, OLR, HRT, etc.), and by performing biochemical methane potential (BMP) test. The substrate ratio has to be optimized for enhancing the methane production and efficient treatment of the waste. During anaerobic co-digestion, the inhibitory effects of refractory organics and other toxic substances on the bioprocess can be mitigated (Siddique and Wahid 2018). Several co-digestion studies on landfill leachate were reported using different substrates and with varying substrate ratios. The substrates used in such studies include cornstalk and vermicompost; Fruit juice/garden waste with sewage sludge and cow dung; straw and food waste; vegetable and fruit waste with cattle manure, sewage sludge with organic household waste; and lignocellulose waste with animal manure; sewage sludge with OFMSW.

In a study to evaluate the co-digestion of landfill leachate and acid mine drainage wastewater, Zhou et al. (2021) used a UASB reactor operated at mesophilic conditions. It was found that COD removal efficiency was 75%, and 69% of sulfate removal was observed when the HRT was 20 h. In another co-digestion study by Berenjkar et al. (2019), sewage sludge and leachate were mixed in 5 different ratios, and the ratio of 95% sewage sludge and 5% leachate had the highest removal of COD (38%) and VS (30%). In another study by de Castro et al. (2020), the effect of the addition of glycerine as co-substrate was evaluated for COD removal and methane content in the biogas. It was observed that the leachate: glycerine ratio of 95.13–4.87% had the highest COD reduction of 92% and methane content of 0.19 L $CH_4$/g VS removed.

#### 6.4.2.3 Algal Bioremediation

Algal bioremediation or phycoremediation is a relatively novel method for landfill leachate treatment. To date, many lab-scale phycoremediation studies have been reported in which leachate is diluted to curtail the toxic effects and inhibition of algal growth. Phycoremediation removes most of the organic contents and some heavy metals such as copper from landfill leachate.

In a study by El Ouaer et al. (2017), the efficiency of microalgae species *Chlorella* for treating leachate from Tunisian landfills was evaluated using lab-scale reactors. It was observed that this species removed 60% of COD and 90% of ammoniacal nitrogen during the operation for about 13 days. In another study by Nordin et al. (2017), biomass content of three *algal* species was checked during the treatment of leachate and yielded: 428 mg/L biomass of *Chlorella sp.*, 805 mg/L of *Oscillatoria sp.*, and 237 mg/L of *Scenedesmus sp.* with an average ammonia removal of 77%, 90% and 66%. A recent study evaluated the efficiency of leachate treatment by a thermophilic alga, *Galdieria sulphuraria*, for its biomass growth and nitrogen removal (Pan et al. 2021). Various dilutions of leachate were used, and it was observed that nitrogen and phosphorous removal and biomass production was higher in 20% leachate concentrations. The nitrogen and phosphorous removals were 99.9% and 34%, whereas the biomass growth was 2.16 g/L, which is two times more than the control.

## 6.5 Conclusions

This chapter is a culmination of conventional and contemporary methods that are used under physicochemical and biological technologies for landfill leachate treatment. In order to meet the standards set by the local regulatory authorities for discharging leachate into the environment, the choice of the appropriate technology or combination of technologies is imperative. The combination of technologies could be either within or between the physicochemical and biological methods. The selection of treatment method depends on several factors such as characteristics of landfill leachate, discharge standards, capital and operational costs. Physicochemical treatment technologies such as filtration and advanced oxidation processes completely remove COD, TDS, BOD, and ions that can seriously impact the receiving environment. The major disadvantage of any physicochemical treatment technology is its cost, as treating a high volume of leachate requires significant capital investment and can increase the operational cost in terms of reagents, manpower, and energy requirements. Even though biological treatment decreases the operational cost, the discharge standards cannot be met due to the recalcitrant characteristic of leachate with high inorganic and low biologically degradable contents. However, certain biological treatment methods such as Annamox and anaerobic digestion can remove ammoniacal nitrogen and COD, respectively, to the expected levels. Hence it is concluded that a combination of physicochemical and biological treatment technologies would be ideal for efficient landfill leachate treatment.

## References


Abuabdou SMA, Teng OW, Bashir MJK, Aun NC, Sethupathi S, Pratt LM (2019) Treatment of tropical stabilised landfill leachate using palm oil fuel ashisothermal and kinetic studies. Desalin Water Treat 144. https://doi.org/10.5004/dwt.2019.23662

Agustina TE, Ang HM, Vareek VK (2005) A review of synergistic effect of photocatalysis and ozonation on wastewater treatment. J Photochem Photobiol C: Photochem Rev 6(4). https://doi.org/10.1016/j.jphotochemrev.2005.12.003

Akinbile CO, Yusoff MS, Ahmad Zuki AZ (2012) Landfill leachate treatment using sub-surface flow constructed wetland by Cyperus haspan. Waste Manag 32(7):1387–1393. https://doi.org/10.1016/j.wasman.2012.03.002

Alfaia RGSM, Nascimento MMP, Bila DM, Campos JC (2019) Coagulation/flocculation as a pretreatment of landfill leachate for minimizing fouling in membrane processes. Desalin Water Treat 159. https://doi.org/10.5004/dwt.2019.24280

Aluko OO, Sridhar M (2014) Evaluation of effluents from bench-scale treatment combinations for landfill leachate in Ibadan, Nigeria. Waste Manag Res 32(1):70–78. https://doi.org/10.1177/0734242X13514624

Aminul Haque M, Hoque MA, Modal MSA, Tauhid-Ur-Rahman M (2013) Characterization of leachate and solid waste of Dhaka City Corporation landfill site for solid waste stabilization. Am J Civ Eng Archit 1(2):39–42. https://doi.org/10.12691/ajcea-1-2-3

Amokrane A, Comel C, Veron J (1997) Landfill leachates pretreatment by coagulation-flocculation. Water Res 31(11). https://doi.org/10.1016/S0043-1354(97)00147-4

Anqi T, Zhang Z, Suhua H, Xia L (2020) Review on landfill leachate treatment methods. IOP Conf Ser Earth Environ Sci 565(1). https://doi.org/10.1088/1755-1315/565/1/012038

Aronsson P, Dahlin T, Dimitriou I (2010) Treatment of landfill leachate by irrigation of willow coppice – plant response and treatment efficiency. Environ Pollut 158(3):795–804. https://doi.org/10.1016/j.envpol.2009.10.003

Ates H, Argun ME (2018) Removal of PAHs from leachate using a combination of chemical precipitation and Fenton and ozone oxidation. Water Sci Technol 78(5):1064–1070. https://doi.org/10.2166/wst.2018.378

Bahmanyar A, Asghari M, Khoobi N (2012) Numerical simulation and theoretical study on simultaneously effects of operating parameters in direct contact membrane distillation. Chem Eng Process Process Intensif 61:42–50. https://doi.org/10.1016/j.cep.2012.06.012

Bardi A, Yuan Q, Tigini V, Spina F, Varese GC, Spennati F, Becarelli S, Di Gregorio S, Petroni G, Munz G (2017) Recalcitrant compounds removal in raw leachate and synthetic effluents using the white-rot fungus Bjerkandera adusta. Water (Switzerland) 9(11):1–14. https://doi.org/10.3390/w9110824

Barnes D, Li X, Chen J (2007) Determination of suitable pretreatment method for old-intermediate landfill leachate. Environ Technol 28(2):195–203. https://doi.org/10.1080/09593332808618782

Bashir MJK, Aziz HA, Yusoff MS, Huqe AAM, Mohajeri S (2010) Effects of ion exchange resins in different mobile ion forms on semi-aerobic landfill leachate treatment. Water Sci Technol 61(3):641–649. https://doi.org/10.2166/wst.2010.867

Bashir MJK, Aziz HA, Amr SSA, Sethupathi S, Ng CA, Lim JW (2015) The competency of various applied strategies in treating tropical municipal landfill leachate. Desalin Water Treat 54(9). https://doi.org/10.1080/19443994.2014.901189

Benoit A, Coqblin M, Nicod JM, Rehn-Sonigo V (2016) Optimizing memory allocation for multistage scheduling including setup times. J Sched 19(6):641–658. https://doi.org/10.1007/s10951-015-0437-x

Berenjkar P, Islam M, Yuan Q (2019) Co-treatment of sewage sludge and mature landfill leachate by anaerobic digestion. Int J Environ Sci Technol 16(5):2465–2474. https://doi.org/10.1007/s13762-018-1889-2

Bilgili MS, Demir A, Ozkaya B (2006) Quality and quantity of leachate in aerobic pilot-scale landfills. Environ Manag 38(2):189–196. https://doi.org/10.1007/s00267-005-0179-1

Bohdziewicz J, Kwarciak A (2008) The application of hybrid system UASB reactor-RO in landfill leachate treatment. Desalination 222(1–3):128–134. https://doi.org/10.1016/j.desal.2007.01.137

Çakmakce M, Kayaalp N, Koyuncu I (2008) Desalination of produced water from oil production fields by membrane processes. Desalination 222(1–3):176–186. https://doi.org/10.1016/j.desal.2007.01.147

Cakmakci M, Baspinar AB, Balaban U, Uyak V, Koyuncu I, Kinaci C (2009) Comparison of nanofiltration and adsorption techniques to remove arsenic from drinking water. Desalin Water Treat 9(1–3):149–154. https://doi.org/10.5004/dwt.2009.765

Camaño Silvestrini NE, Maine MA, Hadad HR, Nocetti E, Campagnoli MA (2019) Effect of feeding strategy on the performance of a pilot scale vertical flow wetland for the treatment of landfill leachate. Sci Total Environ 648:542–549. https://doi.org/10.1016/j.scitotenv.2018.08.132

Campos JC, Moura D, Costa AP, Yokoyama L, Araujo FVDF, Cammarota MC, Cardillo L (2013) Evaluation of pH, alkalinity and temperature during air stripping process for ammonia removal from landfill leachate. J Environ Sci Health – Part A Toxic/Hazard Subst Environ Eng 48(9). https://doi.org/10.1080/10934529.2013.774658

Chaudhari LB, Murthy ZVP (2010) Treatment of landfill leachates by nanofiltration. J Environ Manag 91(5):1209–1217. https://doi.org/10.1016/j.jenvman.2010.01.007

Cheung KC, Chu LM, Wong MH (1997) Ammonia stripping as a pretreatment for landfill leachate. Water Air Soil Pollut 94(1–2). https://doi.org/10.1023/A:1026413031434

Chia Ken L, MJK B, Abu Amr SS, Ching Sim L (2016) Post-treatment of palm oil mill effluent (POME) using combined persulphate with hydrogen peroxide (S 2 O 82-/$H_2O_2$) oxidation. Water Sci Technol J Int Assoc Water Pollut Res 74(11):2675–2682. https://doi.org/10.2166/WST.2016.458

Chong MN, Sharma AK, Burn S, Saint CP (2012) Feasibility study on the application of advanced oxidation technologies for decentralised wastewater treatment. J Clean Prod 35. https://doi.org/10.1016/j.jclepro.2012.06.003

Chung S, Seo CD, Lee H, Choi JH, Chung J (2014) Design strategy for networking membrane module and heat exchanger for direct contact membrane distillation process in seawater desalination. Desalination 349:126–135. https://doi.org/10.1016/j.desal.2014.07.001

Collado S, Núñez D, Oulego P, Riera FA, Díaz M (2020) Effect of landfill leachate ageing on ultrafiltration performance and membrane fouling behaviour. J Water Process Eng 36(March):101291. https://doi.org/10.1016/j.jwpe.2020.101291

Cortez S, Teixeira P, Oliveira R, Mota M (2011a) Denitrification of a landfill leachate with high nitrate concentration in an anoxic rotating biological contactor. Biodegradation 22(3):661–671. https://doi.org/10.1007/s10532-010-9439-8

Cortez S, Teixeira P, Oliveira R, Mota M (2011b) Evaluation of Fenton and ozone- based advanced oxidation processes as mature landfill leachate pre-treatments. J Environ Manag 92(3). https://doi.org/10.1016/j.jenvman.2010.10.035

Dan DA, Fujii D, Soda S, Machimura T, Ike M (2017) Removal of phenol, bisphenol A, and 4-tert-butylphenol from synthetic landfill leachate by vertical flow constructed wetlands. Sci Total Environ 578:566–576. https://doi.org/10.1016/j.scitotenv.2016.10.232

de Almeida R, Moraes Costa A, de Almeida Oroski F, Carbonelli Campos J (2019) Evaluation of coagulation–flocculation and nanofiltration processes in landfill leachate treatment. J Environ Sci Health – Part A Toxic/Hazard Subst Environ Eng 54(11):1091–1098. https://doi.org/10.1080/10934529.2019.1631093

de Castro TM, Torres DGB, Arantes EJ, de Carvalho KQ, Passig FH, Christ D, Gotardo JT, Gomes SD (2020) Anaerobic co-digestion of industrial landfill leachate and glycerin: methanogenic potential, organic matter removal and process optimization. Environ Technol (UK) 41(20):2583–2593. https://doi.org/10.1080/09593330.2019.1575915

Deborde M, von Gunten U (2008) Reactions of chlorine with inorganic and organic compounds during water treatment-kinetics and mechanisms: a critical review. Water Res 42(1–2). https://doi.org/10.1016/j.watres.2007.07.025

Deng Y, Englehardt JD (2006) Treatment of landfill leachate by the Fenton process. Water Res 40(20):3683–3694. https://doi.org/10.1016/j.watres.2006.08.009

Deng Y, Englehardt JD (2007) Electrochemical oxidation for landfill leachate treatment. Waste Manag 27(3). https://doi.org/10.1016/j.wasman.2006.02.004

Deng Y, Feng C, Chen N, Hu W, Kuang P, Liu H, Hu Z, Li R (2018) Research on the treatment of biologically treated landfill leachate by joint electrochemical system. Waste Manag 82. https://doi.org/10.1016/j.wasman.2018.10.028

Deng Y, Chen N, Feng C, Chen F, Wang H, Kuang P, Feng Z, Liu T, Gao Y, Hu W (2019) Treatment of organic wastewater containing nitrogen and chlorine by combinatorial electrochemical system: taking biologically treated landfill leachate treatment as an example. Chem Eng J 364. https://doi.org/10.1016/j.cej.2019.01.176

Dogan S (2018) Optimization of struvite precipitation for landfill leachate treatment. Research 23(1):65–76. https://doi.org/10.17482/uumfd.373302

Dolar D, Košutić K, Strmecky T (2016) Hybrid processes for treatment of landfill leachate: coagulation/UF/NF-RO and adsorption/UF/NF-RO. Sep Purif Technol 168:39–46. https://doi.org/10.1016/j.seppur.2016.05.016

Dong Y, Wang Z, Zhu C, Wang Q, Tang J, Wu Z (2014) A forward osmosis membrane system for the post-treatment of MBR-treated landfill leachate. J Membr Sci 471:192–200. https://doi.org/10.1016/j.memsci.2014.08.023

Drioli E, Ali A, Macedonio F (2015) Membrane distillation: recent developments and perspectives. Desalination 356:56–84. https://doi.org/10.1016/j.desal.2014.10.028

El Ouaer M, Kallel A, Kasmi M, Hassen A, Trabelsi I (2017) Tunisian landfill leachate treatment using Chlorella sp.: effective factors and microalgae strain performance. Arab J Geosci 10(20). https://doi.org/10.1007/s12517-017-3241-4

EL-Bourawi MS, Khayet M, Ma R, Ding Z, Li Z, Zhang X (2007) Application of vacuum membrane distillation for ammonia removal. J Membr Sci 301(1–2):200–209. https://doi.org/10.1016/j.memsci.2007.06.021

EPA (2011) Municipal solid waste generation, recycling, and disposal in the United States: facts and figures for 2010

Essalhi M, Khayet M (2015) Membrane distillation (MD). In: Progress in filtration and separation (Issue Md). Elsevier Ltd. https://doi.org/10.1016/B978-0-12-384746-1.00003-3

Fard NAJH, Jorfi S, Ahmadi M, Mirali S, Kujlu R (2016) Treatment of mature landfill leachate by chemical precipitation and Fenton advanced oxidation process. Environ Health Eng Manag J 3(1):35–40

Fernandes A, Pacheco MJ, Ciríaco L, Lopes A (2015) Review on the electrochemical processes for the treatment of sanitary landfill leachates: present and future. Appl Catal B Environ 176–177. https://doi.org/10.1016/j.apcatb.2015.03.052

Ferraz FM, Povinelli J, Pozzi E, Vieira EM, Trofino JC (2014) Co-treatment of landfill leachate and domestic wastewater using a submerged aerobic biofilter. J Environ Manag 141:9–15. https://doi.org/10.1016/j.jenvman.2014.03.022

Frascari D, Bronzini F, Giordano G, Tedioli G, Nocentini M (2004) Long-term characterization, lagoon treatment and migration potential of landfill leachate: a case study in an active Italian landfill. Chemosphere 54(3):335–343. https://doi.org/10.1016/j.chemosphere.2003.08.013

Geenens D, Bixio B, Thoeye C (2001) Combined ozone-activated sludge treatment of landfill leachate. Water Sci Technol 44(2–3). https://doi.org/10.2166/wst.2001.0790

Ghasemzade R, Pazoki M (2017) Estimation and modeling of gas emissions in municipal landfill (Case study: landfill of Jiroft City). Pollution 3(4):689–700

Gomes LP, Schoenell EK (2018) Application of ozone and ozone + hydrogen peroxide to remove recalcitrant compounds in landfill leachates. Engenharia Sanitaria e Ambiental 23(1). https://doi.org/10.1590/S1413-41522018155758

Grimm J, Bessarabov D, Sanderson R (1998) Review of electro-assisted methods for water purification. Desalination 115(3). https://doi.org/10.1016/S0011-9164(98)00047-2

Guidi Nissim W, Palm E, Pandolfi C, Mancuso S, Azzarello E (2021) Willow and poplar for the phyto-treatment of landfill leachate in Mediterranean climate. J Environ Manag 277(June 2020):111454. https://doi.org/10.1016/j.jenvman.2020.111454

Hasar H, Unsal SA, Ipek U, Karatas S, Cinar O, Yaman C, Kinaci C (2009) Stripping/flocculation/membrane bioreactor/reverse osmosis treatment of municipal landfill leachate. J Hazard Mater 171(1–3):309–317. https://doi.org/10.1016/j.jhazmat.2009.06.003

Hashemi H, Jasemizad T, Derakhshan Z, Ebrahimi A (2017) Determination of Sequencing Batch Reactor (SBR) performance in treatment of composting plant leachate. Health Scope. https://doi.org/10.5812/jhealthscope.13356

Helfferich FG (1995) Ion Exchange. Dover Publications

Hong JG, Zhang B, Glabman S, Uzal N, Dou X, Zhang H, Wei X, Chen Y (2015) Potential ion exchange membranes and system performance in reverse electrodialysis for power generation: a review. J Membr Sci 486(July):71–88. https://doi.org/10.1016/j.memsci.2015.02.039

Ibrar I, Yadav S, Altaee A, Samal AK, Zhou JL, Nguyen TV, Ganbat N (2020) Treatment of biologically treated landfill leachate with forward osmosis: investigating membrane performance and cleaning protocols. Sci Total Environ 744:140901. https://doi.org/10.1016/j.scitotenv.2020.140901

Ince M, Senturk E, Onkal Engin G, Keskinler B (2010) Further treatment of landfill leachate by nanofiltration and microfiltration-PAC hybrid process. Desalination 255(1–3):52–60. https://doi.org/10.1016/j.desal.2010.01.017

Iskander S, Zou S, Brazil B, Novak JT, He Z (2017) Energy consumption by forward osmosis treatment of landfill leachate for water recovery. Waste Manag 1–8. https://doi.org/10.1016/j.wasman.2017.03.026

Islam M, Yuan Q (2020) Fungal treatment of mature landfill leachate utilizing woodchips and wheat-straw as co-substrates. Biodegradation 31(1–2):109–122. https://doi.org/10.1007/s10532-020-09897-9

Ismail S, Tawfik A (2016) Treatment of hazardous landfill leachate using Fenton process followed by a combined (UASB/DHS) system. Water Sci Technol 73(7):1700–1708. https://doi.org/10.2166/wst.2015.655

Jacob P, Phungsai P, Fukushi K, Visvanathan C (2015) Direct contact membrane distillation for anaerobic effluent treatment. J Membr Sci 475:330–339. https://doi.org/10.1016/j.memsci.2014.10.021

Jones DL, Williamson KL, Owen AG (2006) Phytoremediation of landfill leachate. Waste Manag 26(8):825–837. https://doi.org/10.1016/j.wasman.2005.06.014

Jorgensen TC, Weatherley LR (2003) Ammonia removal from wastewater by ion exchange in the presence of organic contaminants. Water Res 37(8):1723–1728. https://doi.org/10.1016/S0043-1354(02)00571-7

Kabdasli I, Tünay O, Öztürk I, Yilmaz S, Arikan O (2000) Ammonia removal from young landfill leachate by magnesium ammonium phosphate precipitation and air stripping. Water Sci Technol 41(1):237–240. https://doi.org/10.2166/wst.2000.0034

Kanyatrakul A, Prakhongsak A, Honda R, Phanwilai S, Treesubsuntorn C, Boonnorat J (2020) Effect of leachate effluent from activated sludge and membrane bioreactor systems with acclimatized sludge on plant seed germination. Sci Total Environ 724:138275. https://doi.org/10.1016/j.scitotenv.2020.138275

Khai NM (2012) Chemical precipitation of ammonia and phosphate from Nam Son Landfill Leachate, Hanoi. Iran J Energy Environ. https://doi.org/10.5829/idosi.ijee.2012.03.05.06

Khayet M, Matsuura T (2011) Membrane distillation: principles and applications. Elsevier

Kim KR, Owens G (2010) Potential for enhanced phytoremediation of landfills using biosolids – a review. J Environ Manag 91(4):791–797. https://doi.org/10.1016/j.jenvman.2009.10.017

Kjeldsen P, Barlaz MA, Rooker AP, Baun A, Ledin A, Christensen TH (2002) Present and long-term composition of MSW landfill leachate: a review. Crit Rev Environ Sci Technol. https://doi.org/10.1080/10643380290813462

Košutić K, Dolar D, Strmecky T (2015) Treatment of landfill leachate by membrane processes of nanofiltration and reverse osmosis. Desalin Water Treat 55(10):2680–2689. https://doi.org/10.1080/19443994.2014.939863

Kulhavý M, Cakl J, Václavík L, Maršálek J (2018) Electrodialysis as part of the integrated membrane process for landfill leachate treatment. Desalination Water Treat 150:13–16. https://doi.org/10.5004/dwt.2019.23511

Kulikowska D, Jóźwiak T, Kowal P, Ciesielski S (2010) Municipal landfill leachate nitrification in RBC biofilm – process efficiency and molecular analysis of microbial structure. Bioresour Technol 101(10):3400–3405. https://doi.org/10.1016/j.biortech.2009.12.050

Kulikowska D, Zielińska M, Konopka K (2019) Treatment of stabilized landfill leachate in an integrated adsorption–fine-ultrafiltration system. Int J Environ Sci Technol 16(1):423–430. https://doi.org/10.1007/s13762-018-1685-z

Kurniawan TA, Lo WH, Chan GYS (2006a) Degradation of recalcitrant compounds from stabilized landfill leachate using a combination of ozone-GAC adsorption treatment. J Hazard Mater 137(1):443–455. https://doi.org/10.1016/j.jhazmat.2006.02.020

Kurniawan TA, Lo WH, Chan GYS (2006b) Physico-chemical treatments for removal of recalcitrant contaminants from landfill leachate. J Hazard Mater 129(1–3). https://doi.org/10.1016/j.jhazmat.2005.08.010

Landfills | US EPA (n.d.)

Levchuk I, Rueda Márquez JJ, Sillanpää M (2018) Removal of natural organic matter (NOM) from water by ion exchange – a review. Chemosphere 192:90–104. https://doi.org/10.1016/j.chemosphere.2017.10.101

Li XZ, Zhao QL (2001) Efficiency of biological treatment affected by high strength of ammonium-nitrogen in leachate and chemical precipitation of ammonium-nitrogen as pretreatment. Chemosphere 44(1):37–43. https://doi.org/10.1016/S0045-6535(00)00382-9

Li XZ, Zhao QL, Hao XD (1999) Ammonium removal from landfill leachate by chemical precipitation. Waste Manag 19(6):409–415. https://doi.org/10.1016/S0956-053X(99)00148-8

Li HS, Zhou SQ, Sun YB, Feng P, Li JD (2009) Advanced treatment of landfill leachate by a new combination process in a full-scale plant. J Hazard Mater 172(1):408–415. https://doi.org/10.1016/j.jhazmat.2009.07.034

Li M, Feng C, Zhang Z, Yang S, Sugiura N (2010a) Treatment of nitrate contaminated water using an electrochemical method. Bioresour Technol 101(16). https://doi.org/10.1016/j.biortech.2010.03.076

Li Y, Lu A, Ding H, Wang X, Wang C, Zeng C, Yan Y (2010b) Microbial fuel cells using natural pyrrhotite as the cathodic heterogeneous Fenton catalyst towards the degradation of biorefractory organics in landfill leachate. Electrochem Commun 12(7). https://doi.org/10.1016/j.elecom.2010.04.027

Li Z, Ren X, Zuo J, Liu Y, Duan E, Yang J, Chen P, Wang Y (2012) Struvite precipitation for ammonia nitrogen removal in 7-aminocephalosporanic acid wastewater. Molecules 17(2):2126–2139. https://doi.org/10.3390/molecules17022126

Lima PR, Mirapalheta A, Henrique dos Santos Andrade M, Vilar EO, e Silva Zanta CL de P, Tonholo J (2010) Energy loss in electrochemical diaphragm process of chlorine and alkali industry – a collateral effect of the undesirable generation of chlorate. Energy 35(5). https://doi.org/10.1016/j.energy.2010.01.039

Lin C, Chang F, Chang C (2000) Co-digestion of leachate with septage using a UASB reactor. Bioresour Technol 73:175–178

Luo H, Zeng Y, Cheng Y, He D, Pan X (2020) Recent advances in municipal landfill leachate: a review focusing on its characteristics, treatment, and toxicity assessment. Sci Total Environ 703:135468. https://doi.org/10.1016/j.scitotenv.2019.135468

Madsen HT (2014) Membrane filtration in water treatment – removal of micropollutants. In: Chemistry of advanced environmental purification processes of water: fundamentals and applications. https://doi.org/10.1016/B978-0-444-53178-0.00006-7

Mariam T, Nghiem LD (2010) Landfill leachate treatment using hybrid coagulation- nanofiltration processes. Desalination 250(2):677–681. https://doi.org/10.1016/j.desal.2009.03.024

Marttinen SK, Kettunen RH, Sormunen KM, Soimasuo RM, Rintala JA (2002) Screening of physical-chemical methods for removal of organic material, nitrogen and toxicity from low strength landfill leachates. Chemosphere 46(6). https://doi.org/10.1016/S0045-6535(01)00150-3

Mehmood MK, Adetutu E, Nedwell DB, Ball AS (2009) In situ microbial treatment of landfill leachate using aerated lagoons. Bioresour Technol 100(10):2741–2744. https://doi.org/10.1016/j.biortech.2008.11.031

Miao L, Wang S, Cao T, Peng Y, Zhang M, Liu Z (2016) Advanced nitrogen removal from landfill leachate via Anammox system based on Sequencing Biofilm Batch Reactor (SBBR): effective protection of biofilm. Bioresour Technol 220:8–16. https://doi.org/10.1016/j.biortech.2016.06.131

Mohammadifakhr M, De Grooth J, Roesink HDW, Kemperman AJB (2020) Forward osmosis: a critical review. PRO 8(4):1–29
Mojiri K (2015) Assessment of various tropical municipal landfill leachate characteristics and treatment opportunities. Global NEST J 17(3):439–450
Mojiri A, Zhou JL, Ratnaweera H, Ohashi A, Ozaki N, Kindaichi T, Asakura H (2021) Treatment of landfill leachate with different techniques: an overview. J Water Reuse Desalin 11(1):66–96. https://doi.org/10.2166/wrd.2020.079
Münch EV, Barr K (2001) Controlled struvite crystallisation for removing phosphorus from anaerobic digester sidestreams. Water Res 35(1):151–159. https://doi.org/10.1016/s0043-1354(00)00236-0
Nagarale RK, Gohil GS, Shahi VK (2006) Recent developments on ion-exchange membranes and electro-membrane processes. Adv Colloid Interf Sci 119(2–3):97–130. https://doi.org/10.1016/j.cis.2005.09.005
Nagy E (2019) Chapter 17 – forward osmosis 2016, pp 447–456. https://doi.org/10.1016/B978-0-12-813722-2.00017-0
Nhat PT, Biec HN, Tuyet Mai NT, Thanh BX, Dan NP (2014) Application of a partial nitritation and anammox system for the old landfill leachate treatment. Int Biodeterior Biodegrad 95(PA):144–150. https://doi.org/10.1016/j.ibiod.2014.05.025
Nordin N, Yusof N, Samsudin S (2017) Biomass production of Chlorella sp., Scenedesmus sp., and Oscillatoria sp. in nitrified landfill leachate. Waste Biomass Valorization 8(7):2301–2311. https://doi.org/10.1007/s12649-016-9709-8
Ozturk I, Altinbas M, Koyuncu I, Arikan O, Gomec-Yangin C (2003) Advanced physico-chemical treatment experiences on young municipal landfill leachates. Waste Manag 23(5). https://doi.org/10.1016/S0956-053X(03)00061-8
Pan S, Dixon KL, Nawaz T, Rahman A, Selvaratnam T (2021) Evaluation of Galdieria sulphuraria for nitrogen removal and biomass production from raw landfill leachate. Algal Res 54(January):102183. https://doi.org/10.1016/j.algal.2021.102183
Peng Y (2017) Perspectives on technology for landfill leachate treatment. Arab J Chem 10:S2567–S2574. https://doi.org/10.1016/j.arabjc.2013.09.031
Pera-Titus M, García-Molina V, Baños MA, Giménez J, Esplugas S (2004) Degradation of chlorophenols by means of advanced oxidation processes: a general review. Appl Catal B Environ 47(4):219–256. https://doi.org/10.1016/j.apcatb.2003.09.010
Pi Y, Zhang L, Wang J (2007) The formation and influence of hydrogen peroxide during ozonation of para-chlorophenol. J Hazard Mater 141(3). https://doi.org/10.1016/j.jhazmat.2006.07.032
Postacchini L, Ciarapica FE, Bevilacqua M (2018) Environmental assessment of a landfill leachate treatment plant: impacts and research for more sustainable chemical alternatives. J Clean Prod 183:1021–1033. https://doi.org/10.1016/j.jclepro.2018.02.219
Pujotomo I, Azzahra S (2019) Electrical energy production process from landfill gas. E3S Web Conf 125(2019):4–7. https://doi.org/10.1051/e3sconf/201912513002
Raghab SM, Abd El Meguid AM, Hegazi HA (2013) Treatment of leachate from municipal solid waste landfill. HBRC J 9(2):187–192. https://doi.org/10.1016/j.hbrcj.2013.05.007
Ranjan K, Chakraborty S, Verma M, Iqbal J, Naresh Kumar R (2016) Co-treatment of old landfill leachate and municipal wastewater in sequencing batch reactor (SBR): effect of landfill leachate concentration. Water Qual Res J Can 51(4):377–387. https://doi.org/10.2166/wqrjc.2016.020
Remmas N, Ntougias S, Chatzopoulou M, Melidis P (2018) Optimization aspects of the biological nitrogen removal process in a full-scale twin sequencing batch reactor (SBR) system in series treating landfill leachate. J Environ Sci Health – Part A Toxic/Hazard Subst Environ Eng 53(9):847–853. https://doi.org/10.1080/10934529.2018.1455375
Ren Y, Ferraz F, Lashkarizadeh M, Yuan Q (2017) Comparing young landfill leachate treatment efficiency and process stability using aerobic granular sludge and suspended growth activated sludge. J Water Process Eng 17(February):161–167. https://doi.org/10.1016/j.jwpe.2017.04.006
Renou S, Givaudan JG, Poulain S, Dirassouyan F, Moulin P (2008) Landfill leachate treatment: review and opportunity. J Hazard Mater 150(3):468–493. https://doi.org/10.1016/j.jhazmat.2007.09.077

Rodrigo MA, Cañizares P, Sánchez-Carretero A, Sáez C (2010) Use of conductive- diamond electrochemical oxidation for wastewater treatment. Catal Today 151(1–2). https://doi.org/10.1016/j.cattod.2010.01.058
Ross TK, Wragg AA (1965) Electrochemical mass transfer studies in annuli. Electrochim Acta 10(11). https://doi.org/10.1016/0013-4686(65)80034-2
Saleem M, Spagni A, Alibardi L, Bertucco A, Lavagnolo MC (2018) Assessment of dynamic membrane filtration for biological treatment of old landfill leachate. J Environ Manag 213(May):27–35. https://doi.org/10.1016/j.jenvman.2018.02.057
Salem Z, Hamouri K, Djemaa R, Allia K (2008) Evaluation of landfill leachate pollution and treatment. Desalination 220(1–3). https://doi.org/10.1016/j.desal.2007.01.026
Saquing JM, Chanton JP, Yazdani R, Barlaz MA, Scheutz C, Blake DR, Imhoff PT (2014) Assessing methods to estimate emissions of non-methane organic compounds from landfills. Waste Manag 34(11):2260–2270. https://doi.org/10.1016/j.wasman.2014.07.007
Sata T (2004) Ion exchange membranes preparation, characterization, modification and application. The Royal Society of Chemistry
Schoeman JJ, Steyn A, Makgae M (2005) Evaluation of electrodialysis for the treatment of an industrial solid waste leachate. Desalination 186(2005):273–289. https://doi.org/10.1016/j.desal.2005.04.061
Shehzad A, Bashir MJK, Sethupathi S, Lim JW (2015) An overview of heavily polluted landfill leachate treatment using food waste as an alternative and renewable source of activated carbon. Process Saf Environ Prot 98. https://doi.org/10.1016/j.psep.2015.09.005
Siddique MNI, Wahid ZA (2018) Achievements and perspectives of anaerobic co-digestion: a review. J Clean Prod 194(1):359–371. https://doi.org/10.1016/j.jclepro.2018.05.155
Siegrist H (1996) Nitrogen removal from digester supernatant – comparison of chemical and biological methods. Water Sci Technol 34(1–2–2 pt 1):399–406. https://doi.org/10.1016/0273-1223(96)00529-X
Silva AC, Dezotti M, Sant'Anna GLSA Jr (2004) Treatment and detoxification of a sanitary landfill leachate. Chemosphere 55:207–214. https://doi.org/10.1016/j.chemosphere.2003.10.013
Silva NCM, Moravia WG, Amaral MCS, Figueiredo KCS (2019) Evaluation of fouling mechanisms in nanofiltration as a polishing step of yeast MBR-treated landfill leachate. Environ Technol (UK) 40(27). https://doi.org/10.1080/09593330.2018.1482568
Singh SK, Townsend TG, Boyer TH (2012) Evaluation of coagulation (FeCl 3) and anion exchange (MIEX) for stabilized landfill leachate treatment and high-pressure membrane pretreatment. Sep Purif Technol 96:98–106. https://doi.org/10.1016/j.seppur.2012.05.014
Siracusa G, Yuan Q, Chicca I, Bardi A, Spennati F, Becarelli S, Levin DB, Munz G, Petroni G, Gregorio SD (2020) Mycoremediation of old and intermediate landfill. Water 12:800
Song U, Waldman B, Park JS, Lee K, Park SJ, Lee EJ (2018) Improving the remediation capacity of a landfill leachate channel by selecting suitable macrophytes. J Hydro Environ Res 20(April):31–37. https://doi.org/10.1016/j.jher.2018.04.005
Sørensen E, Lam KF, Sudhoff D (2014) Special distillation applications. In: Distillation: operation and applications. Elsevier Inc. https://doi.org/10.1016/B978-0-2-386876-3.00009-0
Statista (n.d.)
Sun H, Peng Y, Shi X (2015) Advanced treatment of landfill leachate using anaerobic- aerobic process: organic removal by simultaneous denitritation and methanogenesis and nitrogen removal via nitrite. Bioresour Technol 177:337–345. https://doi.org/10.1016/j.biortech.2014.10.152
Syzdek AC, Ahlert RC (1984) Separation of landfill leachate with polymeric ultrafiltration membranes. J Hazard Mater 9(2):209–220. https://doi.org/10.1016/0304-3894(84)80018-7
Tatsi AA, Zouboulis AI, Matis KA, Samaras P (2003) Coagulation-flocculation pretreatment of sanitary landfill leachates. Chemosphere 53(7). https://doi.org/10.1016/S0045-6535(03)00513-7
Trebouet D, Schlumpf JP, Jaouen P, Quemeneur F (2001) Stabilized landfill leachate treatment by combined physicochemical-nanofiltration processes. Water Res 35(12):2935–2942. https://doi.org/10.1016/S0043-1354(01)00005-7

Visvanathan C, Choudhary MK, Montalbo MT, Jegatheesan V (2007) Landfill leachate treatment using thermophilic membrane bioreactor. Desalination 204(1–3 Special Issue):8–16. https://doi.org/10.1016/j.desal.2006.02.028

Wachinski AM (2016a) Environmental ion exchange: principles and design, 2nd edn. CRC Press

Wachinski AM (2016b) Fundamental principles and concepts of ion exchange. In: Environmental ion exchange. https://doi.org/10.1201/9781315368542-4

Wang Z, Peng Y, Miao L, Cao T, Zhang F, Wang S, Han J (2016) Continuous- flow combined process of nitritation and ANAMMOX for treatment of landfill leachate. Bioresour Technol 214:514–519. https://doi.org/10.1016/j.biortech.2016.04.118

Wang K, Li L, Tan F, Wu D (2018) Treatment of landfill leachate using activated sludge technology: a review. Archaea 2018. https://doi.org/10.1155/2018/1039453

Woo YC, Kim S, Shon HK, Tijing LD (2019) Introduction: membrane desalination today, past, and future. In: Current trends and future developments on (bio-) membranes. Elsevier Inc. https://doi.org/10.1016/B978-0-12-813551-8.00028-0

Wu Y, Zhou S, Qin F, Peng H, Lai Y, Lin Y (2010) Removal of humic substances from landfill leachate by Fenton oxidation and coagulation. Process Saf Environ Prot 88(4):276–284. https://doi.org/10.1016/j.psep.2010.03.002

Xiong J, Zheng Z, Yang X, He J, Luo X, Gao B (2018) Mature landfill leachate treatment by the MBBR inoculated with biocarriers from a municipal wastewater treatment plant. Process Saf Environ Prot 119:304–310. https://doi.org/10.1016/j.psep.2018.08.019

Xu Y, Zhou Y, Wang D, Chen S, Liu J, Wang Z (2008) Occurrence and removal of organic micropollutants in the treatment of landfill leachate by combined anaerobic- membrane bioreactor technology. J Environ Sci 20(11):1281–1287. https://doi.org/10.1016/S1001-0742(08)62222-6

Xu X, Wang G, Zhou L, Yu H, Yang F (2018) Start-up of a full-scale SNAD-MBBR process for treating sludge digester liquor. Chem Eng J 343(January):477–483. https://doi.org/10.1016/j.cej.2018.03.032

Ye J, Mu Y, Cheng X, Sun D (2011) Treatment of fresh leachate with high-strength organics and calcium from municipal solid waste incineration plant using UASB reactor. Bioresour Technol 102(9):5498–5503. https://doi.org/10.1016/j.biortech.2011.01.001

Yuan MH, Chen YH, Tsai JY, Chang CY (2016) Ammonia removal from ammonia-rich wastewater by air stripping using a rotating packed bed. Process Saf Environ Prot 102. https://doi.org/10.1016/j.psep.2016.06.021

Zamri MFMA, Kamaruddin MA, Yusoff MS, Aziz HA, Foo KY (2017) Semi-aerobic stabilized landfill leachate treatment by ion exchange resin: isotherm and kinetic study. Appl Water Sci 7(2):581–590. https://doi.org/10.1007/s13201-015-0266-2

Zhang Q, Xu R, Xu P, Chen R, He Q, Zhong J, Gu X (2014) Performance study of ZrO2 ceramic micro-filtration membranes used in pretreatment of DMF wastewater. Desalination 346:1–8. https://doi.org/10.1016/j.desal.2014.05.006

Zhang C, Ding W, Zeng X, Xu X (2020) Recovery of ammonia nitrogen from landfill leachate using a biopolar membrane equipped electrodialysis system. Water Sci Technol 82(9):1758–1770. https://doi.org/10.2166/wst.2020.438

Zhou Q, Li W, Hua T (2010) Removal of organic matter from landfill leachate by advanced oxidation processes: a review. Int J Chem Eng. https://doi.org/10.1155/2010/270532

Zhou Y, Huang M, Deng Q, Cai T (2017) Combination and performance of forward osmosis and membrane distillation (FO-MD) for treatment of high salinity landfill leachate. Desalination 420(February):99–105. https://doi.org/10.1016/j.desal.2017.06.027

Zhou S, Wang J, Peng S, Chen T, Yue Z (2021) Anaerobic co-digestion of landfill leachate and acid mine drainage using up-flow anaerobic sludge blanket reactor. Environ Sci Pollut Res 28(7):8498–8506. https://doi.org/10.1007/s11356-020-11207-y

Zico MM, Ricci BC, Reis BG, Magalhães NC, Amaral MCS (2021) Sustainable ammonia resource recovery from landfill leachate by solar-driven modified direct contact membrane distillation. Sep Purif Technol 264(September 2020). https://doi.org/10.1016/j.seppur.2021.118356

Zielińska M, Kulikowska D, Stańczak M (2020) Adsorption – membrane process for treatment of stabilized municipal landfill leachate. Waste Manag 114:174–182. https://doi.org/10.1016/j.wasman.2020.07.011

Zolfaghari M, Jardak K, Drogui P, Brar SK, Buelna G, Dubé R (2016) Landfill leachate treatment by sequential membrane bioreactor and electro-oxidation processes. J Environ Manag 184:318–326. https://doi.org/10.1016/j.jenvman.2016.10.010

Zoungrana A, Çakmakci M (2020) From non-renewable energy to renewable by harvesting salinity gradient power by reverse electrodialysis: a review. Int J Energy Res 45(3):3495–3522. https://doi.org/10.1002/er.6062

Zoungrana A, Cakmakci M, Zengin IH, Inoglu O, Elcik H (2016) Treatment of metal-plating waste water by modified direct contact membrane distillation. Chem Pap 70(9):1185–1195. https://doi.org/10.1515/chempap-2016-0066

Zoungrana A, Zengin İH, Elcik H, Özkaya B, Çakmakci M (2017) The treatability of landfill leachate by direct contact membrane distillation and factors influencing the efficiency of the process. Desalin Water Treat 71:233–243. https://doi.org/10.5004/dwt.2017.20494

Zoungrana A, Zengin İH, Türk OK, Çakmakcı M (2020) Ammoniacal nitrogen reclamation by membrane distillation from high ammonia polluted solutions. Chem Pap 74(6):1903–1915. https://doi.org/10.1007/s11696-019-01034-y

Zulaikha S, Lau WJ, Ismail AF, Jaafar J (2014) Treatment of restaurant wastewater using ultrafiltration and nanofiltration membranes. J Water Process Eng 2:58–62. https://doi.org/10.1016/j.jwpe.2014.05.001

# Chapter 7
# Advancements in Operations of Bioreactor Landfills for Enhanced Biodegradation of Municipal Solid Waste

**Abhishek N. Srivastava, Rahul Singh, Sumedha Chakma, and Volker Birke**

**Abstract** Conventional sanitary landfills are of dry tomb type structures with entrapped storage of municipal solid waste (MSW) between the liner and cover layers, which facilitates long-term stabilization, limited gas recovery and unhygienic surroundings. However, various drawbacks pertaining to dry tomb landfilling reinforced the development of essential technologies to intensify biodegradation of MSW disposal in landfills. Bioreactor landfills (BRLs) are well known for overcoming such issues and enhancing the biodegradation of MSW with amplified landfill gas (LFG) generation. However, the efficiency of BRL system is compromised recently based on its shortcomings related to MSW degradation characteristics, inappropriate leachate recirculation and rigorous performance monitoring. Therefore, this chapter explores recent advanced methodologies adopted worldwide for BRL process intensification. Such advancements are classified according to their mode of utilization and stages of digestion phase in BRL system. Incorporation of aeration unit in BRL along with intermittent leachate recirculation has been proved to be advantageous over anaerobic BRL system. However, excessive aeration affects landfill capacity through an imbalance of waste mass settlement. Therefore, hybrid bioreactor system further enhances the MSW decomposition rate, LFG generation

A. N. Srivastava (✉) · S. Chakma
Department of Civil Engineering, Indian Institute of Technology Delhi, New Delhi, India
e-mail: Abhishek.Narayana.Srivastava@civil.iitd.ac.in; cez178533@iitd.ac.in; chakma@iitd.ac.in

R. Singh
Department of Civil Engineering, Indian Institute of Technology Delhi, New Delhi, India

Faculty of Engineering Science, Department of Process, Energy, Environmental and Biotech Engineering, University of Wismar – University of Applied Sciences, Technology, Business, and Design, Wismar, Germany
e-mail: Rahul.Singh@civil.iitd.ac.in

V. Birke
Faculty of Engineering Science, Department of Process, Energy, Environmental and Biotech Engineering, University of Wismar – University of Applied Sciences, Technology, Business, and Design, Wismar, Germany
e-mail: volker.birke@hs-wismar.de



P. Pathak, S. G. Palani (eds.), *Circular Economy in Municipal Solid Waste Landfilling: Biomining & Leachate Treatment*, Radionuclides and Heavy Metals in the Environment, https://doi.org/10.1007/978-3-031-07785-2_7

and leachate decontamination. Several biotechnological applications such as enzymatic amendment, inoculum addition with respective pros and cons are also discussed in detail.

**Keywords** Municipal solid waste · Bioreactor landfill · Hybrid bioreactor landfill · Leachate treatment · Aerated bioreactor landfill

## 7.1 Introduction

Over the decades, extreme consumerism reinforced by whopping population increase has posed high demand of food, energy and goods worldwide that led to the intensification of municipal solid waste (MSW) generation in recent years. In fact, as per World Bank report, its generation will further rise upto 3.4 billion tons (bnt) per year till year 2050 against ~2.2 bnt MSW generated in 2018 (Hoornweg and Bhada-Tata 2012). Globally, approximately 33% of MSW remain uncollected by municipal authorities (Dutta and Jinsart 2020). Out of collected MSW by municipalities, 70% goes for landfill dumping, 19% recyclables are recovered and 11% is utilized for energy recovery through composting or bio methanation (Srivastava and Chakma 2020). MSW comprises of both biodegradables (food waste, kitchen waste, yard waste, paper waste, wood waste) and non-biodegradables (plastics, glass, construction debris, electronic waste, metals etc.). MSW composition varies significantly depending upon distinct continent, countries and their respective lifestyles. The engineered disposal of MSW has always been a paradigm between developing and developed countries. Notably, MSW management is dependent upon the economy of the nation and population pattern. In developed nations, energy recovery infrastructure from MSW is well established unlike developing countries, where actual demand for landfilling is still striving due to poor waste segregation and collection efficiency (Chakma et al. 2014).

Landfilling is the economical and most widely adopted way of disposing nonrecyclable fraction of MSW (Mathur and Chakma 2002). Engineered landfilling of MSW is a storage method of disposal in which waste is confined between liner and cover layers. This practice appears to be popular in developed countries but is lesser adopted in developing countries. In developing countries, most of the MSW is disposed openly without any liner or cover layer due to restricted open space, lack of vigilance and high population density (Nandan et al. 2017). MSW open dumping creates various environmental issues such as greenhouse gas emission, groundwater contamination, particulate emission, slope failure, pests and rodents, fire hazards and unhygienic surroundings (Chakma et al. 2014). Despite of being engineered mode of disposal, landfills pertain several disadvantages such as slow degradation process, uneven degradation caused by heterogeneity of MSW, limited landfill gas (LFG) production, improper utilization of MSW for energy generation and possibilities of groundwater contamination on the breakage of liner and cover layer (Singh et al. 2020). MSW possess substantial amount of organic fraction, thus has high potentials to regain energy through various waste to energy technologies. In conventional dry tomb landfills, it becomes difficult to utilize maximum energy potential due to aforementioned shortcomings and poor leachate management (Srivastava and Chakma 2020).

Bioreactor landfill (BRL) as an alternative for rapid stabilization of MSW with incremented LFG yield as a renewable energy option (Chakma and Mathur 2017). BRL is an engineered modern landfill that works in dual ways of both treating and storing MSW (Grossule et al. 2018). Moreover, as compared to dry tomb landfills, BRLs have several other advantages such as increased landfill capacity due to faster settlement of degrading waste caused by rapid degradation, reduced toxicity of leachate due to associated leachate recirculation mechanism, overall cost and time effectiveness, lower susceptibility of groundwater contamination and greenhouse emission (Gu et al. 2020). Apart from these advantages, BRLs saves annual cost of around \$1.40–2.15 per tonne as compared to dry tomb landfills (Tiseo 2021). The leachate recirculation mechanism equipped with a BRL enhances the moisture content to an appropriate level, which promotes microbial activity for rapid transformation of organic matter unlike conventional dry landfills (Fig. 7.1). Higher biochemical reaction rates in BRLs deliver huge potential for land reclamation (Reinhart and Townsend 2018). Di Maria et al. (2016) reported advantages of BRL over conventional landfill in terms of enhanced LFG yield during initial days of disposal which decreased suddenly and stabilized within 10 years. Improved leachate characteristics with lesser organic strength, limited post closure liabilities are other significant characteristics of BRL (Chakma and Mathur 2013; Srivastava and Chakma 2021a). BRLs pertain sufficient scope of replacing expensive landfill covers to cost-effective covers. Various materials such as native vegetations, bio-covers consisting composts from MSW organic fractions, green covers of native grass and aerobically treated high organic solid sludges have already been recommended as effective BRL cover materials for pilot scale application (He et al. 2012; Perdikea et al. 2008; Wang et al. 2011). Early stabilization of organic matter in BRL reduces the possibilities of generating odorous and noxious gases such as $H_2S$, and $NH_3$, which effectively improves volumetric content of biogas (Ali and Yue 2020; Long et al. 2016).

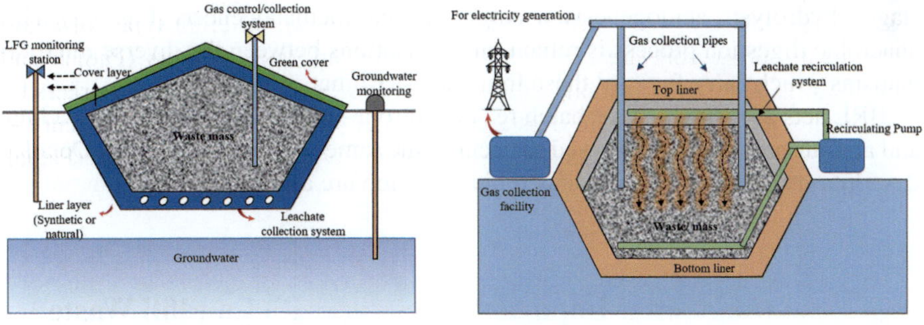


**Fig. 7.1** Bioreactor landfill Vs Conventional dry tomb landfills

BRL technology lags in several aspects such as high risk of fire due to accelerated methane production (Nag et al. 2019). Fire break out in BRL could possibly affect leachate distribution mesh and internal gas collection system, which could further ignite leakage of hazardous elements. There also exist a possibility of slope failure in BRL due to increase in pore water pressure caused by regular leachate recirculation (Giri and Reddy 2014). However, several mathematical models are available for optimizing appropriate leachate injection for maintaining sloe stability as well as adequate biodegradation environment inside BRL system (Byun et al. 2019; Feng et al. 2018). Incorporation of aeration in BRL certainly enhances biodegradation rate, but it affects overall biomethane yield due to inhibition of anaerobic fermentation process (Chu et al. 2021). Inclusion of aeration facility also contributes to raise overall capital investment cost.

Rapid waste generation rate in developing countries is reinforcing the necessity of further enhancement of MSW biodegradation disposed in BRL. Seeding of various degrading microorganism through leachate recirculation during different phases of anaerobic digestion is practiced extensively (Hettiaratchi et al. 2015). Further, aeration and combination of anaerobic-aerobic bioreactors is also being implemented worldwide with focus on rapid stabilization and augmented LFG production (Pellera et al. 2016). Operation of BRL as a hybrid system with sequential anaerobic – aerobic landfilling further accelerates biodegradation rate and LFG production (Ko et al. 2016a). Compaction mechanism equipped with BRL system also assists in accelerating biodegradation through increasing MSW density, even distribution of moisture, nutrients and microorganisms (Qiu et al. 2020). Moreover, co-landfilling of organic substrates for rapid increase in microbial activity is also being practiced worldwide for enhanced biodegradation of co-wastes (Iqbal et al. 2020). In this chapter, MSW degradation mechanism in a typical BRL is discussed in detail along with recent practices of aforementioned advancements for improving biodegradation of MSW.

## 7.2 MSW Degradation Mechanism in BRLs

In anaerobic bioreactor landfill, anaerobic digestion occurs in four successive stages: hydrolysis, acidogenesis, acetogenesis and methanogenesis (Fig. 7.2). This anaerobic digestion process is reliant on the relations between the diverse microorganisms which carry forward these four stages (Kumar et al. 2011).

BRL acts as a single-stage batch reactor for one-time disposable MSW loading, and all four processes are allowed to occur in the same reactor sequentially. Detailed descriptions of these stages of anaerobic digestion are as follows:

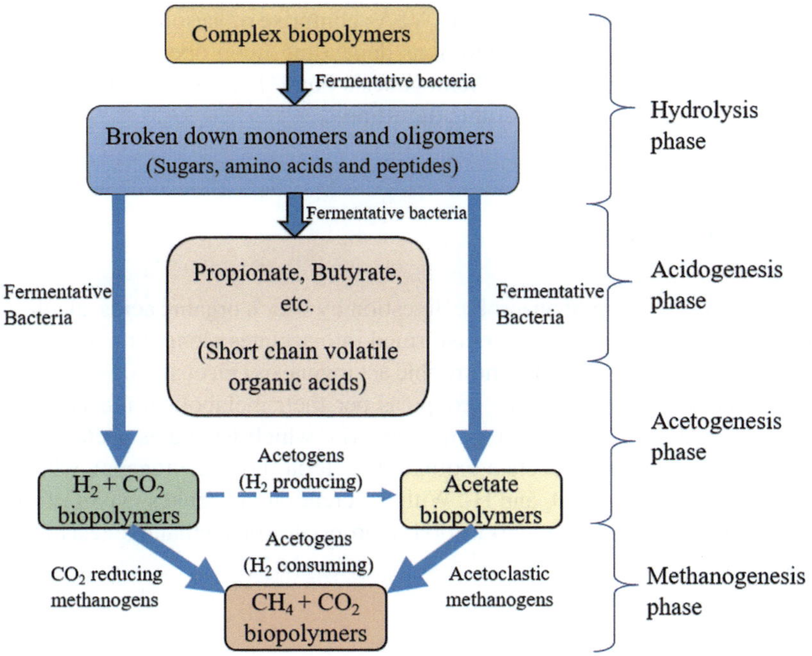


**Fig. 7.2** Phases of anaerobic digestion in landfills

### *7.2.1 Hydrolysis*

Organic fraction of MSW comprises complex bio-polymers, which are unreachable to microbes, is broken down through hydrolysis stage of digestion (Li et al. 2019). Consequently, the process of hydrolysis breaks down the organic macromolecules into its smaller components, which is utilizable by acidogenic bacteria. Hydrolysis, being the rate limiting step of anaerobic digestion, secretes extracellular enzymes that can transform carbohydrates, lipids, and proteins into sugars, long chain fatty acids and amino acids, respectively (Esposito et al. 2012). Certain substrates, having poor biodegradability such as lignin, cellulose, and hemicellulose, can be inaccessible to microbes due to their complex structures. Therefore, degrading enzymes are frequently added to improve the hydrolysis of such substances (Li et al. 2019).

### *7.2.2 Acidogenesis*

Short chained volatile fatty acids (VFAs), monomers of acids, alcohols are produced by acidogenic microorganisms via absorption of hydrolysis products through cell membranes. Post break down monomers and oligomers (sugars, amino acids and peptides) were converted to VFAs, propionates, butyrates and other short chain organic acids by fermentative bacteria. VFAs comprise of various organic acids that are generated in different concentrations depending upon operating conditions of digesters (Vigneron et al. 2007). Acidogenesis phase is considered to be the faster than that of other stages of anaerobic digestion.

### 7.2.3 *Acetogenesis*

Acetogenesis is the step of anaerobic digestion by which organic acids, alcohols and other fermentation products are transformed into acetates along with hydrogen and carbon dioxide through chemolithotrophic acetogenesis (Fu et al. 2019). Acetogenic bacteria can be classified in two groups as per their metabolism mechanism. The first one is proton-reducing acetogenic bacteria, which generates acetates, $H_2$ and $CO_2$. The second group is acetate oxidizing bacteria (homoacetogens), which catalyzes acetates and form $CO_2$ and $H_2$. With the creation of acetates, a part of existing substrate already condenses into a food appropriate for methanogenesis (Yadvika et al. 2004).

### 7.2.4 *Methanogenesis*

Methanogenesis is the final stage of anaerobic digestion, which involves degradation of carbon after exhaustion of all other electron acceptors. Methane is produced via multiple pathways carried out by methanogens (refer Eq.7.1) . Methanogenic bacteria are strict anaerobic bacteria which can only utilize certain organic matters for acetate splitting for methane formation. Moreover, methanogenic microorganisms are very sensitive towards pH, shock loading and temperature (Chen et al. 2008). The methanogenic microbes in BRLs, multiplied under alkaline system with pH levels 6–8 and alkalinity ~2000 mg/L (Xu et al. 2014).

$$4H_2 + CO_2 => CH_4 + 2H_2O \tag{7.1}$$

Depending upon mechanism and nature of biochemical reactions prevailed in these phases, technological advancement is pursued in order to enhance the degradation rate.

## 7.3 Technological Operations for Enhancing MSW Biodegradation in BRLs

There are several process-based stabilization techniques for biodegradation of MSW. All of them control the fundamental parameters that controls the degradation rate in BRL. Some of them are explained in following:

### *7.3.1 Mechanical Pretreatment*

Pulverization of large MSW fraction may assist in maintaining the homogeneity of the waste mix prior to feeding in BRLs and increasing microbial colonies through enhancement of surface area (Carchesio et al. 2020; Nguyen et al. 2021). In addition, mechanically pretreated substrates avoid moisture hindrances caused by large fractions or impermeable substances in commingled waste (Mahar et al. 2016). Pulverized waste favors uniform distribution of leachate without negative head occurrence in waste mix, thus ultimately assisting in improvement of overall landfilling capacity (Srivastava and Chakma 2021b; Tesseme and Chakma 2020). However, there could be negative effect also by shredding the waste as early prevalence of hydrolysis and acidogenesis caused by pretreatment can inhibit the forthcoming methanogenesis stage thereby delaying the overall digestion process with limited energy recovery (Li et al. 2019). Mechanical pretreatment through pulverization of MSW large fractions could only be beneficial only if low pH during acidogenesis and methanogenesis is maintained by pH buffering agents (Pan and Voulvoulis 2007).

### *7.3.2 MSW Compaction*

Compaction may enhance the rate of MSW degradation with appropriate compaction density. Compaction density of waste mix is pivot parameter that influences the landfill stabilization (Fadel et al. 2013). With respect to increase in compaction density the landfill capacity enhances and thereby extends the lifespan of landfill. The compaction also changes the degradation environment through increasing leachate permeability, microbial metabolism and growing microbial colonies (Qiu et al. 2020). In general, compaction is practiced when waste mix is relatively dry, as compaction increases degradation process via enhancement of released moisture. Moreover, compaction may assist in distribution of nutrients and increase the contact time between substrates and bacteria (Ko et al. 2016b). Mechanism of compaction associated with leachate recirculation influences distribution of VFAs concentrations in leachate. The densification of MSW through compaction could cause frequent fluctuation in VFA levels, which alters the methane production

(Fadel et al. 2013). The distribution of nutrients through leachate is directly related to density of waste mass, porosity, saturation and permeability of waste (Qiu et al. 2020). All such parameters are ultimately dependent upon extent of compaction, surface overburden pressure, waste list size and refuse moisture content (Ko et al. 2016b).

### 7.3.3 Co-disposal of Other Organic Wastes

Co-disposal of wastes having substantial organic content with comparatively higher biodegradability can increase the microbial population readily. Even, industrial organic sludges have potential for enhancing co-degradation of MSW and such wastes. Paper mill sludge (PMS) could increase degradation rates and bioenergy production through anaerobic co-disposal in landfill bioreactors (Srivastava and Chakma 2021b). However, maximum bioenergy generation and co-degradation relies on co-disposal ratio of MSW and other organic waste. Above literature recommended equal co-landfilling ratio of effluent treatment sludge from paper mill for maximum biomethane generation and leachate pollutant removal. Another co-disposal simulation of PMS and MSW in lab scale BRL showed considerable reduction in bioavailability of heavy metals (Cu, Cr, Pb, Ni, Cd and Zn) after 300 days of continuous bioreactor operation (Srivastava and Sumedha 2020). Addition of lignocellulosic waste is advisable as it provides extra adsorption sites for seizing mobility of heavy metals (Harman et al. 2007). Transformed humic substances as a result of anaerobic co-digestion of organic matter possess active binding tendency with heavy metals (Qu et al. 2008). Sewage sludge co-landfilling with MSW promotes degradation through moisture enhancement, nutrient supplementation and additional microbial consortia addition (Townsend et al. 2015a). Septage addition with target to enhance methane yield has also been investigated in anaerobic simulated landfill bioreactors, which showed significant rise in biomethane production (Zhao et al. 2005). It is noteworthy that all these sludges offer pH buffering capacity during anaerobic degradation, therefore it is advisable not to use waste having pH in extreme acidic ranges.

### 7.3.4 Enzymatic Amendments

The enzymes produced by fermentative bacteria during anaerobic digestion promote degradation process. Various simulation-based studies have shown positive impacts of waste specific enzyme addition for enhanced degradation of waste mix. In general, paper and cardboard fraction of MSW in landfills capture 40–50% of its space, therefore lignocellulose content also comprises significant fraction in it. Increased lignocellulosic fraction in MSW decreases overall degradation rate and lowers down the energy recovery (Singh et al. 2021). The enzyme addition for degradation of such recalcitrant substances not only can decrease MSW stabilization time but also enhance energy yield (Hettiaratchi et al. 2014). Particularly, the cellulase enzymes are well proven for lignocellulose degradation in bioreactor landfills (Hettiaratchi et al. 2014; Li et al. 2019; Saffira and Kristanto 2018). Similarly, newspaper and softwood were attempted for rapid degradation in landfills using fungal peroxidase and *Agrobacterium,* an isolated lignin-degrading bacterial strain (Muaaz-Us-Salam et al. 2020).

### 7.3.5 Hybrid Operation of BRL

Anaerobic degradation is relatively slow process and limits the full utilization of energy potential of organic fraction. Total aerobic BRL can surely enhance the biostabilization rate with improved leachate characteristics, but it takes out the possibilities of biomethane as energy source. Therefore, combination of aerobic – anaerobic system was proposed as hybrid BRL (Stessel and Murphy 1992). The hybrid BRL operates under series of combinations of aerobic and anaerobic conditions, which allows the landfilling system to attain benefits from both potentially enhanced biochemical reactions and increased landfill gas production (Xu et al. 2014). Hybrid BRL operation assists in buffering of pH, balancing excess moisture content, distribution of microbes and essential nutrients, occurrence of nitrification – denitrification process and controlled leaching of inorganic salts (Erses et al. 2008; Jiang et al. 2021; Shao et al. 2008; Townsend et al. 2015b; Xu et al. 2014). Moreover, intermittent aeration during anerobic operation of bioreactor landfill can significantly improve leachate properties (Nag et al. 2016). In fact, Ko et al. (2016a) reported more than 96% reduction in biochemical oxygen demand (BOD) and chemical oxygen demand (COD) of leachate. Operating BRL as hybrid system can reduce ammonia toxicity as introduction of air supports simultaneous occurrence of nitrification – denitrification (Sun et al. 2013). The facultative conditions prevailed in hybrid BRLs also reduce organic pollutants such as volatile fatty acids and chemical oxygen demand through methanogenesis and denitrification (Xu et al. 2015).

Recently, S.An.A.® (Semi-aerobic, Anaerobic, Aerated) landfill concept was introduced for further advancement of hybrid BRL system (Morello et al. 2017). This approach of hybrid BRL features an initial semi-aerobic phase for increasing biogas yield followed by sequential anaerobic and aerobic phases for the reduction of residual emission (Cossu et al. 2016). Finally, semi-aerobic environment is reestablished with flushing of bioreactor for leaching of inorganic substances. Furthermore, operating BRL as hybrid system also comes with the advantages of mitigation in greenhouse gas concentrations (Jiang et al. 2021). They reported 79.2% reduction in overall greenhouse emission with more than 93% COD removal from leachate after 465 days of operation of hybrid BRL simulator. Apart from these merits, reductions in aromatics and volatile organic compounds (VOCs) were also observed through hybrid BRL operation (Chu et al. 2021). More than 62 VOCs were detected, out of them oxygenated, halogenated, aromatic, alkanes and alkenes were found in abundance. Incorporation of intermittent aeration significantly increased volatilization rates of VOCs (Chu et al. 2021).

## 7.4 Environmental Sustainability of BRL

Sustainability of landfill management and engineered MSW disposal is critical issue for municipal authorities, worldwide. Traditional open landfills possess significant environmental hazards including long-term MSW stabilization, groundwater contamination, greenhouse emissions and unhygienic surroundings; thus, affecting long term sustainability of conventional disposal practice (Singh et al. 2020). On the other side, BRL eliminates such shortcomings through aforementioned advancements in existing technology. Therefore, environmental sustainability of BRLs is specifically designated as minimized short as well as long-term environmental risks upto an acceptable limit through effective design and operation of a BRL system (Grossule et al. 2018). Burton and Watson-Craik (1998) has already recommended criteria for sustainable landfilling that probable pollutive life must be limited to a generation or ~30 years, which signifies that landfilled waste should be stabilized and soluble contaminants flushed out of system within this period. BRL fulfils such criteria to be considered as environmentally sustainable system due to its attributes of early stabilization (10–15 years) of waste mass and containment of possible contaminants. Moreover, biochemical mechanisms and Anammox bacterial activity held in BRLs for nitrogen removal make its operation sustainable by avoiding ammonical toxicity (Shalini and Joseph 2012). Carbon removal through aerobic/anaerobic operation of BRL and upgraded hydraulics accelerate the goal of environmental sustainability of MSW disposal in BRLs. Besides, being advantageous over conventional landfills, BRLs are recommended to follow defined operational protocols and disciplined supervision to undertake landfilling operations for maintaining its long-term sustainability and high efficiency.

## 7.5 Conclusion

The overview of the study concluded that MSW degradation enhancement techniques including pretreatment, increased organic loading and pH maintenance through co-disposal, enzymatic amendments and intermittent aeration could be the paradigm in BRL technology. In recent years, aeration is proven technology that is readily being used in bioreactor landfilling systems, but it pertains the drawbacks of decreased biomethane yield due to interruption in methanogenesis. Hybrid operations of BRL system is recent advancement that significantly enhances the degradation rate and counters need of rapid landfilling technology. Moreover, consecutively sequential anaerobic – aerobic approach and operation is still to be explored in BRLs which could assist in achieving appropriate degradation goals with increased energy recovery. BRL technology could be environmentally sustainable over conventional methods due to its virtues but only if proper operational protocols would be strictly followed.

## References


Ali M, Yue D (2020) Population dynamics of microbial species under high and low ammonia nitrogen in the alternate layer bioreactor landfill (ALBL) approach. Bioresour Technol 315:123787

Burton SAQ, Watson-Craik IA (1998) Ammonia and nitrogen fluxes in landfill sites: applicability to sustainable landfilling. Waste Manag Res 16:41–53

Byun B, Kim I, Kim G, Eun J, Lee J (2019) Stability of bioreactor landfills with leachate injection configuration and landfill material condition. Comput Geotech 108:234–243

Carchesio M, Di Addario M, Tatàno F, de Rosa S, Gambioli A (2020) Evaluation of the biochemical methane potential of residual organic fraction and mechanically-biologically treated organic outputs intended for landfilling. Waste Manag 113:20–31

Chakma S, Mathur S (2017) Modelling gas generation for landfill. Environ Technol 38:1435–1442

Chakma S, Mathur S (2013) Postclosure long-term settlement for MSW landfills. J Hazardous Toxic Radioact Waste 17:81–88

Chakma S, Vaishya RC, Yadav AK (2014) Modeling chemical compositions of municipal solid waste. Environ Geotech 3:65–77

Chen Y, Cheng JJ, Creamer KS (2008) Inhibition of anaerobic digestion process: a review. Bioresour Technol 99:4044–4064

Chu Y-X, Wang J, Tian G, He R (2021) Reduction in VOC emissions by intermittent aeration in bioreactor landfills with gas-water joint regulation. Environ Pollut 118059

Cossu R, Morello L, Raga R, Cerminara G (2016) Biogas production enhancement using semi-aerobic pre-aeration in a hybrid bioreactor landfill. Waste Manag 55:83–92

Di Maria F, Micale C, Sisani L, Rotondi L (2016) Treatment of mechanically sorted organic waste by bioreactor landfill: experimental results and preliminary comparative impact assessment with biostabilization and conventional landfill. Waste Manag 55:49–60

Dutta A, Jinsart W (2020) Waste generation and management status in the fast-expanding Indian cities: a review. J Air Waste Manage Assoc 70:491–503

Erses AS, Onay TT, Yenigun O (2008) Comparison of aerobic and anaerobic degradation of municipal solid waste in bioreactor landfills. Bioresour Technol 99:5418–5426

Esposito G, Frunzo L, Giordano A, Liotta F, Panico A, Pirozzi F (2012) Anaerobic co-digestion of organic wastes. Rev Environ Sci Bio/Technology 11:325–341

El Fadel M, Fayad W, Hashisho J (2013) Enhanced solid waste stabilization in aerobic landfills using low aeration rates and high density compaction. Waste Manag Res 31:30–40

Feng S-J, Chen Z-W, Chen H-X, Zheng Q-T, Liu R (2018) Slope stability of landfills considering leachate recirculation using vertical wells. Eng Geol 241:76–85

Fu B, Jin X, Conrad R, Liu H, Liu H (2019) Competition between chemolithotrophic acetogenesis and hydrogenotrophic methanogenesis for exogenous H2/CO2 in anaerobically digested sludge: impact of temperature. Front Microbiol 10:2418

Giri RK, Reddy KR (2014) Slope stability of bioreactor landfill with leachate recirculation using horizontal trench system. In: Geoenvironmental Engineering. pp 120–129

Grossule V, Morello L, Cossu R, Lavagnolo MC (2018) Bioreactor landfills: comparison and kinetics of the different systems. Detritus In Press, 1. https://doi.org/10.31025/2611-4135/2018.13703

Gu Z, Chen W, Wang F, Li Q (2020) A pilot-scale comparative study of bioreactor landfills for leachate decontamination and municipal solid waste stabilization. Waste Manag 103:113–121

Harman G, Patrick R, Spittler T (2007) Removal of heavy metals from polluted waters using lignocellulosic agricultural waste products. Ind Biotechnol 3:366–374
He R, Xia F-F, Bai Y, Wang J, Shen D-S (2012) Mechanism of H2S removal during landfill stabilization in waste biocover soil, an alterative landfill cover. J Hazard Mater 217:67–75
Hettiaratchi JPA, Jayasinghe PA, Bartholameuz EM, Kumar S (2014) Waste degradation and gas production with enzymatic enhancement in anaerobic and aerobic landfill bioreactors. Bioresour Technol 159:433–436
Hettiaratchi P, Jayasinghe P, Tay J, Yadav S (2015) Recent advances of biomass waste to gas using landfill bioreactor technology – a review. Curr Org Chem 19:413–422
Hoornweg D, Bhada-Tata P (2012) What a waste : a global review of solid waste management. Urban development series. World Bank, pp 1–116
Iqbal A, Ekama GA, Zan F, Liu X, Chui H-K, Chen G-H (2020) Potential for co-disposal and treatment of food waste with sewage: a plant-wide steady-state model evaluation. Water Res 184:116175
Jiang G, Liu D, Chen W, Han Z, Li Q (2021) Greenhouse gas emissions from semi-aerobic bioreactor landfills with different vent-pipe diameters. Environ Sci Pollut Res 28:17563–17572
Ko JH, Ma Z, Jin X, Xu Q (2016a) Effects of aeration frequency on leachate quality and waste in simulated hybrid bioreactor landfills. J Air Waste Manage Assoc 66:1245–1256
Ko JH, Yang F, Xu Q (2016b) The impact of compaction and leachate recirculation on waste degradation in simulated landfills. Bioresour Technol 211:72–79
Kumar S, Chiemchaisri C, Mudhoo A (2011) Bioreactor landfill technology in municipal solid waste treatment: an overview. Crit Rev Biotechnol 31:77–97
Li Y, Chen Y, Wu J (2019) Enhancement of methane production in anaerobic digestion process: a review. Appl Energy 240:120–137. https://doi.org/10.1016/j.apenergy.2019.01.243
Long Y, Fang Y, Shen D, Feng H, Chen T (2016) Hydrogen sulfide (H 2 S) emission control by aerobic sulfate reduction in landfill. Sci Rep 6:1–9
Mahar RB, Sahito AR, Yue D, Khan K (2016) Modeling and simulation of landfill gas production from pretreated MSW landfill simulator. Front Environ Sci Eng 10:159–167
Mathur S, Chakma S (2002) Msw management through bioreactor landfills. In: National seminar on advances in civil engineering: perspectives of developing countries, vol 2. HBTI, Kanpur, pp 546–552
Morello L, Raga R, Lavagnolo MC, Pivato A, Ali M, Yue D, Cossu R (2017) The S. An. A.® concept: semi-aerobic, anaerobic, aerated bioreactor landfill. Waste Manag 67:193–202
Muaaz-Us-Salam S, Cleall PJ, Harbottle MJ (2020) Application of enzymatic and bacterial biodelignification systems for enhanced breakdown of model lignocellulosic wastes. Sci Total Environ 728:138741
Nag M, Shimaoka T, Komiya T (2019) Optimization of operations for nitrous oxide production and mitigation in aerobic and Aerobic-Anaerobic landfill method on MSW degradation. Sci Total Environ 696:134045
Nag M, Shimaoka T, Komiya T (2016) Impact of intermittent aerations on leachate quality and greenhouse gas reduction in the aerobic–anaerobic landfill method. Waste Manag 55:71–82
Nandan A, Yadav BP, Baksi S, Bose D (2017) Recent scenario of solid waste management in India. WSN World Sci News 66:56–74
Nguyen VK, Chaudhary DK, Dahal RH, Trinh NH, Kim J, Chang SW, Hong Y, La DD, Nguyen XC, Ngo HH (2021) Review on pretreatment techniques to improve anaerobic digestion of sewage sludge. Fuel 285:119105
Pan J, Voulvoulis N (2007) The role of mechanical and biological treatment in reducing methane emissions from landfill disposal of municipal solid waste in the United Kingdom. J Air Waste Manage Assoc 57:155–163
Pellera F-M, Pasparakis E, Gidarakos E (2016) Consecutive anaerobic-aerobic treatment of the organic fraction of municipal solid waste and lignocellulosic materials in laboratory-scale landfill-bioreactors. Waste Manag 56:181–189

Perdikea K, Mehrotra AK, Hettiaratchi JPA (2008) Study of thin biocovers (TBC) for oxidizing uncaptured methane emissions in bioreactor landfills. Waste Manag 28:1364–1374
Qiu Z, Li M, Zhang L, Zhao R, Li M (2020) Effect of waste compaction density on stabilization of aerobic bioreactor landfills. Environ Sci Pollut Res 27:4528–4535
Qu X, He PJ, Shao LM, Lee DJ (2008) Heavy metals mobility in full-scale bioreactor landfill: initial stage. Chemosphere 70:769–777
Reinhart DR, Townsend TG (2018) Landfill bioreactor design and operation. Routledge
Saffira N, Kristanto GA (2018) The effect of leachate recirculation with enzyme cellulase addition on waste stability in landfill bioreactor. In: IOP conference series: earth and environmental science. IOP Publishing, p 12119
Shalini SS, Joseph K (2012) Nitrogen management in landfill leachate: application of SHARON. ANAMMOX and combined SHARON–ANAMMOX process Waste Manag 32:2385–2400
Shao L-M, He P-J, Li G-J (2008) In situ nitrogen removal from leachate by bioreactor landfill with limited aeration. Waste Manag 28:1000–1007
Singh D, Chavan D, Pandey AK, Periyaswami L, Kumar S (2021) Determination of landfill gas generation potential from lignocellulose biomass contents of municipal solid waste. Sci Total Environ 785:147243
Singh R, Chakma S, Birke V (2020) Numerical modelling and performance evaluation of multi-permeable reactive barrier system for aquifer remediation susceptible to chloride contamination. Groundw Sustain Dev 10:100317
Srivastava AN, Chakma S (2021a) Dry tomb-bioreactor landfilling approach for enhanced biodegradation and biomethane generation from municipal solid waste co-disposed with sugar mill pressmud. Bioresour Technol 125895
Srivastava AN, Chakma S (2021b) Investigating leachate decontamination and biomethane augmentation through Co-disposal of paper mill sludge with municipal solid waste in simulated anaerobic landfill bioreactors. Bioresour Technol 329:124889
Srivastava AN, Chakma S (2020) Quantification of landfill gas generation and energy recovery estimation from the municipal solid waste landfill sites of Delhi, India. Energy Sources, Part A Recover. Util Environ Eff: 1–14
Srivastava AN, Sumedha C (2020) Heavy metals speciation study revealing merits of anaerobic co-disposal of municipal solid waste with discrete paper mill Sludges: an experimental investigation in simulated landfill bioreactors. J Environ Chem Eng 104337
Stessel RI, Murphy RJ (1992) A lysimeter study of the aerobic landfill concept. Waste Manag Res 10:485–503
Sun Y, Wang Y-N, Sun X, Wu H, Zhang H (2013) Production characteristics of N2O during stabilization of municipal solid waste in an intermittent aerated semi-aerobic bioreactor landfill. Waste Manag 33:2729–2736
Tesseme AT, Chakma S (2020) Small scale bioreactor studies for sustainable municipal solid waste landfilling management in developing countries. Int J Environ Waste Manag 25:194–219
Tiseo I (2021) U.S. price of landfilling municipal waste by region 2019-2020
Townsend TG, Powell J, Jain P, Xu Q, Tolaymat T, Reinhart D (2015a) Moisture supply and conveyance. In: Sustainable practices for landfill design and operation. Springer, pp 133–150
Townsend TG, Powell J, Jain P, Xu Q, Tolaymat T, Reinhart D (2015b) Sustainable practices for landfill design and operation. Springer
Vigneron V, Ponthieu M, Barina G, Audic JM, Duquennoi C, Mazéas L, Bernet N, Bouchez T (2007) Nitrate and nitrite injection during municipal solid waste anaerobic biodegradation. Waste Manag 27:778–791. https://doi.org/10.1016/j.wasman.2006.02.020
Wang J, Xia F-F, Bai Y, Fang C-R, Shen D-S, He R (2011) Methane oxidation in landfill waste biocover soil: kinetics and sensitivity to ambient conditions. Waste Manag 31:864–870
Xu Q, Jin X, Ma Z, Tao H, Ko JH (2014) Methane production in simulated hybrid bioreactor landfill. Bioresour Technol 168:92–96

Xu Q, Tian Y, Wang S, Ko JH (2015) A comparative study of leachate quality and biogas generation in simulated anaerobic and hybrid bioreactors. Waste Manag 41:94–100

Yadvika S, Sreekrishnan TR, Kohli S, Rana V (2004) Enhancement of biogas production from solid substrates using different techniques – a review. Bioresour Technol 95:1–10. https://doi.org/10.1016/j.biortech.2004.02.010

Zhao X, Soong TY, Qian X, O'Keefe L (2005) Enhanced waste decomposition in bioreactor landfill with septage additions. In: Geosynthetics Research and Development in Progress, pp 1–10

# Chapter 8
# Anaerobic Co-digestion of Landfill Leachate with Other Feedstocks

**N. Anand, Sankar Ganesh Palani, Pankaj Pathak, and B. Siva Prasad**

**Abstract** Urbanization and industrialization have led to increase of per capita municipal solid waste (MSW) generation, and the generated waste is treated by landfilling and open dumping or incineration and bioconversion. Landfilling or open dumping is considered to be cost-effective compared to the other treatment methods. Moisture content in MSW and rainfall generates a hazardous liquid waste termed landfill leachate. It is characterized by high chemical oxygen demand, biological oxygen demand, and total dissolved solids, ideal substrate characteristics for the anaerobic digestion process. However, refractory and labile organics in the landfill leachate defies biodegradation, especially during anaerobic digestion. At the same time, high ammoniacal nitrogen and heavy metal content inhibit various pathways involved in anaerobic digestion. However, anaerobic co-digestion of landfill leachate with other co-substrates will enhance its biodegradation. Hence, this chapter focuses on landfill leachate composition and characteristics, recalcitrance mechanisms induced by various organic and inorganic toxic materials, and the anaerobic co-digestibility of landfill leachate with other organic feedstocks.

**Keywords** Landfill leachate · Refractory organics · Ammoniacal nitrogen · Heavy metals · Anaerobic co-digestion

N. Anand · S. G. Palani (✉) · B. Siva Prasad
Environmental Biotechnology Laboratory, Department of Biological Sciences, BITS Pilani, Hyderabad Campus, Hyderabad, Telangana, India
e-mail: sangan@hyderabad.bits-pilani.ac.in

P. Pathak
Department of Environmental Science, SRM University Andhra Pradesh, Guntur, Andhra Pradesh, India



P. Pathak, S. G. Palani (eds.), *Circular Economy in Municipal Solid Waste Landfilling: Biomining & Leachate Treatment*, Radionuclides and Heavy Metals in the Environment, https://doi.org/10.1007/978-3-031-07785-2_8

## 8.1 Introduction

There is a steady increase in municipal solid waste (MSW) generation as the population increases and the affluent lifestyle (Renou et al. 2008). The principal methods followed in India and other third-world countries for the disposal of MSW are landfilling and open dumping (Luo et al. 2020; Renou et al. 2008). The disposal of MSW by landfilling is the most accepted method for its treatment due to its easy maintenance compared to other treatment techniques such as incineration and composting (Renou et al. 2008). Out of 1.49 million tonnes of waste generated in India, 33% is landfilled, 31% openly dumped or left untreated, and 36% is treated by composting, incineration, or other processes (CPCB 2019). Landfilling allows the waste to undergo a series of transformations in controlled conditions until it reaches an inert and stabilized state. Although landfilling is the most economical and financially viable method for MSW disposal, it leads to landfill leachate generation, which, if left untreated, may lead to severe environmental problems. Landfill leachate is generated as the moisture in the waste and rainwater percolates through the waste and in this process extracts the suspended and dissolved solids (both organic & inorganic contents) present in the MSW.

### *8.1.1 Landfill Leachate Generation*

Landfill leachate generation varies due to several factors such as meteorological conditions (rainfall and precipitation), age of the landfill, MSW composition and characteristics, and landfill design and operations. The volume of landfill leachate generated will be higher according to the following reasons: (i) the monsoon or rainy season, (ii) open landfill design, (iii) high moisture content of the waste, and (iv) active operations in landfill (Kjeldsen et al. 2002). Hence, estimating the exact volume of the landfill leachate generated is difficult due to the factors mentioned above.

Various mathematical models have been designed to overcome the issue of calculating the volume of landfill leachate generated. The most basic model to estimate the landfill leachate quantity is the Water Balance Method (WBM), which uses the landfill area and yearly precipitation in that particular region (El-Fadel et al. 1997). Aziz et al. (2012), Ibrahim et al. (2017) have used the above method to determine the quantum of landfill leachate generated in a year and found that it coincides with the landfill data. The equation is as follows:

$$V = 0.15 * R * A$$

Where,

$V$ = volume of landfill leachate generated in a year ($m^3$/year)
$R$ = Annual rainfall (m)

$A$ = Surface area of the landfill ($m^2$)

The other methods for estimating the landfill leachate generation are artificial intelligence (AI) models, such as Hydrologic Evaluation of Landfill Performance (HELP) and Adaptive Neuro-Fuzzy Inference System (ANFIS) (Abunama et al. 2017). The models mentioned earlier are more powerful tools than WBM, but the latter is user-friendly and provides comparable results to the other models. Also, the results obtained via this model is equivalent to the data collected from the landfills. However, the above mentioned method do not include the leachate generation from the moisture content of the waste landfilled. In the study by Anand and Palani (2022), it was observed that by including the leachate generation due to the moisture content of the landfilled waste, the predicted leachate generation rate was closer to the actual leachate generation rate.

### 8.1.2 Landfill Leachate Composition and Characteristics

Landfill leachate composition and characteristics are understood by analyzing various physicochemical parameters. They are Chemical Oxygen Demand (COD), heavy metal content, refractory organics, ammoniacal nitrogen, chlorides, Total Suspended Solids (TSS), Volatile Fatty Acids (VFA) electrical conductivity, alkalinity, pH, total nitrogen, Total Dissolved Solids (TDS), and Biological Oxygen Demand (BOD). These characteristics vary according to the age of the landfill, yearly precipitation, meteorological conditions, type of waste and composition (Kulikowska and Klimiuk 2008; Renou et al. 2008). According to the literature, landfill leachate is usually characterized by high COD, BOD, heavy metals, ammoniacal nitrogen, chlorides, alkalinity, TDS, and refractory organics (Table 8.1). Landfills are usually three types: young landfills, intermediate landfills, and old/legacy landfills according to their ages. As the landfill age increases, there

**Table 8.1** Physicochemical characteristics of landfill leachate from young, old and intermediate landfills

| | Landfill leachate | | |
|---|---|---|---|
| Characteristic | Young | Old | Intermediate |
| COD (mg/L) | 677–65,333 | 839–34,480 | 1352–12,000 |
| BOD (mg/L) | 405–48,000 | 90–990 | 76–1250 |
| Ammoniacal nitrogen (mg/L) | 7.4–1574 | 26–2720 | 451–2700 |
| Lead (ppm) | 0.09–1.9 | 0.18–1.6 | 0.004–3 |
| Nickel (ppm) | 0.02–0.22 | 0.09–4.63 | 0.1–1.339 |
| Copper (ppm) | 0.001–0.46 | 0.05–0.44 | 0.015–3.62 |
| Zinc (ppm) | 0.09–5.6 | 0.064–5.84 | 0.017–7.1 |
| Iron (ppm) | 0.8–208 | 3.1–53 | 0.27–166 |
| Cadmium (ppm) | <0.05 | 0.032 | 0.259 |
| Chromium (ppm) | 1.3 | 0.92 | 0.182 |

will be significant changes in the COD, BOD, ammoniacal nitrogen, heavy metal content, pH, and total solids content (Kulikowska and Klimiuk 2008). Young landfills receiving fresh waste have high amounts of readily biodegradable components such as volatile fatty acids and low amounts of inorganic compounds in their landfill leachate. In contrast, old landfills which are not operational still generate landfill leachate with high amounts of non-biodegradable or ‘difficult-to-biodegrade’ organic contents and inorganic compounds. Young landfill leachate has very high COD & BOD, low heavy metals and other inorganic compounds, whereas old landfill leachate will have high amounts of refractory organics and inorganic substances such as heavy metals, with lower COD & BOD concentrations (Tränkler et al. 2005; Vaccari et al. 2019; Anand and Palani. 2022)). Comparison of COD, BOD, ammoniacal nitrogen and heavy metals of leachate from young, old and intermediate landfills are summarized in Table 8.1. From the study by Anand and Palani (2022), it was observed that segregation of waste prior to landfilling reduced the concentration of above mentioned components in young landfill, as compared to other young landfills.

The contamination potential of landfill leachate is determined by calculating the Landfill Leachate Pollution Index (LPI) based on following parameters such as COD, BOD, ammoniacal nitrogen, TKN, pH, TDS, phenolic compounds, chlorides, cyanide, total coliforms, iron, lead, copper, nickel, zinc, chromium, mercury, and arsenic (Kumar and Alappat 2004). The equation used for calculating LPI is as follows:

$$LPI = \sum_{i=1}^{n} w_i p_i$$

Where,

$w_i$ = $i^{th}$ pollutants weight
$p_i$ = $i^{th}$ pollutants sub-index score
$n$ = number of variables

Usually, in uncontrolled landfills and open dumpsites LPI values are usually above 33, and in old landfills it will be less than 24. However, in the study by Anand and Palani (2022), the LPI values of the young and old landfills were 26.65 and 33.20 because of segregation and landfilling mixed waste, respectively. Also, the ground water quality can be monitored using water quality index (WQI), which can be determined by analysing the following water quality parameters: pH, electrical conductivity, total hardness, total dissolved solids, alkalinity, sodium, potassium, calcium, magnesium, nitrates, bicarbonate, chloride, sulphates, fluorides, chromium, zinc, lead, copper, manganese, iron, nickel and cadmium. Various locations around the landfill has to be considered and the sample should be taken for analysing the WQI around the landfill (Akter et al. 2016; Chaurasia et al. 2018). The WQI is calculated on the basis of the following set of equations.

$$q_i = \left(\frac{C_i}{S_i}\right) \times 100$$

Where, $q_i$, $C_i$, and $S_i$ indicates quality rating scale, concentration of $i$ parameter and standard value of $i$ parameter, respectively.Relative weight scale was calculated with the following equation,

$$w_i = \frac{1}{S_i}$$

The overall WQI is calculated by the following equation,

$$WQI = \frac{\sum q_i w_i}{\sum w_i}$$

Biological characterization is an integral part of landfill leachate characteristics as it provides us with data on the toxicity of landfill leachate. Although LPI and WQI provides information regarding landfill leachate toxicity, biological characterization is crucial since it sheds light on the actual effects of landfill leachate toxicity. Toxicity studies using various organisms have been carried out to determine the toxic effects of landfill leachate. Some of the toxicity studies involve microorganisms, aquatic organisms, plants, and animals. In the toxicity studies reported by Anand and Palani (2022), the germination index was lower but the root inhibition and DNA damage were higher in samples treated with landfill leachates.

## 8.2 Landfill Leachate Treatment

Landfill leachate with a high concentration of contaminants must be treated before its discharge into the environment. EPA sets the discharge standards for landfill leachate, and most of the treatment facilities have to follow these standards for the safe disposal of landfill leachate into the environment. The treatment method will be decided according to the characteristics of the landfill leachate generated. The primary treatment techniques used for treating landfill leachate are biological and physicochemical treatment methods. Physicochemical treatment methods include adsorption, coagulation-flocculation, ion exchange, filtration techniques, Oxidation process, air stripping (Renou et al. 2008). Biological treatment methods include activated sludge process, MBBR, UASB and anaerobic digestion (Luo et al. 2020). Various combinations of physicochemical and biological treatment techniques are used to overcome the difficulties posed by high amounts of organic content and ammoniacal nitrogen.

### 8.2.1 Physicochemical Treatment

Landfill leachate with high concentrations of refractory organics such as humic and fulvic acids, ammoniacal nitrogen and adsorbable organic halides (AOXs), polychlorinated biphenyls (PCBs), and heavy metals is treated using physicochemical methods. Majorly used techniques for treating landfill leachate are coagulation-flocculation, membrane filtration, advanced oxidation process (Torretta et al. 2017). Electrochemical treatment methods include electrocoagulation, electro-Fenton, and electro-oxidation. These methods are majorly used to remove organic content and ammoniacal nitrogen from landfill leachate (Zhang et al. 2014).

### 8.2.2 Biological Treatment

The biological treatment of landfill leachate is majorly followed due to its efficiency and cost-effectiveness compared to physicochemical treatment techniques. The biological treatment uses microorganisms to remove organic and inorganic substances such as ammoniacal nitrogen, phosphorous, heavy metals, and other refractory organics. Primary biological treatment methods are classified into aerobic and anaerobic techniques. In aerobic treatment, organic compounds are converted to carbon-di-oxide and biomass/sludge. During anaerobic treatment, these compounds are converted to biogas containing primarily methane and carbon-di-oxide and biomass/sludge, the volume of which is much lesser than that is generated from the aerobic system (Goli et al. 2019).

Aerobic treatment is majorly used to remove the ammonia content, and in this process, a part of readily biodegradable substances is removed (Wiszniowski et al. 2006). Aerobic treatment mainly consists of the activated sludge process, moving bed biofilm reactor, sequencing batch reactor, rotating biological contactor, fungal treatment, and phytoremediation.

Anaerobic treatment of landfill leachate is preferred over aerobic treatment due to the following advantages: lower sludge production, lower energy consumption, and energy recovery by methane generation. Primary methods used in the anaerobic digestion of landfill leachate are UASB and ANAMMOX processes (Kheradmand et al. 2010). Anaerobic digestion converts organic contaminants such as carbohydrates, proteins or fats to biogas, containing mainly methane and carbon-di-oxide. Hence, landfill leachate contains high organic content and can be considered a valuable substrate for anaerobic digestion. The significant advantage of this process is that the biogas produced can be used to maintain the operating temperature, reducing the operational cost. High strength wastewaters, including landfill leachate, are commonly treated using UASB due to their high organic loading rates. UASB has been widely studied to treat landfill leachate, and it was observed that COD removal is high, but the ammoniacal nitrogen removal is low. It has also been observed that when used in combination with other techniques such as aerobic SBR or membrane filtration or conjunction with chemical treatment methods, COD and ammoniacal nitrogen removal got enhanced (Marttinen et al. 2002). The ANAMMOX process is commonly used to remove ammoniacal nitrogen by utilizing ammonium and nitrite as electron donors and acceptors. Either by using the ANAMMOX process solely or combined with other treatment methods, it has been observed that nitrogen removal reached more than 90% and COD removal of more than 80% (Wu et al. 2018). It has been suggested that the ANAMMOX process is better when treating mature landfill leachate with lower COD and higher ammoniacal nitrogen contents than other anaerobic digestion techniques. Anaerobic membrane reactors are used to reduce the start-up time of anaerobic digesters by combining the MBBR technology with anaerobic digestion process. Many researchers have utilized this technology to remove more than 90% of the COD present in leachate at much shorter HRT compared to other anaerobic digestion technologies. The theoretical methane yield is lesser in anaerobic membrane reactor (0.23 to 0.33 $LCH_4$ / g COD removed) compared to conventional anaerobic treatment methods (0.382 $LCH_4$/g COD removed) (Abuabdou et al. 2020).

## 8.3 Anaerobic Digestion of Landfill Leachate

Anaerobic mono-digestion or single-stage AD has not been studied extensively due to the inhibitory effects of contaminants in the landfill leachate. Various studies reported that high concentrations of ammoniacal nitrogen, heavy metals, high salinity, and refractory organics in landfill leachate affect the AD process (Özgür Akta 2001). Each of the contaminants mentioned above has its mechanism or process to inhibit the AD process. In order to comprehensively understand the process by which landfill leachate inhibits AD, it is imperative to study the mechanism of inhibition of the above contaminants, which will aid in designing processes to overcome their inhibition.

### *8.3.1 Inhibition of Anaerobic Digestion of Landfill Leachate*

High concentrations of ammoniacal nitrogen, heavy metals, high salt content, and refractory organics inhibit the landfill leachate anaerobic digestion process. While high COD and low ammoniacal nitrogen generally characterize young landfill leachate, they may contain higher concentrations of heavy metals, salinity, or refractory organics, that inhibit the AD process. Mature landfill leachate is known to have high concentrations of ammoniacal nitrogen, refractory organics, and heavy metals, which inhibit the anaerobic digestion process (Luo et al. 2020; Renou et al. 2008). Inhibition exerted on anaerobic digestion by these compounds is discussed below.

#### 8.3.1.1 Ammoniacal Nitrogen

Ammoniacal nitrogen is an important compound that can be used as a nitrogen source for microbial growth, but it can inhibit microorganisms' growth in higher concentrations. Predominantly ammoniacal nitrogen is present as free ammonia ($NH_3$) or unionized ammonia and ammonium ion ($NH_4^+$) or ionized ammonia, which can inhibit the AD and even cause the failure of the process (Angelidaki et al. 1993). Total ammoniacal nitrogen (TAN) refers to the combination of these forms of ammonia, and their concentration may vary depending on the operating conditions, such as pH and temperature of the process (Dong and Sartaj 2016). A study performed to determine the inhibitive effect of TAN on microorganisms was found to be in the range of 1500–5000 mg/L. In the form of free ammoniacal nitrogen (FAN), ammoniacal nitrogen directly inhibits biogas production at high pH and temperature by affecting the microorganisms (Sung and Liu 2003). FAN can penetrate the cell membrane and cause proton imbalance, which changes the cell's intracellular pH and affects the microorganism's growth. Due to intracellular pH change, the activities of the microorganisms' intracellular enzymes are inhibited, affecting AD. Methanogenic microorganisms are susceptible to ammonium concentration and are readily inhibited than hydrolytic and acidogenic microorganisms. A study by Yuzer et al. (2012), found that as the free ammonia concentration increased, COD and BOD removal decreased. The above study also observed that ammonia concentrations above 1500 mg/L completely inhibited AD. In comparison, concentrations from 200–1000 mg/L had adverse effects on AD, which decreased the COD and BOD removal efficiencies to less than 20%. In another study by Baraza et al. (2019), it was observed that ammoniacal nitrogen concentrations above 1350 mg/L affected methane production. This concentration had only an inhibitory effect rather than causing a complete failure of the process.

#### 8.3.1.2 Refractory Organics

The major challenge in the anaerobic digestion of landfill leachate is refractory organics such as humic and fulvic acids. Low hydrolysis rate and substrate biodegradability are directly related to low biogas production in AD (Vavilin et al. 2008). Refractory organics such as humic and fulvic acids can affect hydrolysis rates due to their lower biodegradability. Humic acid can affect the hydrolysis stage by binding to the enzymes that are involved in cellulose degradation. In addition to inhibition of hydrolysis, the methanogenic stage inhibition was also observed. To investigate the effect of HA on methanogens, Khadem et al. (2017) observed that HA concentrations of 5 kg/m$^3$ affected the methanogens. Yap et al. (2018) studied the effects of HA on the hydrolysis of cellulose and acetoclastic methanogens. It was observed that hydrolysis rate slowed down at HA concentrations of 2 g/L, and complete inhibition of hydrolysis and the activity of acetoclastic methanogens were affected at HA concentrations above 5 g/L. Partial inhibition of methanogens was observed at 10 g/L HA, and at 20 g/L complete inhibition of methanogens was seen.

Another study by Dang et al. (2016) investigated the effects of fulvic acid on methanogenic and syntrophic microorganisms. It was observed that increasing fulvic acid concentration inhibited the methanogenic process as well as syntrophic bacteria adversely.

#### 8.3.1.3 Heavy Metals

Landfill leachate is characterized by the higher concentrations of heavy metals such as Fe, Cu, Cr, Zn, Ni, Pb, and Al. Other heavy metals such as As, Hg, and Cd are present at lower concentrations (Carvajal-Flórez & Santiago-Alonso Cardona-Gallo 2019). Heavy metals affect AD by binding to thiol (and other) groups on enzyme and protein molecules, disrupting their normal functions or replacing usually occurring metals in enzymes (Guo et al. 2019). Some studies performed to determine the inhibitory effects of heavy metals on AD are presented below: Cu affected the activities of methanogens and cellulase at 130 mg/L, Ni at 32 mg/L, Fe inhibited cellulase activity at 20 g/L, Cd and Zn affected methanogenic activity at 1 mg/L and 50 mg/L, respectively (Altaş 2009).

### *8.3.2 Mitigation of Inhibitory Components of Landfill Leachate*

#### 8.3.2.1 Pretreatment to Enhance AD

Humic acid inhibition of cellulolytic enzymes can be mitigated by two approaches: (i) removing humic acid from the substrate and (ii) reducing the binding capacity of cellulolytic enzymes to humic acid by adding compounds. By adding coagulants and flocculants, humic acid was removed from contaminated aquatic sites. Alternatively, the addition of $Ca^{2+}$ ions can reverse the inhibition of potato protein hydrolysis by humate. Adding calcium, iron, and magnesium salts alleviated the hydrolysis inhibition by humic acid, and the efficiencies of each salt reached up to 65%, 72%, and 75%, respectively, when compared to control. Azman et al. (2015) studied the effects of salts' addition in mitigating the impact of HA on AD. The study observed the HA mitigation effects of various salts of Na, K, Ca, Mg and Fe. The observed results showed that adding Ca, Mg, and Fe salts increased methane production by 60%, and no visible difference was seen with Na and K salts. It was proposed that the Ca, Mg, and Fe salts were bound to the active sites of humic acid and rendered it unavailable for the binding of enzymes. Li et al. (2021) studied the addition of metal ions such as $Ca^{2+}$ and $Al^{3+}$ and observed that only 40% of HA inhibition was mitigated when these metals were added separately. When used in combination, it mitigated 80% HA inhibition.

It was proposed that excess hydrolysis enzyme addition may overcome the inhibitory effects of humic acid. Humic acid extraction methods have been successfully used to pre-treat the landfill leachate. In a recent study, alkali pretreatment was employed to extract humic acid, and the total methane yield increased by 50%. Castrillón et al. (2010) investigated the effects of the addition of iron trichloride or aluminium polychloride to remove refractory and non-biodegradable contaminants from landfill leachate before anaerobic treatment. It was observed that the process's efficiency increased after including the pretreatment step during the treatment (Castrillón et al. 2010).

Struvite precipitation is a crucial method to remove ammonia from wastewater. This method utilizes $Mg^{2+}$ salts such as magnesium hydroxide and magnesium chloride for precipitating ammonia. Huang et al. (2014) studied magnesium and phosphate salts' effects on ammonia removal from landfill leachate. It was observed that the addition of these salts mitigated ammonia inhibition by 68%. Other studies state that adding adsorption compounds such as activated carbon, alumina and silica gel can minimize ammonia's inhibitory effect. Abdul Aziz et al. (2004) studied activated carbon and limestone addition for ammonia removal from landfill leachate. It was found that the addition of activated carbon and limestone aided in reducing 40% of ammonia from landfill leachate.

A pretreatment step using the ammonia stripping technique can be used to remove ammonia before AD. In this process, air or steam is passed through the substrate, and the ammonia removed can be recovered using the adsorption process. A study by Marttinen et al. (2002) suggested that at pH 11 and temperature 20 °C, 94% of ammoniacal nitrogen removal was observed. Xiong et al. (2018) studied the effects of using vacuum steam for ammonia removal, and this process can achieve a removal rate ranging from 74–90%. (Silva et al. 2015) observed that ammonium carbonate and ammonium sulfate could be synthesized in this process as a by-product of ammonia removal. The significant advantage of ammonia stripping over struvite precipitation is that the pH and temperature need not be altered and can be performed at room temperature for efficient ammonia removal. Hence, ammonia stripping is a practical pretreatment step before anaerobic digestion (Renou et al. 2008). Figure 8.1 displays the different inhibitory agents in the anaerobic digestion of landfill leachate and their mechanism of action.

## 8.4 Anaerobic Co-digestion as an Alternative to Enhance AD

Anaerobic co-digestion is a proven technology where two or more substrates are treated simultaneously, which provides stability and increases the treatment process's efficiency. Stabilization of the AD process provides proper nutrient balance, ensures syntrophism between microorganisms, increases OLR and C/N ratio. These processes can be maintained easily using co-digestion of substrates which will eventually enable better methane production. The co-digestion process needs to be optimized by focussing on the following aspects: characteristics of wastewater, functional parameters (such as pH, OLR, C/N ratio, temperature and HRT), and biochemical methane potential (BMP) of the substrates (Siddique and Wahid 2018).

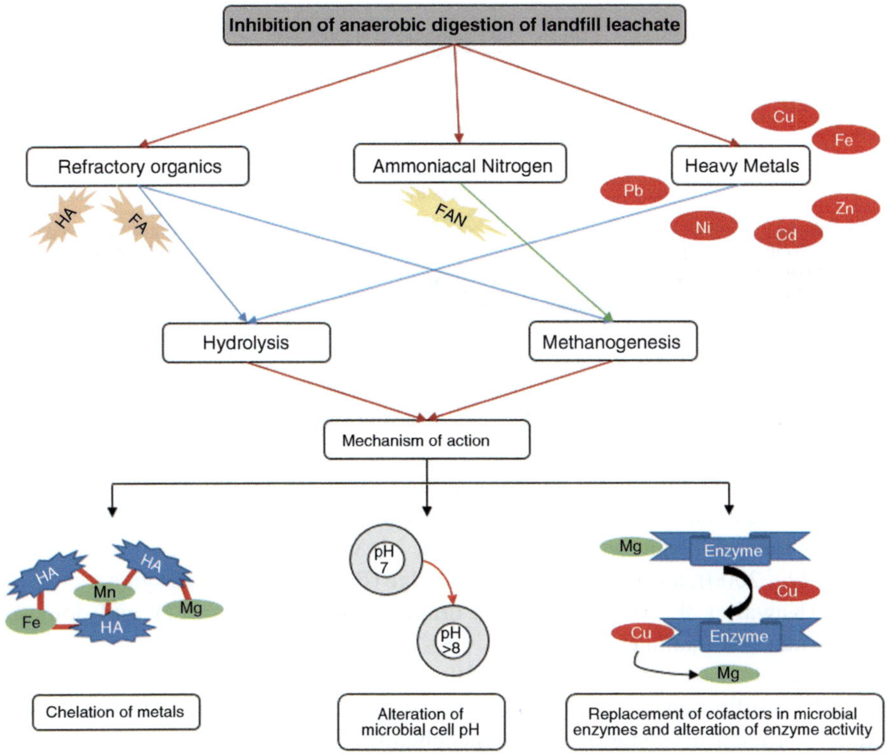


**Fig. 8.1** Inhibitory agents of AD of landfill leachate and their mechanism of action

The substrate ratio has to be optimized for enhancing the methane production and efficient treatment of the waste. According to Pecorini et al. (2016), BMP tests help design the treatment process and understand the biodegradability and bioavailability of the substrates. During this process, the inhibitory effects of refractory organics and other toxic substances can be diluted and mitigated from inhibiting the process (Siddique and Wahid 2018). Various co-digestion studies have been investigated with different substrates at varying compositions to understand the co-digestion process. Few such substrate combinations are as follows: Cornstalk with vermicompost (Chen et al. 2010); Sewage sludge with cow dung and garden waste/fruit juice waste (Rao and Baral 2011); Food waste with straw (Yong et al. 2015); Cattle manure with fruit & vegetable waste, organic household waste and sewage sludge (Tufaner and Avşar 2016); Animal manure with lignocellulose material (Neshat et al. 2017).

According to the SCOPUS survey, co-digestion of landfill leachate with other substrates has not been studied extensively, and the total number of publications is 39 as of August 2021. The co-substrates used along with landfill leachate include sewage sludge (Berenjkar et al. 2019; Hombach et al. 2003; Montusiewicz and Lebiocka 2011; Montusiewicz et al. 2018); diluted municipal sludge cake and fruit/

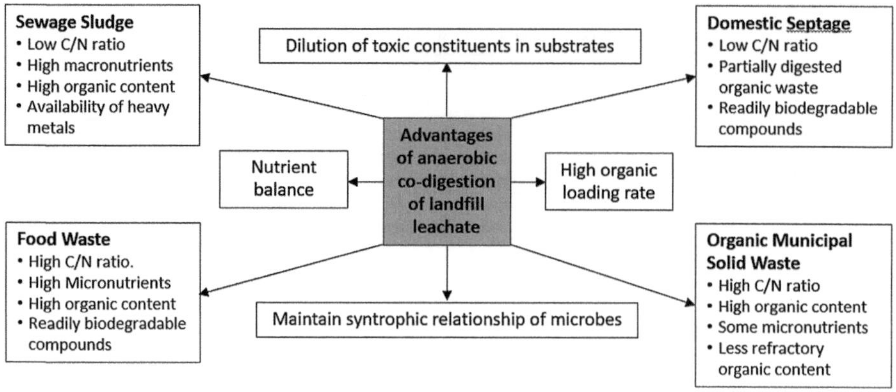


**Fig. 8.2** Advantages of anaerobic co-digestion of landfill leachate

wine juice production waste (Barrantes Leiva et al. 2014); food waste (Liao et al. 2014; Zhang et al. 2015); glycerine (de Castro et al. 2019, 2021); the organic fraction of municipal solid waste (Guven et al. 2018; T.A.S et al. 2020); septage (Lin et al. 1999, 2000); used oils and sewage sludge (Pastor et al. 2013); synthetic wastewater (Kawai et al. 2012); acid mine drainage (Zhou et al. 2021); and domestic wastewater (Moujanni et al. 2019).

Anaerobic co-digestion of the landfill leachate can enhance the anaerobic digestion of landfill leachate by diluting inhibitory agents' toxic effects. This process is equally efficient to some of the physicochemical or pretreatment processes mentioned above, and it is cost-effective compared to the above techniques. Figure 8.2 explains the advantages of co-digestion with landfill leachate and the different substrates used for co-digestion.

## *8.4.1 Anaerobic Co-digestion with Sewage Sludge*

Sewage Sludge (SS) is the most commonly used co-substrate in the anaerobic digestion of landfill leachate. It is used due to the high C/N ratio, readily biodegradable organic compounds, availability of micronutrients, and low heavy metals, enhancing the anaerobic co-digestion of landfill leachate. Berenjkar et al. (2019) studied the influence of landfill leachate during the anaerobic digestion of sewage sludge. Sewage sludge and landfill leachate were mixed in 5 different ratios, with the landfill leachate percentage being 0, 5, 10, 20, and 40% v/v. The experiment was performed in Wheaton glass bottles of 600 mL and maintained at 37 °C using a water bath. The observed results showed decreased COD, VS, and TS removal efficiencies as the landfill leachate concentration increased. The biogas production decreased with increasing landfill leachate volume. Although the parameters were affected, it was observed that TN removal efficiency was not affected. The results indicate that landfill leachate can be added at lower doses for treatment of SS, enhancing the efficiency of the process.

Montusiewicz et al. (2018) studied the effects of pretreatment on co-digestion of SS and mature LL. Hydrodynamic cavitation was used for pretreating landfill leachate before anaerobic co-digestion. Landfill leachate was pretreated using a 30 L reactor setup, and the co-digestion study was performed using 40 L working volume digesters operated at 35 °C. The co-substrates ratio was 95% SS and 5% pretreated landfill leachate, and each run was maintained at an HRT of 20 days. The results indicated that COD, TS, and VS removal efficiencies of pretreated landfill leachate and un-treated landfill leachate were similar, indicating that SS and landfill leachate can be treated in the co-digestion process without pretreatment. It was concluded that the addition of landfill leachate in low doses enhanced the treatment of SS.

Co-digestion of sewage sludge and intermediate landfill leachate was studied by (Montusiewicz and Lebiocka 2011). The experiment was performed in a 40 L reactor operated at 35 °C with mixed sludge and landfill leachate. In Run 1, mixed sludge was fed to the reactor, and landfill leachate was added in Runs 2 and 3. In Run 2, where the ratio of sludge to landfill leachate was 20:1, COD, TS, and VS removal was low than Run 3 (10:1), but biogas yield and methane yield were higher in Run 2 than Run 3. It was concluded that the addition of low volumes of landfill leachate enhanced biogas production and methane yield.

Hombach et al. (2003) researched the co-digestion of sewage sludge and landfill leachate using lab-scale reactors of working volume 500 mL, and it was operated at mesophilic conditions. The HRT of the reactor was 1 day during the initial stages, and as the landfill leachate volume increased, the HRT was decreased. The reactors were operated in two phases, with phase 1 receiving landfill leachate less than 12% sewage sludge volume, and in phase 2, the landfill leachate volume was increased to 20%. During phase 1, the methane yields and biogas production increased. Biogas yield and the VS removal decreased as the landfill leachate volume was increased above 20% in phase 2.

### *8.4.2 Anaerobic Co-digestion with Domestic Sewage*

Domestic sewage is used as a co-substrate, usually to dilute the toxic compounds present in landfill leachate. Domestic sewage has a low C/N ratio but has high amounts of micronutrients. Domestic sewage has less organic content than landfill leachate, by which it can dilute the toxic effects of ammoniacal nitrogen, heavy metals, and refractory organics, thereby enhancing landfill leachate treatment (Ferraz et al. 2014).

Moujanni et al. (2019) investigated the co-digestion of landfill leachate and domestic wastewater (DWW). The experiment was performed in 5 different ratios of LL:DWW (100%:0%; 75%:25%; 50%:50%; 25%:75%; 0%:100%) in 1.3 L working volume anaerobic digester operated at mesophilic conditions. According to the results, the LL: DWW of 50%:50% had the highest cumulative biogas production, followed by a ratio of 25%:75%. The addition of DWW diluted the toxic compounds in landfill leachate and enhanced the biogas production, which can be confirmed with the cumulative biogas production with a 100% landfill leachate ratio.

### 8.4.3 Anaerobic Co-digestion with Food Waste

Food waste is the second commonly used co-substrate for landfill leachate co-digestion in an anaerobic digestion system. General characteristics of food waste are high amounts of readily biodegradable compounds and availability of micronutrients. Nevertheless, during anaerobic digestion of food waste, VFA influx can cause process failure. When landfill leachate is added as a co-substrate, it will be used as a buffer system since landfill leachate has refractory organic compounds whose biodegradation is slower than food waste (Renou et al. 2008).

Liao et al. (2014) studied the effect of raw landfill leachate on the anaerobic digestion of food waste. Landfill leachate was added in 7 different ratios to an anaerobic digester containing food waste in a digester of 1.5 L working volume, and the digester was maintained at 35C with an HRT of 30 days. The reactor with 50% landfill leachate volume had a higher gas yield compared to other ratios. The VFA concentration dropped in ratios 5 to 7, indicating that the VFA produced during the treatment process is utilized for biogas production. In other ratios, the VFA concentration over days did not drop to the expected level, indicated by lower gas yields.

Zhang et al. (2015) investigated the effect of the addition of landfill leachate on food waste anaerobic digestion. In this study, 300 mL reactors were used, and the volumes of landfill leachate added were 0%, 5.8%, 11.6%, and 22.7% in four reactors. The HRT of the experiment was 20 days, and the temperature of operation was 35 °C. VS removal and methane yields were higher in ratio, having 22.7% landfill leachate than the other ratios, and total VFA production was in accordance with methane yield for ratio having 22.7% landfill leachate.

### 8.4.4 Anaerobic Co-digestion with the Organic Fraction of Municipal Solid Waste

The organic fraction of MSW (OFMSW) is used as a co-substrate in landfill leachate treatment to balance the C/N ratio (T.A.S et al. 2020). T.A.S et al. (2020) studied the effects of co-digestion of organic fraction of MSW and landfill leachate in a pilot plant using an anaerobic gas-lift reactor. The study was performed using a 42m$^3$ biogas plant for 47 weeks with an HRT of 14.5 days. Three reactors were operated with OFMSW, LL, and a combination of OFMSW and LL, with a biogas production of 73.8, 42, and 63.8 m$^3$/day, respectively.

Guven et al. (2018) co-digested landfill leachate with three different substrates (RO concentrate, SS, and OFMSW) in 7 different ratios along with mono-digestion of all four substrates. The experiments were performed in 300 mL serum bottles and for seven weeks. The results observed were as follows: in mono-digestion of the substrates, OFMSW had the highest methane yield, followed by sewage sludge, RO concentrate, and finally landfill leachate. Whereas in co-digestion, ratio 7 with landfill leachate (32.5%), RO concentrate (0%), SS (16.5%) and OFMSW (16.5%) had highest methane yield. The ratios having higher concentrations of landfill leachate or RO concentrate had lower methane yield. Hence, it was concluded that landfill leachate, SS, and OFMSW co-digestion was better at lower landfill leachate concentrations.

### *8.4.5 Anaerobic Co-digestion with Septage*

Domestic septage co-digestion with landfill leachate has not been studied extensively. According to the literature, the addition of septage can dilute the toxic compounds in landfill leachate, thereby enhancing landfill leachate treatment (Lin et al. 1999). Treatment efficiency of co-digesting landfill leachate and septage was studied by (Lin et al. 1999). Four continuously stirred tank reactors were used for this study, and the operating temperature was 35 ± 1 °C. Landfill leachate and septage were mixed in four ratios (0:1, 1:1, 1:2 and 1:3) and operated at three different HRTs 20, 10 and 5.3 days. The results showed that the increase in septage volume increased the COD removal, phosphorous and ammoniacal nitrogen removal. Methane yield also increased with an increase in septage volume. UASB reactor was used to evaluate the co-digestion of septage and landfill leachate in a study by (Lin et al. 2000). The reactors were operated at an HRT of 1.5 days at 35 ± 1 °C as the operating temperature. Landfill leachate and septage were mixed in 3 ratios 1:1, 1:2 and 1:3 and the removal efficiencies are as follows: Removal of total and soluble COD, total and volatile solids, total phosphorous, and ammoniacal nitrogen were higher in the experiments with higher septage volume. The results obtained from treating landfill leachate and septage were comparable with the treatment of landfill leachate with sewage.

### *8.4.6 Anaerobic Co-digestion with Wine/Fruit Juice Waste*

Four different substrates, such as landfill leachate, municipal sludge cake, screen cake from the treatment plant of fruit juice and thickened waste activated sludge, were used in this study (Barrantes Leiva et al. 2014). The experiment was evaluated using eight digesters operated in single and co-digestion modes for an HRT of 10 and 20 days. Co-digestion of all these substrates increased the efficiency of the treatment process in comparison to single-stage digesters. The operating conditions of the digesters were mesophilic, and results suggested that there were no significant differences in the production of biogas and removal of organic content between the eight digesters. It was concluded that co-digestion enhanced the treatment of landfill leachate and other co-substrates (Barrantes Leiva et al. 2014).

### *8.4.7 Anaerobic Co-digestion with Acid Mine Drainage*

Anaerobic co-digestion of landfill leachate and acid mine drainage was explored by Zhou et al. (2021) in a UASB reactor of working volume 3.2 L at 35 °C. The effect of different HRT (30, 20, 12, 8 h) was investigated in this study by analyzing the COD and sulfate removal efficiencies. With 20 h HRT, 75% COD removal and 69% sulfate removal was found to be the best compared to other HRTs. From the microbial analysis, the most abundant microorganisms were *Desulfobulbus* and *Methanobacterium* of sulfate-reducing bacteria and methanogen groups, respectively.

### *8.4.8 Anaerobic Co-digestion with Glycerin/Glycerol*

de Castro et al. (2021) investigated the process efficiency of co-digestion of industrial landfill leachate with crude residual glycerine in different ratios. Glycerin was mixed in 5 different ratios, i.e., 0%, 1.5%, 5%, 8%, and 10%, and the remaining being landfill leachate. The experiments were performed in 400 mL working volume reactors operated at 30 °C and 30 days of incubation. According to the results, the ideal mixing ratio was 95% landfill leachate and 5% crude residual glycerine, with the highest biogas production and methane yield.

### *8.4.9 Anaerobic Co-digestion with Synthetic Wastewater*

In a study by Kawai et al. (2012), landfill leachate is co-digested with synthetic wastewater, a labile substrate, in different ratios to understand the biodegradation of refractory and labile substrates. The digester was started with 100% synthetic wastewater (phase 1), and slowly landfill leachate proportion was increased to 100% (phase 4), and 100% synthetic wastewater was reintroduced in phase 7. The highest COD removal was observed for 100% synthetic wastewater, and the lowest COD removal and specific methane yields were observed for 100% landfill leachate. During phases 2 and 3, COD removal efficiency and methane yields decreased when landfill leachate concentration was increased. Landfill leachate being highly refractory and toxic, affected the microorganisms, resulting in digester upset. It was concluded that mono-digestion of landfill leachate is not recommended because of its toxicity eliciting potential. This study also suggested that a labile co-substrate be utilized to degrade refractory landfill leachate in a co-digestion experiment. The treatment efficiencies of the studies mentioned above are given in the following Table 8.2.

**Table 8.2** Co-digestion of Landfill leachate with different substrates

| Optimum substrate ratio | Role of LL | Temperature °C | COD removal (%) | VS removal (%) | Biogas production $m^3$/Kg Vs | Reference |
|---|---|---|---|---|---|---|
| 1:1 (DS:LL) | Main co-substrate | 35 ± 1 | 81 | 41 | 0.62 $m^3$/kg vs | Lin et al. (1999) |
| 1: 1 DS: LL | Main co-substrate | 35 ± 1 | 48 | 68 | NA | Lin et al. (2000) |
| 10:1 SS:LL | Pretreated landfill leachate as co-substrate | 35 | ~39 | ~40 | 1.24 $m^3$/kg VS | Montusiewicz et al. (2018) |
| 95% SS and 5% LL | Potential use of LL as co-substrate | 35 | 38 | 30 | NA | Berenjkar et al. (2019) |
| 33% LL and 67% SS | Main co-substrate | 37 ± 1 | 68 | NA | 0.11 L CH4/ gCOD | Kawai et al. (2012) |
| 95.13% LL and 4.87% Gly | Main co-substrate | 30 ± 1 | 92 | NA | 0.19 LCH4/ gVS | de Castro et al. (2021) |
| 77.3% FW: 22.7% FL | To provide process stability and improve biogas production during FW anaerobic digestion | 37 | NA | 78 ± 0.8 | 500 mL/ gVS $CH_4$ | Zhang et al. (2015) |
| LL 12% of sludge volume | Landfill leachate impact on sewage sludge digestion | 35 | NA | 49 | NA | Hombach et al. (2003) |
| LLL and AMD | Addition of AMD to enhance LL digestion | 35 | 83 | NA | NA | Zhou et al. (2021) |
| 50% LL:50% WW | Main co-substrate | 35 | NA | NA | 430 mL/ gVS | Moujanni et al. (2019) |
| 1:1 LL:OFMSW | Addition of landfill leachate on OFMSW digestion | 35 | NA | 49 | 0.58 m3/ KgVS | T.A.S et al. (2020) |

*LL* landfill leachate, *SS* sewage sludge, *DS* domestic septage, *FW* food waste, *FL* Fresh Landfill leachate, *OFMSW* Organic Fraction of Municipal Solid Waste, *AMD* Acid Mine Drainage, *WW* Waste water

## 8.5 Future Prospects

Anaerobic co-digestion of landfill leachate has not been studied extensively due to the inhibitory role of various components present in landfill leachate. Moreover, the reported studies identified the optimum ratio to be used for the process and the landfill leachate percentage is less in most of the studies. Co-digestion studies were performed only at mesophilic conditions, and the thermophilic treatment of landfill leachate has not been studied. The behaviour of landfill leachate as a co-substrate at higher temperatures and the mechanism of ameliorating the inhibitory effects of the compounds mentioned above during co-digestion of landfill leachate has to be understood thoroughly to conclude the success of anaerobic co-digestion for the treatment of landfill leachate. It is concluded that anaerobic co-digestion of landfill leachate is an un-identified grey area that has to be understood better to enhance the treatment of landfill leachate.

## 8.6 Conclusion

Variation in landfill leachate characteristics and its relation to landfill age and seasonal changes have been discussed briefly, along with the calculation of LPI to assess the contamination potential of landfill leachate. Landfill leachate treatment by anaerobic digestion is inhibited by high amounts of refractory organics such as humic & fulvic acids, ammoniacal nitrogen and heavy metals, and the mechanism of inhibition has been discussed in detail in this chapter. Also, the use of chemical additives and pretreatment techniques/methods to ameliorate these components' effects during anaerobic digestion was explained. The use of anaerobic co-digestion as an alternative method for treating landfill leachate and the effect of landfill leachate concentration during co-digestion has been studied extensively. The majority of the co-digestion studies have been performed with optimum concentration of landfill leachate in order to avoid the inhibition of the process, except for few studies performed by Lin et al. (1999, 2000), de Castro et al. (2021), Moujanni et al. (2019), and T.A.S et al. (2020).

**Acknowledgement** N. Anand and P. Sankar Ganesh thank BITS Pilani, Hyderabad Campus, India, for providing fellowship and extending the required support to conduct this work, respectively.

## References

Abdul Aziz H, Adlan MN, Zahari MSM, Alias S (2004) Removal of ammoniacal nitrogen (N-NH3) from municipal solid waste leachate by using activated carbon and limestone. Waste Manag Res 22(5):371–375. https://doi.org/10.1177/0734242X04047661

Abuabdou SMA, Ahmad W, Aun NC, Bashir MJK (2020) A review of anaerobic membrane bioreactors (AnMBR) for the treatment of highly contaminated landfill leachate and biogas production: effectiveness, limitations and future perspectives. J Clean Prod 255:120215. https://doi.org/10.1016/j.jclepro.2020.120215

Abunama T, Othman F, Alslaibi T, Abualqumboz M (2017) Quantifying the generated and percolated leachate through a landfill's lining system in Gaza strip, Palestine. Polish J Environ Stud 26(6):2455–2461. https://doi.org/10.15244/pjoes/73803

Akter T, Jhohura FT, Akter F, Chowdhury TR, Mistry SK, Dey D, Barua MK, Islam MA, Rahman M (2016) Water quality index for measuring drinking water quality in rural Bangladesh: A cross-sectional study. J Health Popul Nutr 35(1):1–12. https://doi.org/10.1186/s41043-016-0041-5

Altaş L (2009) Inhibitory effect of heavy metals on methane-producing anaerobic granular sludge. J Hazard Mater 162(2–3):1551–1556. https://doi.org/10.1016/j.jhazmat.2008.06.048

Anand N, Palani SG (2022) A comprehensive investigation of toxicity and pollution potential of municipal solid waste landfill leachate. Science of The Total Environment https://doi.org/838155891-10.1016/j.scitotenv.2022.155891

Angelidaki I, Ellegaard L, Ahring BK (1993) A mathematical model for dynamic simulation of anaerobic digestion of complex substrates: focusing on ammonia inhibition. Biotechnol Bioeng 42(2):159–166. https://doi.org/10.1002/bit.260420203

Aziz HA, Adlan MN, Amilin K, Yusoff MS, Ramly NH, Umar M (2012) Quantification of leachate generation rate from a semi-aerobic landfill in Malaysia. Environ Eng Manag J 11(9):1581–1585. https://doi.org/10.30638/eemj.2012.198

Azman S, Khadem AF, Zeeman G, van Lier JB, Plugge CM (2015) Mitigation of humic acid inhibition in anaerobic digestion of cellulose by addition of various salts. Bioengineering 2(2):54–65. https://doi.org/10.3390/bioengineering2020054

Baraza X, Sáez-Navarrete C, Torres-Castillo R (2019) Anaerobic biodegradability of leachate from MSW intermediate landfill. Afinidad 76(585):3–11

Barrantes Leiva M, Hosseini Koupaie E, Eskicioglu C (2014) Anaerobic co-digestion of wine/fruit-juice production waste with landfill leachate diluted municipal sludge cake under semi-continuous flow operation. Waste Manag 34(10):1860–1870. https://doi.org/10.1016/j.wasman.2014.06.027

Berenjkar P, Islam M, Yuan Q (2019) Co-treatment of sewage sludge and mature landfill leachate by anaerobic digestion. Int J Environ Sci Technol 16(5):2465–2474. https://doi.org/10.1007/s13762-018-1889-2

Carvajal-Flórez E, Cardona-Gallo S-A (2019) Technologies applicable to the removal of heavy metals from landfill leachate. Environ Sci Pollut Res 26(16):15725–15753. https://doi.org/10.1007/s11356-019-04888-7

Castrillón L, Fernández-Nava Y, Ulmanu M, Anger I, Marañón E (2010) Physico-chemical and biological treatment of MSW landfill leachate. Waste Manag 30(2):228–235. https://doi.org/10.1016/j.wasman.2009.09.013

Chaurasia AK, Pandey HK, Tiwari SK, Prakash R, Pandey P, Ram A (2018) Groundwater quality assessment using Water Quality Index (WQI) in parts of Varanasi District, Uttar Pradesh, India. J Geol Soc India 92(1):76–82. https://doi.org/10.1007/s12594-018-0955-1

Chen G, Zheng Z, Yang S, Fang C, Zou X, Luo Y (2010) Experimental co-digestion of corn stalk and vermicompost to improve biogas production. Waste Manag 30(10):1834–1840. https://doi.org/10.1016/j.wasman.2010.03.014

CPCB, 2019 (2019) CPCB Annual Report Annual Report. CPCB Website, December, 2–2. https://www.rtda.gov.rw/fileadmin/templates/publications/RWANDA_Annual_Report_2018-2019_SHARING.pdf

Dang Y, Lei Y, Liu Z, Xue Y, Sun D, Wang LY, Holmes DE (2016) Impact of fulvic acids on bio-methanogenic treatment of municipal solid waste incineration leachate. Water Res 106:71–78. https://doi.org/10.1016/j.watres.2016.09.044

de Castro TM, Arantes EJ, de Mendonça Costa MSS, Gotardo JT, Passig FH, de Carvalho KQ, Gomes SD (2021) Anaerobic co-digestion of industrial waste landfill leachate and glycerin in a continuous anaerobic bioreactor with a fixed-structured bed (ABFSB): effects of volumetric organic loading rate and alkaline supplementation. Renew Energy 164:1436–1446. https://doi.org/10.1016/j.renene.2020.10.100

de Castro TM, Torres DGB, Arantes EJ, de Carvalho KQ, Passig FH, Christ D, Gotardo JT, Gomes SD (2019) Anaerobic co-digestion of industrial landfill leachate and glycerin: methanogenic potential, organic matter removal and process optimization. Environ Technol (UK) 0:1–11. https://doi.org/10.1080/09593330.2019.1575915

Dong S, Sartaj M (2016) Statistical analysis and optimization of ammonia removal from landfill leachate by sequential microwave/aeration process using factorial design and response surface methodology. J Environ Chem Eng 4(1):100–108. https://doi.org/10.1016/j.jece.2015.10.029

El-Fadel M, Findikakis AN, Leckie JO (1997) Modeling leachate generation and transport in solid waste landfills. Environ Technol (UK) 18(7):669–686. https://doi.org/10.1080/09593331808616586

Ferraz FM, Povinelli J, Pozzi E, Vieira EM, Trofino JC (2014) Co-treatment of landfill leachate and domestic wastewater using a submerged aerobic biofilter. J Environ Manage 141:9–15. https://doi.org/10.1016/j.jenvman.2014.03.022

Goli A, Shamiri A, Khosroyar S, Talaiekhozani A, Sanaye R, Azizi K (2019) A review on different aerobic and anaerobic treatment methods in dairy industry wastewater. J Environ Treat Tech 7(1):113–141

Guo Q, Majeed S, Xu R, Zhang K, Kakade A, Khan A, Hafeez FY, Mao C, Liu P, Li X (2019) Heavy metals interact with the microbial community and affect biogas production in anaerobic digestion: A review. J Environ Manage 240(February):266–272. https://doi.org/10.1016/j.jenvman.2019.03.104

Guven H, Akca MS, Iren E, Keles F, Ozturk I, Altinbas M (2018) Co-digestion performance of organic fraction of municipal solid waste with leachate: preliminary studies. Waste Manag 71:775–784. https://doi.org/10.1016/j.wasman.2017.04.039

Hombach ST, Oleszkiewicz JA, Lagasse P, Amy LB, Zaleski AA, Smyrski K (2003) Impact of landfill leachate on anaerobic digestion of sewage sludge. Environ Technol (UK) 24(5):553–560. https://doi.org/10.1080/09593330309385590

Huang H, Xiao D, Zhang Q, Ding L (2014) Removal of ammonia from landfill leachate by struvite precipitation with the use of low-cost phosphate and magnesium sources. J Environ Manage 145:191–198. https://doi.org/10.1016/j.jenvman.2014.06.021

Ibrahim TNT, Mahmood NZ, Othman F (2017) Estimation of leachate generation from MSW landfills in Selangor, Malaysia. Asian J Microbiol Biotechnol Environ Sci 19(1):44–49

Kawai M, Kishi M, Hamersley MR, Nagao N, Hermana J, Toda T (2012) Biodegradability and methane productivity during anaerobic co-digestion of refractory leachate. Int Biodeter Biodegr 72:46–51. https://doi.org/10.1016/j.ibiod.2012.04.003

Khadem AF, Azman S, Plugge CM, Zeeman G, van Lier JB, Stams AJM (2017) Effect of humic acids on the activity of pure and mixed methanogenic cultures. Biomass Bioenergy 99:21–30. https://doi.org/10.1016/j.biombioe.2017.02.012

Kheradmand S, Karimi-Jashni A, Sartaj M (2010) Treatment of municipal landfill leachate using a combined anaerobic digester and activated sludge system. Waste Manag 30(6):1025–1031. https://doi.org/10.1016/j.wasman.2010.01.021

Kjeldsen P, Barlaz MA, Rooker AP, Baun A, Ledin A, Christensen TH (2002) Critical reviews in environmental science and technology present and long-term composition of MSW landfill leachate: a review present and long-term composition of MSW landfill leachate: a review. Crit Rev Environ Sci Technol 32(324):37–41. http://www.tandfonline.com/action/journalInformation?journalCode=best20%5Cn. https://doi.org/10.1080/10643380290813462

Kulikowska D, Klimiuk E (2008) The effect of landfill age on municipal leachate composition. Bioresour Technol 99(13):5981–5985. https://doi.org/10.1016/j.biortech.2007.10.015

Kumar D, Alappat BJ (2004) Selection of the appropriate aggregation function for calculating Leachate Pollution Index. Pract Periodic Hazard Toxic Radioact Waste Manag 8(4):253–264. https://doi.org/10.1061/(ASCE)1090-025X(2004)8:4(253)

Li J, Hao X, van Loosdrecht MCM, Liu R (2021) Relieving the inhibition of humic acid on anaerobic digestion of excess sludge by metal ions. Water Res 188. https://doi.org/10.1016/j.watres.2020.116541

Liao X, Zhu S, Zhong D, Zhu J, Liao L (2014) Anaerobic co-digestion of food waste and landfill leachate in single-phase batch reactors. Waste Manag 34(11):2278–2284. https://doi.org/10.1016/j.wasman.2014.06.014

Lin C, Chang F, Chang C (1999) Anaerobic co-digestion of septage and landfill leachate. Bioresour Technol 68:275–282

Lin C, Chang F, Chang C (2000) Co-digestion of leachate with septage using a UASB reactor. Bioresource Technol 73:175–178

Luo H, Zeng Y, Cheng Y, He D, Pan X (2020) Recent advances in municipal landfill leachate: a review focusing on its characteristics, treatment, and toxicity assessment. Sci Total Environ 703:135468. https://doi.org/10.1016/j.scitotenv.2019.135468

Marttinen SK, Kettunen RH, Sormunen KM, Soimasuo RM, Rintala JA (2002) Screening of physical-chemical methods for removal of organic material, nitrogen and toxicity from low strength landfill leachates. Chemosphere 46(6). https://doi.org/10.1016/S0045-6535(01)00150-3

Montusiewicz A, Lebiocka M (2011) Co-digestion of intermediate landfill leachate and sewage sludge as a method of leachate utilization. Bioresour Technol 102(3):2563–2571. https://doi.org/10.1016/j.biortech.2010.11.105

Montusiewicz A, Bis M, Pasieczna-Patkowska S, Majerek D (2018) Mature landfill leachate utilization using a cost-effective hybrid method. Waste Manag 76:652–662. https://doi.org/10.1016/j.wasman.2018.03.012

Moujanni AE, Qarraey I, Ouatmane A (2019) Anaerobic codigestion of urban solid waste fresh leachate and domestic wastewaters: biogas production potential and kinetic. Environ Eng Res 24(1):38–44. https://doi.org/10.4491/eer.2018.082

Neshat SA, Mohammadi M, Najafpour GD, Lahijani P (2017) Anaerobic co-digestion of animal manures and lignocellulosic residues as a potent approach for sustainable biogas production. Renew Sustain Energy Rev 79(July 2016):308–322. https://doi.org/10.1016/j.rser.2017.05.137

Özgür Akta FÇ (2001) Addition of activated carbon to batch activated sludge reactors in the treatment of landfill leachate and domestic wastewater. J Chem Technol Biotechnol 76(8):793–802. https://doi.org/10.1002/jctb.450

Pastor L, Ruiz L, Pascual A, Ruiz B (2013) Co-digestion of used oils and urban landfill leachates with sewage sludge and the effect on the biogas production. Appl Energy 107:438–445. https://doi.org/10.1016/j.apenergy.2013.02.055

Pecorini I, Baldi F, Carnevale EA, Corti A (2016) Biochemical methane potential tests of different autoclaved and microwaved lignocellulosic organic fractions of municipal solid waste. Waste Manag 56:143–150. https://doi.org/10.1016/j.wasman.2016.07.006

Rao PV, Baral SS (2011) Experimental design of mixture for the anaerobic co-digestion of sewage sludge. Chem Eng J 172(2–3):977–986. https://doi.org/10.1016/j.cej.2011.07.010

Renou S, Givaudan JG, Poulain S, Dirassouyan F, Moulin P (2008) Landfill leachate treatment: review and opportunity. J Hazard Mater 150(3):468–493. https://doi.org/10.1016/j.jhazmat.2007.09.077

Siddique MNI, Wahid ZA (2018) Achievements and perspectives of anaerobic co-digestion: a review. J Clean Prod 194(1):359–371. https://doi.org/10.1016/j.jclepro.2018.05.155

Silva NFP, Gonçalves AL, Moreira FC, Silva TFCV, Martins FG, Alvim-Ferraz MCM, Boaventura RAR, Vilar VJP, Pires JCM (2015) Towards sustainable microalgal biomass production by phycoremediation of a synthetic wastewater: a kinetic study. Algal Res 11:350–358. https://doi.org/10.1016/j.algal.2015.07.014

Sung S, Liu T (2003) Ammonia inhibition on thermophilic anaerobic digestion. Chemosphere 53(1):43–52. https://doi.org/10.1016/S0045-6535(03)00434-X

T.A.S J, Mamindlapelli NK, Begum S, Arelli V, Juntupally S, Ahuja S, Dugyala SK, Anupoju GR (2020) Anaerobic mono and co-digestion of organic fraction of municipal solid waste and landfill leachate at industrial scale: impact of volatile organic loading rate on reaction kinetics, biogas yield and microbial diversity. Sci Total Environ 748:142462. https://doi.org/10.1016/j.scitotenv.2020.142462

Torretta V, Ferronato N, Katsoyiannis IA, Tolkou AK, Airoldi M (2017) Novel and conventional technologies for landfill leachates treatment: a review. Sustainability (Switzerland) 9(1):1–39. https://doi.org/10.3390/su9010009

Tränkler J, Visvanathan C, Kuruparan P, Tubtimthai O (2005) Influence of tropical seasonal variations on landfill leachate characteristics – results from lysimeter studies. Waste Manag 25(10):1013–1020. https://doi.org/10.1016/j.wasman.2005.05.004

Tufaner F, Avşar Y (2016) Effects of co-substrate on biogas production from cattle manure: a review. 2303–2312. https://doi.org/10.1007/s13762-016-1069-1

Vaccari M, Tudor T, Vinti G (2019) Characteristics of leachate from landfills and dumpsites in Asia, Africa and Latin America: an overview. Waste Manag 95:416–431. https://doi.org/10.1016/j.wasman.2019.06.032

Vavilin VA, Fernandez B, Palatsi J, Flotats X (2008) Hydrolysis kinetics in anaerobic degradation of particulate organic material: an overview. Waste Manag 28(6):939–951. https://doi.org/10.1016/j.wasman.2007.03.028

Wiszniowski J, Robert D, Surmacz-Gorska J, Miksch K, Weber JV (2006) Landfill leachate treatment methods: a review. Environ Chem Lett 4(1):51–61. https://doi.org/10.1007/s10311-005-0016-z

Wu L, Li Z, Zhao C, Liang D, Peng Y (2018) A novel partial-denitrification strategy for post-anammox to effectively remove nitrogen from landfill leachate. Sci Total Environ 633:745–751. https://doi.org/10.1016/j.scitotenv.2018.03.213

Xiong J, Zheng Z, Yang X, Dai X, Zhou T, He J, Luo X (2018) Recovery of NH3-N from mature leachate via negative pressure steam-stripping pretreatment and its benefits on MBR systems: a pilot scale study. J Clean Prod 203:918–925. https://doi.org/10.1016/j.jclepro.2018.08.285

Yap SD, Astals S, Lu Y, Peces M, Jensen PD, Batstone DJ, Tait S (2018) Humic acid inhibition of hydrolysis and methanogenesis with different anaerobic inocula. Waste Manag 80:130–136. https://doi.org/10.1016/j.wasman.2018.09.001

Yong Z, Dong Y, Zhang X, Tan T (2015) Anaerobic co-digestion of food waste and straw for biogas production. Renew Energy 78:527–530. https://doi.org/10.1016/j.renene.2015.01.033

Yuzer B, Akgul D, Mertoglu B (2012) Effect of high ammonia concentration on UASB reactor treating sanitary landfill leachate. Fen Bilimleri Dergisi 24(2):59–67

Zhang H, Wang Z, Liu C, Guo Y, Shan N, Meng C, Sun L (2014) Removal of COD from landfill leachate by an electro/Fe2+/peroxydisulfate process. Chem Eng J 250:76–82. https://doi.org/10.1016/j.cej.2014.03.114

Zhang W, Zhang L, Li A (2015) Anaerobic co-digestion of food waste with MSW incineration plant fresh leachate: process performance and synergistic effects. Chem Eng J 259:795–805. https://doi.org/10.1016/j.cej.2014.08.039

Zhou S, Wang J, Peng S, Chen T, Yue Z (2021) Anaerobic co-digestion of landfill leachate and acid mine drainage using up-flow anaerobic sludge blanket reactor. Environ Sci Pollut Res 28(7):8498–8506. https://doi.org/10.1007/s11356-020-11207-y

# Chapter 9
# Anaerobic Ammonia Oxidation Enrichment to Enhance Landfill Leachate Treatment

**S. Sri Shalini and Kurian Joseph**

**Abstract** Ammonia nitrogen concentrations in landfill leachate varies around 100–3000 mg/L, considered to be a persistent pollutant in landfill. The nitrogen profile and its mechanism involved in the landfill waste decomposition is explored in detail in this chapter. Various physical, chemical and biological nitrogen removal systems for their merits and demerits are addressed. Among them the novel anaerobic ammonium oxidation (ANAMMOX) process has benefits for nitrogen removal than conventional biological process with lower power requirements and operational expenditure. The foremost challenge in scale-up is the need for large-scale enrichment of AnAOB (or ANAMMOX) bacteria. This chapter presents the study results of successful enrichment of AnAOBs using mined municipal solid waste (MSW) from landfill. Large-scale anaerobic fed batch reactor loaded with mined MSW as seed at a solids loading rate of 1739 g of TS operated for 250 days with nutrient supplement addition. AnAOB enrichment was confirmed based on specific ANAMMOX activity with 0.005–0.10 mg Amm-N/mg MLVSS/d, hydrazine (0.02 mg/L) and hydroxylamine (0.003 mg/L) formation, biomass accumulation – 8200 mg/L with free nitrous acid and free ammonia concentrations of 0.0–0.002 and 0.5–34.0 mg/L, respectively. Ammonia removal efficiency of 87 ± 6 mg/L, nitrite utilization (98%) and nitrate formation (4.6–24.1 mg/L) revealed the AnAOB enrichment using mined MSW for enhanced landfill leachate treatment using ANAMMOX process.

S. S. Shalini (✉)
Centre for Climate Change and Disaster Management, Department of Civil Engineering, Anna University, College of Engineering Guindy Campus, Chennai, Tamil Nadu, India
e-mail: srishalini10@gmail.com

K. Joseph
Centre for Climate Change and Disaster Management, Department of Civil Engineering, Anna University, College of Engineering Guindy Campus, Chennai, Tamil Nadu, India

Centre for Environmental Studies, Department of Civil Engineering, Anna University, College of Engineering Guindy Campus, Chennai, Tamil Nadu, India



P. Pathak, S. G. Palani (eds.), *Circular Economy in Municipal Solid Waste Landfilling: Biomining & Leachate Treatment*, Radionuclides and Heavy Metals in the Environment, https://doi.org/10.1007/978-3-031-07785-2_9

**Keywords** Landfill leachate · Ammonia nitrogen removal · Municipal solid waste · ANAMMOX process · AnAOB enrichment

## 9.1 Nitrogen Profile During Landfill Waste Decomposition

Managing the municipal solid waste (MSW) is a main urban services offered by the government to people, due to their increasing waste generation in the recent decades. The rapid urbanisation trend, lifestyle changes and population surge has led the world to produce waste around 2.01 BT in 2016 and it will upsurge to 2.59 BT in 2030 and 3.40 BT in 2050 (Silpa et al. 2018). World's MSW composition contained 44% of organic waste (food and green waste), 17% of paper and cardboard, 12% of plastic, 5% of glass, 4% of metal, 2% of wood, 2% of rubber and leather and 14% of other components. The major MSW treatment and disposal method is landfilling in most of the countries. The contaminated liquid from the landfills are landfill leachate that is produced due to various processes inside the landfill environment such as rainwater infiltration, water holding capacity and biological decomposition (Renou et al. 2008). Proper landfill treatment is decided upon the extent of its leachate treatment. The landfill leachate composed of high contents of organic and nitrogen compounds. These are the primary pollutants that pose major impact on the environment and makes the leachate treatment more complicated (Berge et al. 2005; Renou et al. 2008). The heavy metals and xenobiotic compounds are other common pollutants in the leachate (Kjeldsen et al. 2002). Leachate composition vary widely in different landfills and its characteristics fluctuate based on factors of age of landfill, site hydrogeology, seasonal weather conditions, precipitation, moisture passing into the landfill, waste type and composition, landfill design, and landfill operational conditions (Show et al. 2019).

Landfill age is the main factor responsible for the waste composition during the entire landfill operational period. The landfill leachate can be classified based on their landfill age as young or fresh, intermediate or middle and mature or old age leachates. The landfill leachate compositions based on age of the landfill as given in the Table 9.1 (Miao et al. 2014; Luo et al. 2015; Show et al. 2019).

Leachate characteristics varies based on the landfill waste decomposition phase period. Initially the waste decomposition within the landfill proceeds aerobically as initial adjustment phase of lag or aerobic phase is typically short. The oxygen available is also exhausted quickly and it becomes the limiting factor. The next, transition phase involves shifting from aerobic to anaerobic environment due to depletion of oxygen and moisture content of the waste, where facultative anaerobic microorganisms becomes active. The total volatile acids found in leachate with higher concentration of chemical oxygen demand (COD) reveals the greater anaerobic bacterial activity. In addition, nitrogen compounds detected due to hydrolysis and fermentation of protein compounds. In early acidic phase of the landfill, young leachate contains maximum content of all the compounds due to their strong waste decomposition and higher leaching condition. Rapid hydrolysis of biodegradable fraction

**Table 9.1** Landfill leachate characteristics based on landfill age

| S. N. | Parameters | Young or fresh | | Intermediate or middle | Mature or old |
|---|---|---|---|---|---|
| 1. | Period | Transition | Acid formation | Methane formation | Final maturation |
| 2. | Age (years) | 0–5 | 5–10 | 10–20 | ≥20 |
| 3. | Phase | Acidogenic | Acidogenic | Methanogenic | Stabilization/ methanogenic |
| 4. | Stability | Unstable | Unstable | Moderately | Stable |
| 5. | pH | ≤6.5 | 6.5–7.5 | 6.6–7.5 | 7.5–9 |
| 6. | BOD (mg/L) | 100–11,000 | 1000–57,000 | 100–3500 | 4–120 |
| 7. | COD (mg/L) | ≥10,000 | ≥10,000 | 4000–10,000 | ≤4000 |
| 8. | $BOD_5$/COD | ≥0.3 | 0.5–1 | 0.1–0.5 | ≤0.1 |
| 9. | $NH_4$-N (mg/L) | ≤400 | ≤400 | 6–430 | 3000–5000 |
| 10. | Nitrite (mg/L) | 0.1–500 | 0.1–20 | 0.1–1.5 | 0.5–0.6 |
| 11. | Level of heavy metals | Low to medium | Low to medium | Low | Low |

of the solid waste by acidogenic bacteria results in volatile organic acids. Consequently, a decrease in pH occurs with concomitant mobilization and possible complexation of metal species. Nutrients such as nitrogen and phosphorous are released and utilized in support of microbial biomass growth. Hydrogen gas might be detected and may affect the nature and type of intermediary metabolism and product formation. As there is no free oxygen in the waste, maximum levels of COD and biochemical oxygen demand (BOD) are observed in young leachate. This $BOD_5$ and COD then consequently reduces with the increasing landfill age of more than ten years.

In the methane fermentation phase, intermediary products, volatile organic acids and hydrogen gas produced in the acid formation are resulted into methane ($CH_4$) and carbon dioxide ($CO_2$) by strict anaerobes. Hydrophilic methanogenic bacteria convert hydrogen ($H_2$) and $CO_2$ into $CH_4$. Acetophilic methanogenic bacteria convert acetic acid ($CH_3COOH$) into $CH_4$ and $CO_2$. Sulphate-reducing bacteria convert hydrogen, acetic acid and higher volatile fatty acids (VFAs) into $CO_2$ and $H_2S$. As a result, VFA consumption by methanogens, the pH drifts to neutrality. Oxidation reduction potential (ORP) in methane fermentation phase is at their lowest reducing levels with the accumulation of reduced sulphur and nitrogen species. The leachate organic strength characterized by low BOD is drastically decreased in correspondence with the increase in gas production. Leachate containing heavy metals decrease due to high pH value that creates complexation and precipitation, and transport it to solid phase. The final phase of solid waste decomposition is characterized by a lower rate of biological activity with a significant drop in landfill gas production. Hence, in the prolonged methanogenic phase of a landfill, produces a stable composition of leachate i.e. old or mature landfill leachate with lower content of compounds. However, the ammonia ($NH_4$-N) composition did not decrease in leachate, it increases with the increasing landfill age and becomes the persistent pollutant (Kjeldsen et al. 2002; Berg et al. 2005). The ammonia nitrogen in the leachate can vary between 10 to 10,000 mg/L (Tatsi and Zouboulis 2002; Statom et al. 2004; Berge et al. 2005). The ammonia composition is based on the nitrogen content in the waste. The nitrogen profile and mechanisms involved during the landfill waste decomposition varies widely.

The nitrogen concentrations during the landfill waste decomposition for a conventional landfill leachate and landfill bioreactor leachate (in landfill bioreactors (LFBR) where landfill leachate is recirculated in the landfill) are summarized in the Table 9.2 (Boonchai 2004; Berge et al. 2005).

In the conventional landfill, leachate nitrogen concentration greatly depends on waste decomposition. Initially, total kjeldahl nitrogen (TKN) are in higher concentrations in transition phase (180–860 mg/L) and reduces in acid formation and methanogenic because of the microbial assimilation of nitrogenous compounds. Similar trend is seen in nitrate ($NO_3$) concentrations. It increases in transition phase (0.1–5.1 mg/L) due to oxidation of ammonia nitrogen and reduces in acid formation and completely absent in methanogenic phase due to conversion to ammonia or nitrogen gas. The ammonia nitrogen follows a different trend during waste decomposition. In transition phase, ammonia concentration is about 120–125 mg/L. It increases in acid formation phase due to release of ammonia mainly from nitrate reduction and protein breakdown (2–1030 mg/L). The ammonia nitrogen in conventional landfills does not reduce in methanogenic and maturation phase present in the range of 6–430 mg/L. The ammonia concentrations are even higher (above 1800 mg/L) in landfill bioreactor in different degradation phases, constituting it as persistent pollutants (Kjeldesen et al. 2002; Boonchai 2004; Berge et al. 2005). The ammonia nitrogen content causes aquatic toxicity and eutrophication in the natural water ecosystem that impairs the water quality of water ecosystems especially drinking water, toxic algal blooms, water anoxia and loss of habitat or biodiversity (Tchobanoglous et al. 2003; Weigelhofer et al. 2018). Hence, the landfill leachate treatment should be efficient to remove the ammonia nitrogen content to the level of recommended discharge limits. The leachate treatment using various physical, chemical and biological nitrogen removal systems are explored.

**Table 9.2** Nitrogen content in landfill leachate during landfill waste decomposition

| | | Conventional landfill (mg/L) | | | Landfill bioreactor (mg/L) |
|---|---|---|---|---|---|
| S.No | Degradation phases | TKN | Nitrate | Ammonia | Ammonia |
| 1. | Transition | 180–860 | 0.1–5.1 | 120–125 | 76–125 |
| 2. | Acid formation | 14–1970 | 0.05–19 | 2–1030 | 0–1800 |
| 3. | Methane fermentation | 25–82 | Absent | 6–430 | 32–1850 |
| 4. | Final maturation | 7–490 | 0.5–0.6 | 6–430 | 420–580 |

## 9.2 Nitrogen Removal Techniques for Landfill Leachate Treatment

Nitrogen removal methods can be classified as physical, chemical and biological techniques.

### *9.2.1 Physical and Chemical Nitrogen Removal Methods for Landfill Leachate Treatment*

Several physical and chemical techniques for leachate nitrogen removal available are air/steam stripping, vacuum distillation, ammonia precipitation, ion exchange, breakpoint chlorination, activated carbon, ammonia adsorption, chemical reduction and advanced oxidation processes (AOPs). Some of the AOPs for leachate treatment for ammonia and organic nitrogen conversion into $N_2$ or $NO_3$ are ozonation, peroxon oxidation, photo-catalytic oxidation, Fenton and electrochemical oxidation (Kim et al. 2007; Zhang et al. 2009; Capodaglio et al. 2015). The conventional techniques of air stripping and, coagulation and flocculation usually expensive due to their primary expense of operation, power and chemicals. The AOPs are not economically feasible for large-scale application (Wiszniowski et al. 2006). The coagulation-flocculation process for leachate treatment are expensive and generates waste that requires further secondary treatment. Several combination of processes could be used for different aged landfill leachates according to their characteristics. Physico-chemical leachate treatment generally applied as a pre-treatment or as an advanced post treatment (Amokrane et al. 1997; Ozturk et al. 2003; Zhang et al. 2016). However, physico-chemical methods for leachate treatment has advantages such as simple, not affected by temperature variations, application for high concentrated leachates, its disadvantages such as expensive due to operational cost, post-treatment for sludge treatment, fouling problems and energy intensive makes it not a feasible method for highly nitrogen polluted leachate (Kurniawan et al. 2006; Peng et al. 2008). The biological treatment processes are commonly used for leachate nitrogen removal (Liang and Liu 2007; Mulder 2003; Janus and Vander 1997).

### *9.2.2 Biological Nitrogen Removal Methods for Landfill Leachate Treatment*

The biological nitrogen removal is highly recommended due to its effectiveness, economical and feasible technology (Abelling and Seyfried 1992; Hoilijoki et al. 2000; Peng et al. 2008; Ahemd and Lan 2012). Some of the different biological methods for leachate treatment processes include aerobic, anaerobic, combined anaerobic–aerobic by various systems such as activated sludge process (ASP), sequencing batch reactor (SBR), Single reactor for high activity ammonia removal

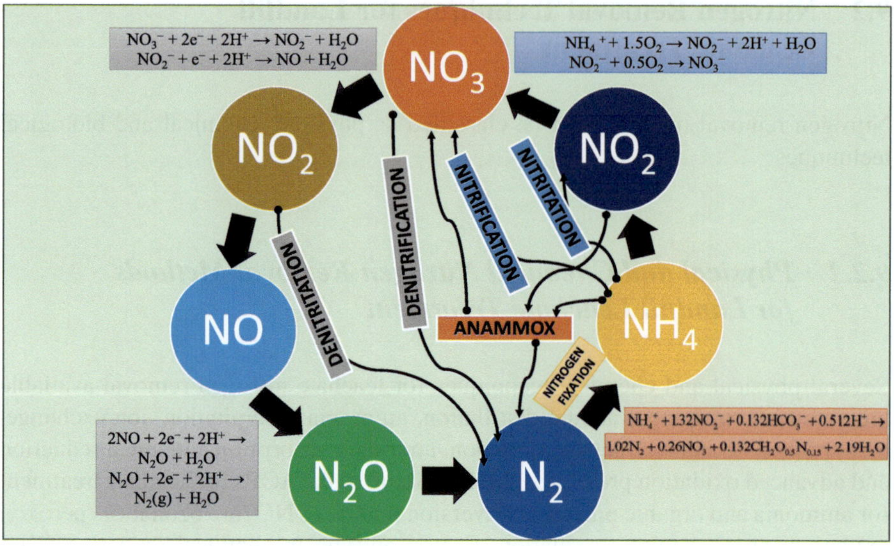


**Fig. 9.1** Nitrogen transformations in the nitrogen cycle with conventional and other novel nitrogen removal processes

over nitrite (SHARON), anaerobic ammonium oxidation (ANAMMOX), membrane bioreactor (MBR), trickling filters, rotating biological contactors (RBC), landfill bioreactors, microbial electrochemical cells, phototrophic systems, etc. (Tengrui et al. 2007; Long et al. 2008; Renou et al. 2008; Sri Shalini and Kurian Joseph 2012; Winkler and Straka 2019). The nitrogen transformations in the nitrogen cycle with various nitrogen removal processes are depicted in the Fig. 9.1.

#### 9.2.2.1 Conventional Biological Nitrogen Removal Methods for Landfill Leachate Treatment

The conventional biological nitrogen removal are by nitrification and denitrification processes. The nitrification is an aerobic process by autotrophs where, the ammonium oxidizing bacteria (AOB) (e.g. *Nitrosomonas*) convert ammonia to nitrite ($NO_2$) known as nitritation and nitrite oxidizing bacteria (NOB) (e.g. *Nitrobacter*) convert nitrite to nitrate called as nitratation using ammonia or $NO_2$ as electron donor and $O_2$ as an electron acceptor (as shown in Fig. 9.1). The denitrification is usually an anaerobic process by heterotrophs where, denitrifiers subsequently convert $NO_3$ to $N_2$ gas with organic carbon as an electron donor. Sometimes it can also takes place in aerobic condition (Metcalf and Eddy 2003). Hence, denitrification can takes place as autotrophic or heterotrophic process. Similarly, some heterotrophic nitrifying bacteria (such as *Bacillus* strains, *Alcaligenes faecalis, Acinetobacter* sp., *Paracoccus denitrificans, Thiosphaera pantotropha,* etc.) can do both heterotrophic nitrification and aerobic denitrification (Hong et al. 2017). Different combinations of nitrification and denitrification in different reactors for leachate treatment are tabulated in Table 9.3.

**Table 9.3** Different combinations of nitrification and denitrification in different reactors for landfill leachate treatment

| S. No | Biological Treatment | Leachate characteristics | Removal efficiency | Reference |
|---|---|---|---|---|
| 1. | Anaerobic sequencing batch reactor (ASBR) combining pulsed sequencing batch reactor (PSBR) | Raw leachate from MSW sanitation landfill site (Beijing, China) $NH_4^+$-N – 1154 mg/L COD – 8528 mg/L pH – 8.0; $BOD_5$/COD – 0.5–0.7 | TN removal – 97% and COD removal – 89%; effluent TN – 40 mg $L^{-1}$ | Zhu et al. (2013) |
| 2. | Upflow anaerobic biofilm reactor, an aerobic-activated sludge reactor and a clarifier | Kyungjoo City municipal wastes landfill site, South Korea $NH_4^+$-N – 1682 mg/L; COD – 24,400 mg/L pH – 7.3; $BOD_5$/COD – 0.44 | Maximum $NH_4^+$-N removal rate 0.84 kg $NH_4^+$-N/$m^3$ /d i.e. 50% in anaerobic reactor and the maximum nitrification rate was 0.5 kg $NO_3$ -N/$m^3$/d in aerobic reactor | Im et al. (2001) |
| 3. | Two-stage UASB and anoxic–oxic reactor | MSW landfill of Haidian District of Beijing city (PRC) $NH_4^+$-N – 1100–2000 mg/L; COD – 8550–12,500 mg/L pH – 7.2–7.9; $BOD_5$/COD – 0.47 | Maximum $NH_4^+$-N removal – 0.68 kg $NH_4^+$-N/$m^3$/d and about 99% of $NH_4^+$-N removal and 81–93% total nitrogen removal | Peng et al. (2008) |
| 4. | Sequencing batch reactor (SBR) and the effluent to an upflow anaerobic sludge blanket | Liulitun landfill of Beijing city $NH_4^+$-N – 738–22,500 mg/L; COD – 7856–22,500 mg/L TN – 839–1390 mg/L pH – 7.1–8.5 | $NH_4^+$-N – 99.7%, total nitrogen – 98.3% and COD – 96.7%. | Sun et al. (2010) |
| 5. | Upflow anaerobic sludge blanket and anoxic/oxic (A/O) | MSW landfill from Beijing $NH_4^+$-N – 1200–1800 mg/L; COD – 2000–4500 mg/L pH – 8.0–8.3 | $NH_4^+$-N: 90–100% (<15 mg/L); Total inorganic nitrogen: 67–80% | Zhang et al. (2007) |
| 6. | Landfill bioreactor | MSW from mined dumpsite Simulated leachate with maximum $NH_4^+$-N – 652 mg/L; maximum COD – 2780 mg/L pH – 7.0–8.3 | $NH_4^+$-N – 49%; COD removal – 68% | Sri Shalini and Kurian Joseph (2018) |

Nitrogen removal in different reactors showed more than 90% in most of the reactors. Recently, landfill bioreactors used for ammonia nitrogen removal from leachate achieved 49 percent in 336 days at nitrogen loading rate (NLR) of 1.0 kg $NH_4$-N/$m^3$/d (Sri Shalini and Kurian Joseph 2018). Membrane bioreactors are also used for landfill treatment with anoxic or oxic (A/O) process studied for a duration of one hundred and thirteen days. The results revealed that COD, $NH_4$-N and TN removal was 80.60%, 99.04% and 74.87%, respectively (Liu et al. 2017).

The biological nitrogen removal processes are activated sludge processes are common because of their advantages such as simple process and economical (Zhu et al. 2013). The nitrogen removal technique for young age or intermediate age landfill leachate containing higher carbon to nitrogen ratio use conventional nitrification and denitrification method or other methods. They are combining the two-stage UASB and A/O, or endogenous denitrification method (Peng et al. 2008; Miao et al. 2015). Whereas, the nitrogen removal of lower carbon to nitrogen ratio of mature landfill leachate using conventional nitrogen removal techniques requires higher carbon sources leads to increase the cost of the treatment and difficult to manage the process (Kulikowska et al. 2008). Despite the conventional nitrogen removal processes having efficient nitrogen removal rate, the drawbacks such as demand of higher oxygen requirement and carbon sources, pollutant emissions such as nitrous oxide, inadequate sludge settleability (especially in suspended growth system), extended period for complete removal and frequent changes in the leachate composition all makes it complicate (Burton and Watson-Craik 1998; Hoilijoki et al. 2000; Loukidou and Zouboulis 2001; Henze et al. 2008; Ganigue et al. 2009). Conventional biological nitrogen removal methods is not feasible to reduce the nitrogen content completely to meet the new discharge requirements of below 10 mg of total nitrogen (Khin and Annachhatre 2004). The other new methods for nitrogen removal such as phototrophic systems, microbial electrochemical technologies and novel methods are as follows.

#### 9.2.2.2 Phototrophic Systems for Landfill Leachate Treatment

In phototrophic systems, the microorganisms such as algae and bacteria use the sunlight as energy to decrease the nitrogen content (Hulsen et al. 2016; Hong et al. 2017; Winkler and Straka 2019). The different phototrophic systems are suspended growth systems such as waste stabilisation ponds (WSPs), High rate algal pond (HRAP), stirred tank reactors, and tubular photobioreactors (PBRs) and, as attached growth system is algal turf scrubber (ATS). The comparison of these systems for their energy yield, energy consumption, treatment efficacy and operational cost were demonstrated by several researchers (Shoener et al. 2014; Winkler and Straka 2019). Phototrophs or photosynthetic bacteria can be further classified as two types of oxygenic and anoxygenic phototrophs. The phototrophs that oxidises the water and produces the molecular oxygen as a by-product are oxygenic phototrophic bacteria such as single celled algae and microalgae (e.g. cyanobacteria). The phototrophs that uses hydrogen sulphide and thiosulfate as a reducing agent instead of $O_2$

called anoxygenic phototrophic bacteria (APB) such as purple sulfur bacteria (PPB), purple nonsulfur bacteria, green sulfur bacteria, and green nonsulfur bacteria. The phototrophic PPB can grow as photo-autotrophic or heterotrophic and chemo-heterotrophic (Ghosh et al. 2017). APBs conduct nitrogen fixation, ammonification, ammonia assimilation, nitrification, and denitrification with their preferable nitrogen source as ammonium, rarely uses nitrate and nitrite (Madigan 1995; Hong et al. 2017). Hong et al. (2017) showed *Marichromatium gracile: YL28* performed simultaneous heterotrophic nitrification and denitrification at low dissolved oxygen (DO) and utilise $NO_2$ as nitrogen supplement for growth. These APBs are widespread use in aquaculture for inorganic nitrogen removal (Qi et al. 2009) and also applied in the highly concentrated wastewater treatment (Hulsen et al. 2016). Some microalgae studies focussed on the application for leachate treatment (Lin et al. 2007; Tsarpali et al. 2012; Zhao et al. 2014). The landfill leachate composition impact on the tolerant microalgae species and the leachate toxicity were assessed by Tsarpali et al. (2012). The landfill leachate toxicity decreased by addition of *Chlorella pyrenoidosa* and *Chlamydomonas snowiae* (Lin et al. 2007). A study by Martins et al. (2013) revealed 75–99% nitrogen removal from the leachate using mixed microalgae strains especially dominated by Chlamydia. Mustafa et al. (2012) conducted leachate experiments with five types of microalgae mixed with natural algae. The ammonia removal achieved was around 99.9%. In another study by Zhao et al. (2014) showed the nitrogen removal from landfill leachate using microalgae and bacteria with the culture conditions containing 221.6 of mg/L of TN spiked with 10% leachate. The leachate results revealed ammonia removal of 95% and TN of 90%, respectively. The phototrophic systems are good for nitrogen removal, however they are energy intensive process such as PBRs and some passive phototrophic systems of WSPs, HRAPs and ATSs does not consume energy but requires larger land areas (Shoener et al. 2014) makes it not convenient for efficient leachate treatment.

#### 9.2.2.3 Microbial Electrochemical Technologies (METs) for Landfill Leachate Treatment

The microbial electrochemical systems utilises the catalytic reactions of microorganisms capable of extracellular electron transfer for generating electrical power, by converting the energy found in organic compounds. METs, consists of an anode whereas in the electrons flow from it through an external circuit to a cathode and the electrical current is consumed in a reduction reaction (Rozendal et al. 2008). The two types of METs are microbial fuel cells (MFCs) that utilise the microbial decomposition of organic compounds to produce a current and in microbial electrolysis cell (MEC), the hydrogen or methane production from organic compound by applying a current. MFCs recently are used for leachate nitrogen treatment. Hassan et al. (2018) showed landfill leachate treatment in a MFC with an outcome of 66% ammonia removal and 86% of nitrite removal. The removal is lower when compared to other previous nitrogen removal studies. The cathode cost is also high in MFCs.

More research is required for full scale applications of MFCs for leachate treatment with cost-effective cathodic substances. Hence, novel nitrogen removal methods for less energy consumption and cost-effective methods were explored.

#### 9.2.2.4 Novel Nitrogen Removal Methods for Landfill Leachate Treatment

Different novel methods of biological nitrogen removal are nitritation/denitritation, nitritation-endogenous denitritation (ED) and deammonification techniques such as partial nitrification/nitritation/SHARON and ANAMMOX processes (depicted in Fig. 9.1). In the nitritation process, the aerobic oxidation of $NH_4$-N to $NO_2$-N is conducted using autotrophic aerobic ammonium oxidising bacteria (first step of the nitrification process). In the denitritation process, further the nitrite nitrogen in anoxic condition reduced to nitrogen gas using heterotrophic bacteria. These bacteria takes up the organic substrate. A study by Chen et al. (2016) on landfill leachate treatment using nitritation and denitritation process with operating condition of DO at 0.1 to 0.5 mg/L resulted in $NO_2$ build-up of ninety percent and $NH_4$-N removed more than ninety five percent. In nitritation-endogenous denitritation, it uses the intracellular stored polymers, where the organic compounds from the leachate is utilised with no further addition of carbon for the nitrogen removal process. The main key organisms for this ED process are glycogen accumulating organisms (Winkler et al. 2011) that develops well in the leachate composition of low phosphorus to COD ratio with lesser value of 0.002 (Miao et al. 2015). The leachate nitrogen removed using glycogen accumulating organisms in endogenous denitrification process, but takes longer duration than the conventional biological removal processes.

The deammonification process is by conversion of ammonia nitrogen to nitrogen gas. SHARON and anaerobic ammonium oxidation processes are deammonification processes (as shown in Fig. 9.1). SHARON/ANAMMOX is advantageous than conventional nitrification-denitrification and nitritation/denitritation due to the low energy, lower oxygen requirements (during aerobic conditions), no carbon source requirements, low sludge production and works in high ammonia-nitrogen levels and high temperature (30–40 °C) (Sri Shalini and Kurian Joseph 2012; Malamis et al. 2014; Shi et al. 2013). In deammonification, the 55% raw ammonia nitrogen converted to nitrite nitrogen using aerobic SHARON process and further ammonia nitrogen anaerobically oxidized using nitrite nitrogen to nitrogen gas in ANAMMOX process.

## 9.3 ANAMMOX Process for Landfill Leachate Treatment

ANAMMOX is a novel and autotrophic process (Mulder et al. 1995). It is highly efficient for good nitrogen removal from wastewaters (Kartal et al. 2010; Li et al. 2019). Mature leachate nitrogen concentration is effectively removed by ANAMMOX (Anfruns et al. 2013). It is conducted using *Planctomycetales.* The main organism involved in this process are anaerobic ammonia-oxidizing bacteria (AnAOB). It uses nitrite nitrogen as electron acceptor. The process is an anaerobic oxidization of ammonia nitrogen to nitrogen gas (Strous et al. 1999). It optimally requires $NO_2$ to $NH_4$-N ratio of 1.32 (Fig. 9.1) (Strous et al. 1998) provided by partial nitrification process. In the theoretical aspects, the highest nitrogen removal capacity is around 89%, as the remaining eleven percent nitrogen converted as nitrate (Lackner et al. 2014).

There are two arrangements in ANAMMOX process, a single step ANAMMOX process where SHARON and ANAMMOX conducted in one reactor and a 2-step ANAMMOX in two individual reactors, where SHARON process, a stable, partial nitrification conducted separately (Hellinga et al. 1998) and followed by ANAMMOX process in another separate reactor. Even though single reactor are advantageous in terms of compactness, it requires proper controllers for two processes. The two stage is beneficial as better control of the two processes could be achieved. Many studies were addressed for combining these two systems in different reactors for leachate application. Partial nitrification-ANAMMOX conducted in SBR systems for leachate treatment (Miao et al. 2014). Three SBR system with prior pre-treatment resulted in more than ninety percent of nitrogen removal. Continuous flow system of nitritation and ANAMMOX system gave TN removal of ninety four percent (Wang et al. 2016). Efficient operation of ANAMMOX process is required for attaining good removal outcomes. The ANAMMOX is most suitable to the leachate composition having low carbon to nitrogen ratio. The different reactor configurations available for ANAMMOX landfill leachate treatment are SBRs, granular sludge blanket reactors (GSBRs), moving bed biofilm reactors (MBBRs), rotating biological contactors (RBCs), fluidised bed biofilm reactor (FBBR), UASB, etc. The different ANAMMOX reactors for landfill leachate treatment with their nitrogen loading and removal are tabulated in Table 9.4. Bio-carrier bed for mature leachate treatment by ANAMMOX process gave a total removal efficiency of around 93% (Phan et al. 2014) and a biofilm system gave 95% (Miao et al. 2016). The studies showed that for higher nitrogen removal efficiency with higher nitrogen loading rate, it required a longer operational period. If factors influencing the ANAMMOX is controlled, then in shorter duration a higher nitrogen removal rates can be achieved.

**Table 9.4** Different ANAMMOX reactors for landfill leachate treatment

| Sl No | Reactor type | Working volume (L) | Operation period (d) | Nitrogen loading rate (kg N/d) | Nitrogen removal (%) | Reference |
|---|---|---|---|---|---|---|
| 1. | UASB | 4.46 | 222 | 0.40 | 88 | Zhang and Zhou (2006) |
| 2. | FBBR | 36 | 100 | 0.06–0.11 | 60 | Liang and Liu (2008) |
| 3. | RBC | NA | NA | 1.50 | 40 | Hippen et al. (2001) |
| 4. | SBR | 2 | 360 | 0.24 | 89 | Ruscalleda et al. (2008) |

*NA* Not Available

**Table 9.5** Main factors influencing the ANAMMOX process

| S. No | Factors | Concentration | Impact on ANAMMOX activity | Reference |
|---|---|---|---|---|
| 1. | Dissolved oxygen | 0.06 mg/L | Reversibly inhibition | Paredes et al. (2007) |
| 2. | Temperature | 30–37 °C | Good activity | Dongen et al. (2001) |
| 3. | pH | 7.0–8.5 | Good activity | Dongen et al. (2001) and Reginatto et al. (2005) |
| 4. | Organic matter | 300 mg COD/L | Inactivation | Chamchoi et al. (2008) |
| 5. | Nitrate nitrogen | 45 mM | 50% inhibition | Bettani et al. (2010) |
| 6. | Nitrite nitrogen | 100 mg $NO_2$-N/L | Complete inhibition | Bettani et al. (2010) |
| 7. | Ammonium | >70 mg N/L | Free ammonia inhibition | Bettani et al. (2010) |
| 8. | C/N ratio | >1 | Complete inhibition | Guven et al. (2005) |

## *9.3.1 Factors Influencing the ANAMMOX Process*

Several factors influence the ANAMMOX process, among them some of the major parameters that has the highest impact are dissolved oxygen, temperature, pH, alkalinity and organic matter (Dongen et al. 2001; Bettazi et al. 2010; Tang et al. 2010). The concentration of main parameters influence the ANAMMOX process are summarised in Table 9.5.

The ANAMMOX bacteria is severely affected by DO, even in the range of 0 to 0.2 mM it is inhibited (Jetten et al. 1999). The AnAOB bacterial growth metabolism is strongly influenced by pH and temperature. The optimum pH and temperature was 6.7–8.5 and 30–37 °C, respectively, sometimes vary from twenty to forty-three degree Celsius (Strous et al. 1999; Dongen et al. 2001; Reginatto et al. 2005). pH decides free ammonia (FA) and free nitrous acid (FNA) concentrations (Anthonisen et al. 1976). Mainly ANAMMOX activity is affected by pH rather than FA. The ammonium and nitrite concentrations affect the ANAMMOX activity as their affinity constants are <0.1 mg/L (Strous et al. 1999). The ANAMMOX affected by ammonium greater than 70 mg N/L is inhibited by free ammonia (Bettazi et al. 2010). It is also affected by 100 mg/L of $NO_2$ concentration (Strous et al. 1999).

The high nitrite concentrations are inhibitors for ANAMMOX process. Recently, nitrate/nitrite - dependent anaerobic methane oxidation (n-DAMO) coupling to ANAMMOX was used for leachate treatment (Ni et al. 2020). It is mainly for both $NH_4$-N and $CH_4$ removal conducted in a membrane biofilm reactor (MBfR). The study achieved approximately six kilograms of nitrogen removed per cubic metre per day. Even, higher organic matter concentration affects ANAMMOX bacteria, as the organic matter presence accelerates the growth of denitrifying bacteria. If COD is present in lower concentration (upto 150 mg/L), it is not affected (Ni et al. 2012; Miao et al. 2014), it can co-exist with denitrifiers and have beneficial for nitrogen removal. However, more than 300 mg COD/L affects AnAOB population and decrease it (Chamchoi et al. 2008; Molinuevo et al. 2009; Ni et al. 2012; Miao et al. 2019). Influent COD has to be monitored effectively and controlled properly for effective ANAMMOX activity. If the organic content is completed used by other combined process such as dentirification, then good removal is achieved. When compared to other anaerobic bacteria (such as denitrifiers), the AnAOB bacteria grows slow and necessary for enriching it to conduct the ANAMMOX process rapidly.

### *9.3.2 AnAOB Enrichment Studies for ANAMMOX Process*

Different seeds are used for enriching the AnAOB bacteria. They are soil and activated sludge (Fux et al. 2002; Lopez et al. 2008; Wang et al. 2009; Lotti et al. 2015), anaerobic sludge digester (Bagchi et al. 2010), anaerobic/aerobic granular sludge (Tang et al. 2009), UASB (Ahn et al. 2004; Tran et al. 2006) and denitrifying reactor sludge (Graaf et al. 1996). Nitrogen loading rate and their specific ANAMMOX activity (SAA) in different AnAOB enrichment studies are given in Table 9.6.

The AnAOB bacteria enrichment from a denitrifying sludge of FBR showed a nitrogen removal efficiency of 65% with SAA of 0.02 g Amm/g MLVSS/d using nitrate addition and sand as carrier (Graaf et al. 1995). Higher ANAMMOX activity with 0.06 to 0.08 g Amm/g MLVSS/d was achieved using the granular sludge of UASB from brewery wastewater treatment plant as seed for AnAOB bacteria which gave 60–75% nitrogen removal (Ahn et al. 2004) and granules from FBR as seed for AnAOB bacteria growth in SBR gave 78% (Strous et al. 1998). The different seed had varying start-up duration for the enrichment. The simultaneous partial nitrification, anaerobic ammonium oxidation and denitrification (SNAD) sludge from a full-scale leachate treatment unit with start-up of 561 days gave total nitrogen removal efficiency of 82% (Kwak et al. 2020). The denitrification sludge which had start-up of 100 and 105 days removed 0.609 and 2.090 kg $N/m^3/d$, respectively. The UASB sludge of 150 days removed 0.090 kg $N/m^3/d$ (Chamchoi and Nitisoravut 2007;

**Table 9.6** AnAOB enrichment studies in various ANAMMOX processes

| S. No | AnAOB seed for ANAMMOX process | Nitrogen loading rate (kg N/m$^3$/d) | Specific ANAMMOX activity (g N/g VSS/d) | Reference |
|---|---|---|---|---|
| 1. | Activated sludge | 2.6 | 0.30 | Fux et al. (2002) |
| 2. | Activated sludge | 1.6 | 0.92 | Lopez et al. (2008) |
| 3. | Activated sludge | 0.08 | 0.35 | Wang et al. (2009) |
| 4. | Denitrifying sludge | 58.5 | 1.6 | Tsushima et al. (2007) |
| 5. | Denitrifying sludge | 1.2 | 150 | Strous et al. (1997) |
| 6. | Denitrifying sludge | 4.8 | 0.02 | Graaf et al. (1995) |
| 7. | UASB | 0.015 | 0.64 | Tran et al. (2006) |
| 8. | UASB sludge | 2.2 | 0.08 | Ahn et al. (2004) |
| 9. | Anaerobic digester sludge | 0.231 | 0.43 | Bagchi et al. (2010) |
| 10. | Granules from FBR | 3.2 | 0.05 | Strous et al. (1998) |

Zhang et al. 2008). The main challenge in scale-up of the ANAMMOX process is the need for large-scale enrichment of AnAOB bacteria. A case study is presented for the AnAOB enrichment using municipal solid waste in large-scale reactors by controlling the influencing factors of ANAMMOX process for effective nitrogen removal.

## 9.4 Case Study on AnAOBs Enrichment Using MSW for ANAMMOX Process

The AnAOB enrichment using mined municipal solid waste as seed was conducted in large-scale anaerobic reactors for 250 days. The different factors affecting AnAOB population was controlled for good enrichment and for initiating the ANAMMOX process.

### 9.4.1 Materials and Methods

The AnAOB/ANAMMOX bacterial enrichment from municipal solid waste conducted in large-scale anaerobic batch reactors, run in duplicates, operated for 250 days. Mined MSW was utilised as seed. The operational features of AnAOB enrichment reactors are feed to seed ratio with 60/40, enrichment medium as feed (weight: 2472 g or volume: 3000 mL), mined MSW as seed (wet weight: 2112 g and dry weight of 1739 g of TS) and monitored every week. 40% reactor volume was left as headspace. Nitrogen gas was purged into the reactors for maintaining the anaerobic condition at the flow rate of 1 mL/min for half an hour. Biogas production measured using water displacement technique. Nutrients such as ammonium sulphate (1 mM) and sodium nitrite (1 mM) were added as ammonium and nitrite source, respectively. The enrichment units by fed batch mode. Intermittent addition of medium (without any nitrogen supplement) to maintain the reactor liquid level. Enrichment medium composition used in the study was as per the Graaf et al. (1996). The experiments were carried out at ambient temperature. The enrichment of AnAOB was monitored based on the nitrogen transformations, concentrations of hydrazine and hydroxylamine, FA, FNA and bacterial biomass development. The MLVSS and MLSS as per the standard methods (APHA 1998). Presence of AnAOB (or ANAMMOX) confirmed using specific ANAMMOX activity and bacterial morphology studies by scanning electron microscopy (SEM).

**Analytical procedure for characterization of seed and samples**

**Free ammonia and Free nitrous acid**

The FA and FNA in the reactors quantified as per Anthonisen et al. (1976) as given in the Eqs. (9.1) and (9.2).

$$FA\left(NH_3,\text{mg}/\text{L}\right)=17/14\times\left(\left(NH_4\right)\times10^{pH}\right)/\,e^{(6344/(273+t))}+10^{pH}) \tag{9.1}$$

$$FNA\left(HNO_2,\text{mg}/\text{L}\right)=46/14\times\left(\left(\text{NO}_2\right)/\,e^{(-2300/(273+t))}\times10^{pH}\right) \tag{9.2}$$

Where,

17/14 – molecular weight of the ammonia/atomic weight of nitrogen
46/14 – molecular weight of the nitrous acid/atomic weight of nitrogen

**Bacterial Morphology Studies** Surface morphologies of bacteria from reactors were observed by SEM (SEM S-3400 N, Manufactured by Hitachi/Quanta 200F, Manufactured by FEI, Germany). The moisture content of reactor samples was completely removed and treated with ethanol and coated with gold before subjecting it to SEM.

**Statistical Analyses** SPSS software package (version 20.0) was used for the statistical studies. Duplicability of the enrichment reactors was validated by Pearson's correlation analysis. The results were considered significant, when the significance coefficient lesser than 0.05.

### 9.4.2 *Results and Discussion*

AnAOB biomass enrichment for a large production volume with sustainable longer duration of 250 days was investigated in large-scale reactors (AnAOB reactor – 5 L capacity). The characteristics of the mined waste used as seed for ANAMMOX is given in Table 9.7. The moisture content of mined wastes was in the range of 20.8%, it was little higher due to rainy season sampling (July–September). Mined wastes had solids content of 79.2% of total solids containing 13.3% of VS, TOC of 5.3% and TIN of 340.8 mg/kg. Mined waste characteristics are comparable with other studies (Scheu and Bhattacharya 1997; Karthikeyan et al. 2007). Presence of macroelements such as carbon, nitrogen and oxygen, and microelements such as silicon, iron, aluminium, calcium, sodium, magnesium, phosphorus, sulphate, chlorine, potassium, zirconium and tellurium was found in mined MSW. These elements of mined MSW and its inorganic content support the growth of AOBs and AnAOBs (Erses et al. 2008; Bi et al. 2014). Mined MSW proved to be the most favourable seed and substrate for the enrichment (Sri Shalini et al. 2015).

Figure 9.2a depicts pH trend and temperature changes in the reactors. Alkaline pH of 7.0 to 8.6 and temperature was around 28 to 34 °C was observed. It showed optimum growth parameters of AnAOBs prevailed in reactors (Dongen et al. 2001; Strous et al. 2002; Guven et al. 2004; Jin et al. 2008; Li et al. 2009). Optimum pH condition was maintained in the reactor without any alkali or acid addition. Anoxic state was maintained in the reactor, however the initial DO (<0.5 mg/L) present was consumed by AOBs for nitrite production. Similar findings demonstrated by Tang et al. (2010) and Li et al. (2009). Conductivity and salinity used as monitoring

**Table 9.7** Physico-chemical characteristics of mined MSW (seed and substrate)

| Sl No | Parameters | Mined MSW samples |
|---|---|---|
| 1. | Moisture content (%) | 20.8 |
| 2. | Total solids (%) | 79.2 |
| 3. | Volatile solids (%) | 13.3 |
| 4. | Organic carbon (%) | 5.3 |
| 5. | pH | 7.0 |
| 6. | Conductivity (μS/cm) | 1561 |
| 7. | Water soluble COD (mg/kg) | 4042 |
| 8. | Water soluble ammonia-N (mg/kg) | 98 |
| 9. | Water soluble nitrite-N (mg/kg) | 31.0 |
| 10. | Water soluble nitrate-N (mg/kg) | 211.8 |
| 11. | Water soluble hydroxylamine (mg/kg) | 0.003 |

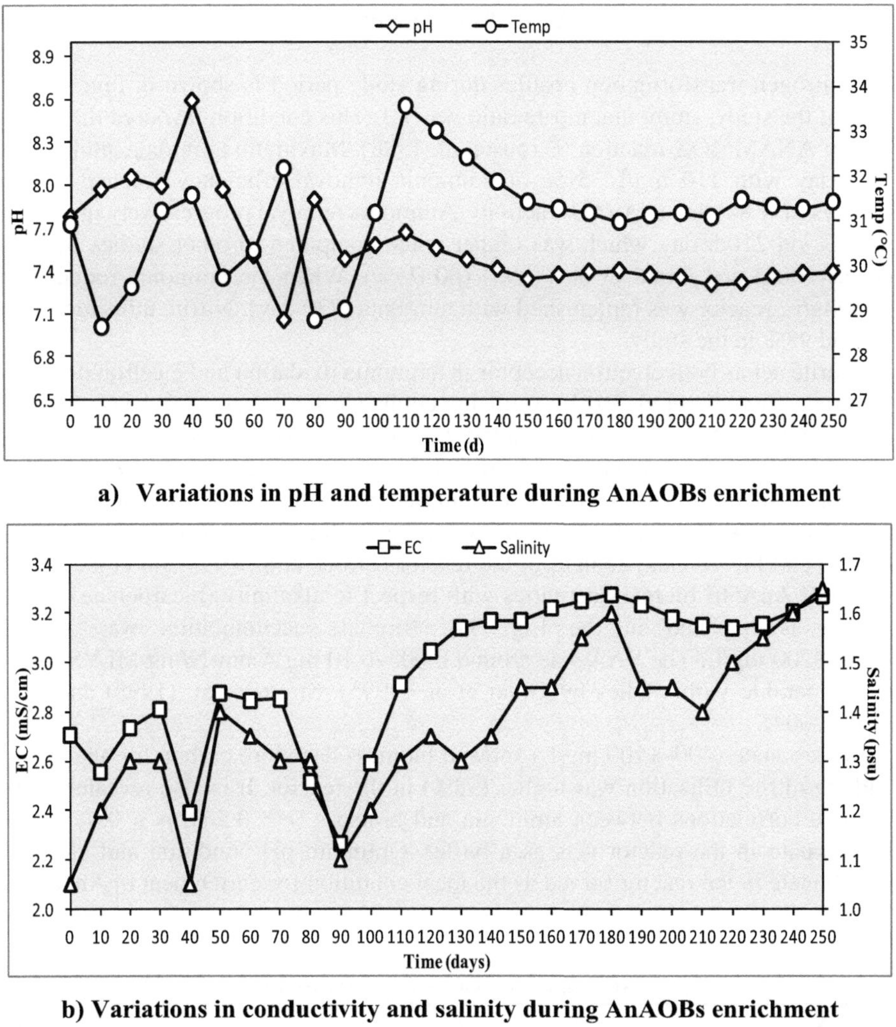


**Fig. 9.2** Variations in (**a**) pH and temperature, (**b**) conductivity and salinity during AnAOBs enrichment in large-scale reactors

parameters for AnAOB activity. The conductivity and salinity during the study period varied in the range of 2.3–3.3 mS/cm and 1.1 to 1.7 psu, respectively as depicted in the Fig. 9.2b. These conditions were favourable for AnAOB development (Terada et al. 2011; Fernandez et al. 2014). Positive correlations was seen between conductivity and salinity ($r = 0.882$, $p \leq 0.01$).

#### 9.4.2.1 Performance of AnAOBs Enrichment in Large-Scale Reactors

The nitrogen transformation profiles during study period is shown in Fig. 9.3. At start of the study, ammonia: nitrite ratio was 1:1. This condition favoured the initiation of ANAMMOX reaction (Strous et al. 1998). Maximum ammonia attained on 70th day with 110 mg/L. 51% of ammonia removal efficiency reached within 10 days (day 80) due to AnAOB activity. Ammonia removal progressively increased to 87% on 210th day which was higher when compared to other studies by Ahn et al. (2004) and Graaf et al. (1995) (60–75%). When the ammonia reduced to <25 mg/L, reactor was replenished with nutrients (220 day). Nitrite utilisation was around 98% in the study.

Nitrite act as both electron acceptor in ammonia oxidation and electron donor in $CO_2$ reduction (Jin et al. 2014). Reducing equivalents necessary for the $O_2$ fixation obtained from $NO_2$ to $NO_3$ oxidation (Graaf et al. 1996; Kartal et al. 2012). ANAMMOX development demonstrated by the simultaneous reduction with ammonia and nitrite with increase in the nitrate production. Nitrate during the study period was 4.6–24.1 mg/L. AnAOB biomass growth revealed from the nitrate formation and bicarbonate content in the reactor (Strous et al. 1998; Jin et al. 2014). Trend of AnAOB biomass activities with respect to alkalinity (bicarbonate) in the reactor is depicted in the Fig. 9.4. Biomass accumulation was around 1120–8200 mg/L. The SAA was around 0.005–0.10 mg Amm-N/mg MLVSS/d. It is comparable with studies by Graaf et al. (1995), Strous et al. (1998) and Ahn et al. (2004).

Bicarbonate (400–1100 mg/L) showed biomass has more carbon for anabolism and the nitrite utilisation was higher (98%) in the reactor. It is also revealed from positive correlations between ammonia and biomass ($r = 0.569$, $p \leq 0.01$). This bicarbonate in the reactor acts as a buffer. Optimum pH condition and elevated bicarbonate in the reactor served as the ideal condition for enrichment of AnAOBs.

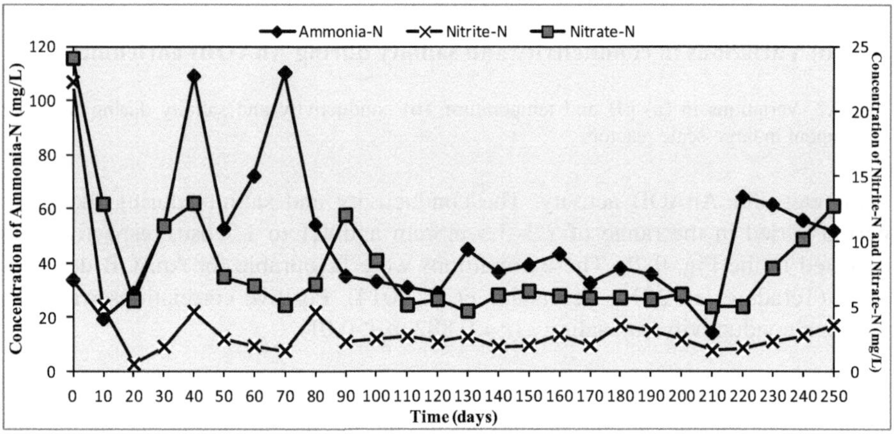


**Fig. 9.3** Nitrogen transformation profiles during large-scale enrichment of AnAOBs

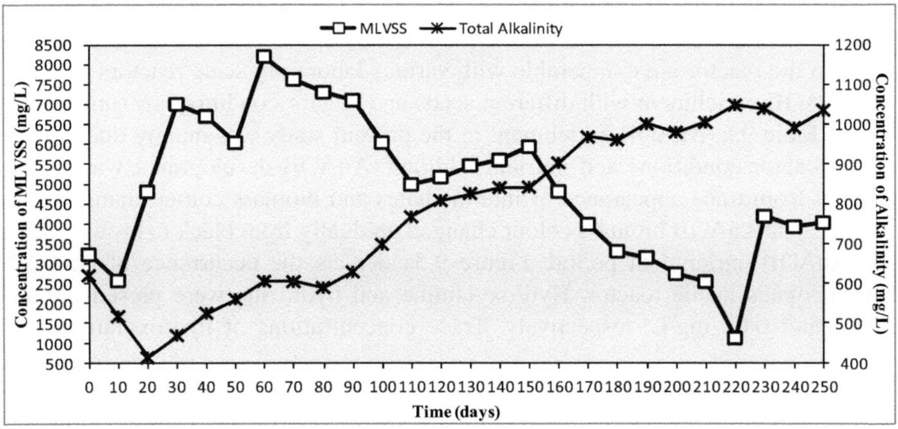


**Fig. 9.4** MLVSS and alkalinity variations during large-scale enrichment of AnAOBs

Hence, pH and alkalinity are considered as indicator for AnAOB development (Zhang and Zhou 2006; Yang et al. 2010; Jin et al. 2014). Bicarbonate never exceeded inhibitory concentration of 2000 mg/L for AnAOB when analysed in the reactor. This inhibitory concentration for AnAOB was reported by Liao et al. (2008). Organic content (COD <250 mg/L) originated from seed (mined MSW) led to the coexistence of AnAOB and other heterotrophic bacteria. Under anaerobic atmosphere with presence of nitrate and COD led to denitrifying bacteria in the reactor which removed COD (60.4%). Similar coexistence of AnAOB with heterotrophic population was observed in preliminary and bench-scale reactors. Such interactions of AnAOB with heterotrophic bacteria were focussed recently by many researchers (Tsushima et al. 2007; Molinuevo et al. 2009; Anjali and Sabumon 2014; Du et al. 2014; Li et al. 2014).

AnAOB coexistence with heterotrophs exhibited a positive effect in nitrogen removal. In the start-up of the reactor there existed a competition between them for nitrite. Since, feed medium in the reactor contained only inorganic carbon that favoured AnAOB existence rather than heterotrophs. This was illustrated from the growth of AnAOB population with 8200 mg/L. As the study progressed, lower population of heterotrophs survived due to unsuitable reactor conditions to grow. Hence, heterotrophs switched the main electron acceptor from COD to nitrate (due to deficiency of organic content) and consequently reduced nitrate to nitrite. This nitrite was further used as electron acceptor by AnAOB which was revealed from its increased AnAOB activity in the reactor (Figs. 9.3 and 9.4). Nitrate content controls the proliferation of heterotrophic population in the reactor. This was supported by Sahinkaya et al. (2011).

Environmental conditions (COD, nitrite, nitrate, ammonium, pH and temperature) have to be controlled to get a stability among the ANAMMOX and denitrification population (Molinuevo et al. 2009) in reactor. The optimum reactor conditions (pH 7.0 to 8.6; temperature 28 to 34 °C), higher ammonia removal efficiency

(87.3%) and nitrate accumulation (4.6–24.1 mg/L) in the reactor are indication of AnAOB biomass development in the reactors. These results of the AnAOB enrichment in the reactor are comparable with various laboratory scale reactors operated for AnAOB enrichment with different seeds and reactor conditions as summarised in the Table 9.6. AnAOB enrichment in the present study was mainly due to optimum reactor conditions and nutrient addition. AnAOB development was further evident from trace appearance of intermediates and biomass colour change in the reactor. The AnAOB biomass colour changed gradually from black to brown during the AnAOB enrichment period. Figure 9.5a depicts the occurrence of AnAOBs intermediates in the reactor. Hydroxylamine and hydrazine were present around 0.003 and 0.02 mg/L, respectively. Trace concentrations of hydroxylamine and

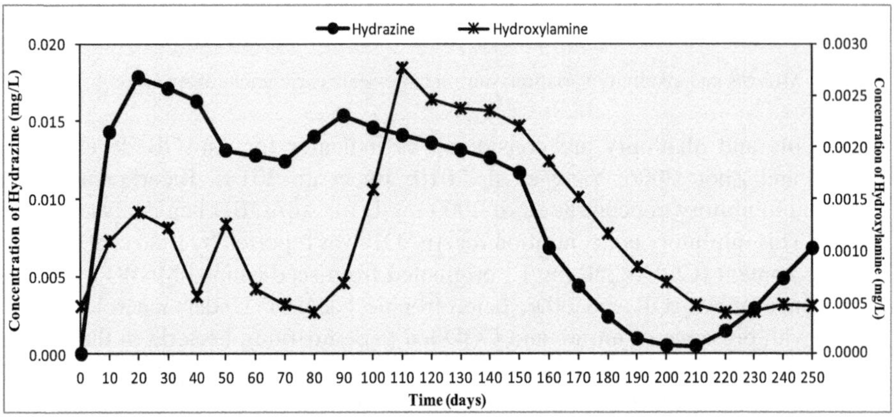


**a) Hydrazine and Hydroxylamine profile during AnAOBs enrichment**

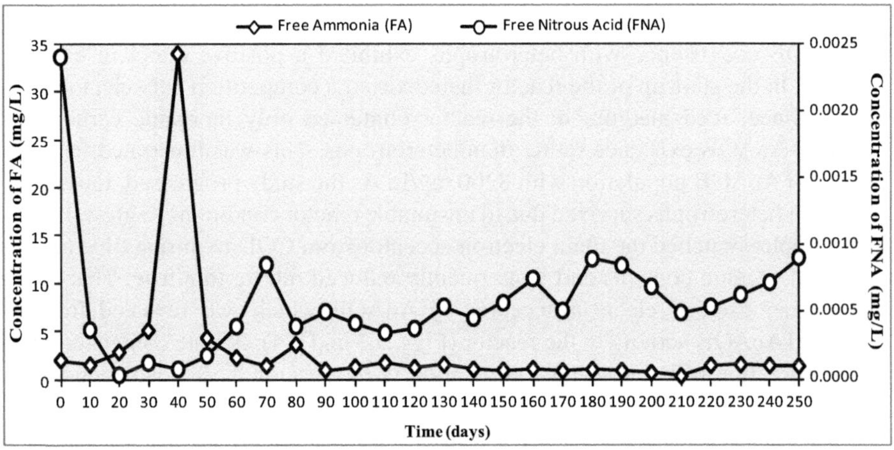


**b) Free Ammonia and Free Nitrous Acid profile during AnAOBs enrichment**

**Fig. 9.5** (**a**) Hydrazine and Hydroxylamine and (**b**) Free Ammonia and Free Nitrous Acid profile during AnAOBs enrichment in large-scale reactors hydrazine confirms the AnAOB growth in the reactor (Shivaraman and Shivaraman 2003; Third et al. 2005). Positive correlations exhibited between the intermediates ($r = 0.397, p \leq 0.05$).

Negative correlations exhibited between ammonia and hydroxylamine ($r = -0.407, p \leq 0.01$). This illustrated that oxidation of ammonia led to the formation of hydroxylamine. AnAOB biomass activity showed positive correlations with the intermediates (hydrazine: $r = 0.728, p \leq 0.01$). Optimum pH and temperature were key parameters for AnAOB population which was revealed from positive correlations of them with hydrazine ($r = 0.543, p \leq 0.01$) and hydroxylamine ($r = 0.476, p \leq 0.05$), respectively. The trend in the FA and FNA is as depicted in Fig. 9.5b. FA and FNA were around 0.5–34.0 mg/L and 0–0.002 mg/L, respectively (Anthonisen et al. 1976; Furukawa et al. 2009). Whereas, FNA was well below the inhibiting range of nitrifying bacteria (0.2 mg/L) (Anthonisen et al. 1976; Furukawa et al. 2009) and AnAOB (0.1 mg/L) (Puyol et al. 2014). NOB inhibition was observed only due to FA above 0.1 mg/L throughout the study period. Nitrifying population survival was according to the inhibitions of FA and FNA. Existence of AOBs and NOBs are required for AnAOB to grow in mixed cultures (Guven et al. 2004).

Optimistic relationship existed between AOB and AnAOB for maintaining reactor condition (consumption of oxygen by AOB) (Li et al. 2009; Tang et al. 2010). Likewise, they share common features like metabolising ammonia and hydroxylamine (Kartal et al. 2011). SAA of 0.10 mg Amm-N/mg MLVSS/d confirmed AnAOB as dominant population in the reactor.

#### 9.4.2.2 Confirmation Studies of AnAOBs Enrichment in Large-Scale Reactors

The Fig. 9.6 displays SEM images of AnAOB biomass from reactor on 250th day at different magnifications for observation of the biomass population dispersion. It demonstrated that biomass contained single and aggregates of *coccoid*-shaped cells with little interspaces between them. It also showed high compactness with a cauliflower-like feature as the dominant microorganism that was AnAOB organism. Similar findings were reported by several authors (Jetten et al. 1999; Ni et al. 2010). Presence of other microorganisms in the biomass along with spherical shaped or short-rod shaped bacteria with some filamentous bacteria was observed. Similar observation was reported by Liao et al. (2007).

SEM observation together with reactor performance authenticated the enrichment of AnAOBs using mined MSW as seed. The key issues addressed during the AnAOB enrichment studies are addition of nitrogen load and media, maintenance of anaerobic condition, control of FA and FNA below inhibitory range, continuous monitoring of nitrogen content and tracking the trace appearances of AnAOB intermediates (hydrazine and hydroxylamine). Enriched AnAOB biomass (2570–10,200 mg/L) from mined MSW was available as seed for conducting the ANAMMOX process for leachate treatment.

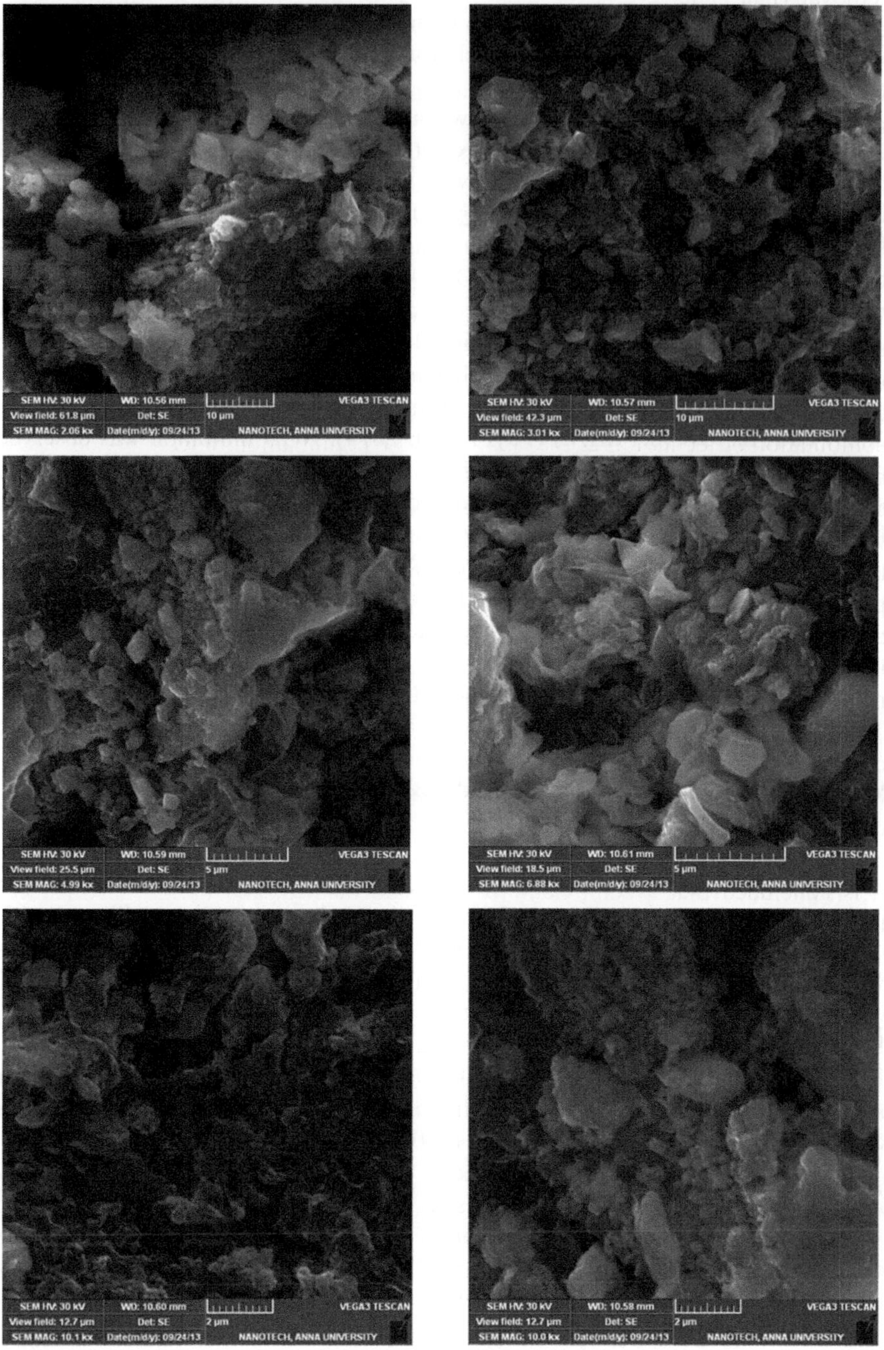


**Fig. 9.6** SEM photographs of enriched AnAOB biomass

## 9.5 Conclusions

The ammonia nitrogen persistence is remaining as ever-lasting pollutant, which is a critical issue in landfill treatment. Extensive studies on ammonia removal from leachate using conventional nitrification and denitrification, new methods of phototrophic and microbial electrochemical systems and other novel nitrogen removal methods of SHARON and ANAMMOX processes were explored in this chapter. Application of novel method of ANAMMOX process for the ammonia nitrogen removal from leachate is advantageous than the conventional processes based on environmental and economic aspects. The AnAOB (or ANAMMOX) bacterial enrichment for ANAMMOX process is essential for large-scale applications of landfill leachate treatment. In this chapter, a case study on AnAOB enrichment using mined municipal solid waste for ANAMMOX process was demonstrated. AnAOB biomass development was revealed based on nitrogen conversions of ammonia, nitrite and nitrate nitrogen, trace appearance in the intermediates of hydrazine and hydroxylamine, free ammonia, free nitrous acid, alkalinity and biomass build-up with MLVSS and MLSS content. AnAOB activity was demonstrated by the specific ANAMMOX activity (0.005–0.10 mg Amm-N/mg MLVSS/d) and, ammonia removal (87% ± 6) with simultaneous nitrite utilisation (98–100%) and subsequent nitrate production in the reactors.

**Acknowledgements** The corresponding author (Dr. S. Sri Shalini) gratefully acknowledges the financial support provided by the All India Council for Technical Education (AICTE-NDF), Government of India and Department of Science and Technology (DST), Ministry of Science and Technology, Government of India under Women Scientists Scheme-A (WOS-A).

## References

Abelling U, Seyfried CF (1992) Anaerobic–aerobic treatment of high strength ammonium wastewater-nitrogen removal using nitrite. Water Sci Technol 26:1007–1015

Ahemd FN, Lan CQ (2012) Treatment of landfill leachate using membrane bioreactors: a review. Desalination 287:41–54

Ahn YH, Hwang IS, Min KS (2004) Anammox and partial denitrification in anaerobic nitrogen removal from piggery waste. Water Sci Technol 49(5–6):145–153

Amokrane A, Comel C, Veron J (1997) Landfill leachate pretreatment by coagulation-flocculation. Water Res 31:2775–2782

Anfruns A, Gabarró J, Gonzalez-Olmos R, Puig S, Balaguer MD, Colprim J (2013) Coupling anammox and advanced oxidation-based technologies for mature landfill leachate treatment. J Hazard Mater 258–259:27–34

Anjali G, Sabumon PC (2014) Unprecedented development of anammox in presence of organic carbon using seed biomass from a tannery Common Effluent Treatment Plant (CETP). Bioresour Technol 153:30–38

Anthonisen AC, Loehr RC, Prakasam TBS, Srinath EG (1976) Inhibition of nitrification by ammonia and nitrous acid. J Water Pollut Control Fed 46:835–852

APHA (1998) Standard methods for the examination of water and wastewater, 20th edn. American Public Health Association/American Water Works Association/Water Environment Federation, Washington, DC. ISBN 0-87553-235-7

Bagchi S, Biswas R, Nandy T (2010) Alkalinity and DO as controlling parameters for ammonia removal through partial Nitritation and ANAMMOX in a single-stage bioreactor. J Ind Microbiol Biotechnol 37:871–876

Berge ND, Reinhart DR, Townsend TG (2005) The fate of nitrogen in bioreactor landfills. Crit Rev Environ Sci Technol 35(4):365–399

Bettazzi E, Caffaz S, Vannini C, Lubello C (2010) Nitrite inhibition and intermediates effects on Anammox bacteria: a batch-scale experimental study. Process Biochem 45(4):573–580

Bi Z, Qiao S, Zhou J, Tang X, Cheng Y (2014) Inhibition and recovery of Anammox biomass subjected to short-term exposure of Cd, Ag, Hg and Pb. Chem Eng J 244:89–96

Boonchai W (2004) Application of membrane bioreactor systems for landfill leachate treatment. Ph.D. thesis, Asian Institute of Technology, Thailand

Burton SAQ, Watson-Craik IA (1998) Ammonia and nitrogen fluxes in landfill sites: applicability to sustainable landfilling. Waste Manag Res 16:41–43

Capodaglio AG, Hlavínek P, Raboni M (2015) Physico-chemical technologies for nitrogen removal from wastewaters: a review. Rev Ambient Água 10(3):481–498

Chamchoi N, Nitisoravut S (2007) Anammox enrichment from different conventional sludges. Chemosphere 66:2225–2232

Chamchoi N, Nitisoravut S, Schmidt JE (2008) Inactivation of ANAMMOX communities under concurrent operation of anaerobic ammonium-N oxidation 7 (ANAMMOX) and denitrification. Bioresour Technol 99:3331–3336

Chen ZG, Wang XJ, Yang YY, Mirino MW, Yuan YL (2016) Partial nitrification and denitrification of mature landfill leachate u.sing a pilot-scale continuous activated sludge process at low dissolved oxygen. Bioresour Technol 218:580–588

Dongen U, Jetten MSM, van Loosdrecht MCM (2001) The SHARON-ANAMMOX process for treatment of ammonium rich wastewater. Water Sci Technol 44:153–160

Du R, Peng Y, Cao S, Wu C, Weng D, Wang S, He J (2014) Advanced nitrogen removal with simulataneous anammox and denitrification in sequencing batch reactor. Bioresour Technol 162:316–322

Erses AS, Onay TT, Yenigun O (2008) Comparison of aerobic and anaerobic degradation of municipal solid waste in bioreactor landfills. Bioresour Technol 99:5418–5426

Fernandez I, Bravo JI, Mosquera-Corral A, Pereira A, Campos JL, Mendez R, Melo LF (2014) Influence of the shear stress and salinity on Anammox biofilms formation: modelling results. Bioprocess Biosyst Eng 10:1955–1961

Furukawa K, Inatomi Y, Qiao S, Quan L, Yamamoto T, Isaka K (2009) Innovative treatment system for digester liquor using anammox process. Bioresour Technol 100(22):5437–5443

Fux C, Boehler M, Huber P, Brunner I, Siegrist H (2002) Biological treatment of ammonium-rich wastewater by partial nitritation and subsequent anaerobic ammonium oxidation (Anammox) in a pilot plant. J Biotechnol 99:295–306

Ganigue R, Gabarro J, Sanchez-Melsio A, Ruscalleda M, Lopez H, Vila X, Colprim J, Balaguer MD (2009) Long-term operation of a partial pilot plant treating leachate with extremel high ammonium concentration prior to an anammox process. Bioresour Technol 100:5624–5632

Ghosh S, Dairkee UK, Chowdhury R, Bhattacharya P (2017) Hydrogen from food processing wastes via photofermentation using Purple Non-sulfur Bacteria (PNSB) – a review. Energ Conver Manage 141:299–314

Graaf AA, Mulder A, Debruijn P, Jetten MSM, Robertson LA, Kuenen JG (1995) Anaerobic oxidation of ammonium is a biologically mediated process. Appl Environ Microbiol 61(4):1246–1251

Graaf AA, Brujin PD, Robertson LA, Jetten MSM, Kuenen JG (1996) Autotrophic growth of anaerobic ammonium-oxidising micro-organisms in a fluidized bed reactor. Microbiology 142:2187–2196

Guven D, van de Pas-Schoonen K, Schmidt MC, Strous M, Jetten MSM, Sozen S, Orhon D, Schmidt I (2004) Implementation of the anammox process for improved nitrogen removal. J Environ Sci Health - Toxic/Hazard Subst Environ Eng A39:1729–1738

Guven D, Dapena A, Kartal B, Schmid MC, Maas B, Pas-Schoonen K, Sozen S, Mendez R, Op den Camp HJM, MSM J-T, Strous M, Schmidt I (2005) Propionate oxidation by and methanol inhibition of anaerobic ammonium-oxidizing bacteria. Appl Environ Microbiol 71(2):1066–1071

Hassan M, Wei H, Qiu H, Su Y, Jaafry SWH, Zhan L, Xie B (2018) Power generation and pollutants removal from landfill leachate in microbial fuel cell: variation and influence of anodic microbiomes. Bioresour Technol 247:434–442

Hellinga C, Schellen AAJC, Mulder JW, Van Lossdrecht MCM, Heijnen JJ (1998) The Sharon process: an innovative method for nitrogen removal from ammonium-rich wastewater. Water Sci Technol 34:135–142

Henze M, Loosdrecht MCMV, Ekama G, Bridjanovic D (2008) Biological wastewater treatment. Modelling and Design, IWA Publishing, Principles, pp 139–151

Hippen A, Helmer C, Kunset S, Rosenwinkel KH, Seyfried CF (2001) Six years practical experience with aerobic/anoxic deammonification in biofilm systems. Water Sci Technol 44:39–48

Hoilijoki TH, Kettunen RH, Rintala JA (2000) Nitrification of anaerobically pretreated municipal landfill leachate at low temperature. Water Res 34:1435–1446

Hong X, Chen Z, Zhao C, Yang S (2017) Nitrogen transformation under different dissolved oxygen levels by the anoxygenic phototrophic bacterium Marichromatium gracile. World J Microbiol Biotechnol 33(6):113

Hülsen T, Barry EM, Lu Y, Puyol D, Keller J, Batstone DJ (2016) Domestic wastewater treatment with purple phototrophic bacteria using a novel continuous photo anaerobic membrane bioreactor. Water Res 100:486–495

Im JH, Woo HJ, Choi MW, Han KB, Kim CW (2001) Simultaneous organic and nitrogen removal from municipal landfill leachate using an anaerobic–aerobic system. Water Res 35:2403–2410

Janus HM, Vander RHF (1997) Don't reject the idea of treating reject water. Water Sci Technol 34(3–4):87–94

Jetten MSM, Strous M, van de Pas-Schoonen KT, Schalk J, van Dongen UGJM, van de Graaf AA, Logemann S, Muyzer G, van Loosdrecht MSM, Kuenena JG (1999) The anaerobic oxidation of ammonium. FEMS Microbiol Rev 22:421–437

Jin RC, Hu BL, Zheng P, Qaisar M, Hu AH, Islam E (2008) Quantitative comparison of stability of ANAMMOX process in different reactor configurations. Bioresour Technol 99:1603–1609

Jin R, Yu J, Ma C, Yang G, Zhang J, Chen H, Zhang Q, Ji Y, Hu B (2014) Transient and long-term effects of bicarbonate on the ANAMMOX process. Appl Microbiol Biotechnol 98:1377–1388

Kartal B, Kuenen JG, van Loosdrecht MCM (2010) Sewage treatment with anammox. Science 328:702–703

Kartal B et al (2011) Molecular mechanism of anaerobic ammonium oxidation. Nature 479:127–130

Kartal B, Niftrik L, Keltjens JT, Op den Camp HJ, Jetten MS (2012) Anammox – growth physiology, cell biology, and metabolism. Adv Microb Physiol 60:211–262

Karthikeyan OP, Swati M, Nagendran R, Joseph K (2007) Performance of bioreactor landfill with waste mined from a dumpsite. Environ Monit Assess 135:141–151

Khin T, Annachhatre AP (2004) Novel microbial nitrogen removal processes. Biotechnol Adv 22:519–532

Kim D, Ryu HD, Kim MS, Kim J, Lee SI (2007) Enhancing struvite precipitation potential for ammonia nitrogen removal in municipal landfill leachate. J Hazard Mater 146(1–2):81–85

Kjeldsen P, Barlaz MA, Rooker AP, Baun A, Ledin A, Christensen TH (2002) Present and long-term composition of MSW landfill leachate: a review. Crit Rev Environ Sci Technol 32(4):297–336

Kulikowska D, Klimiuk E (2008) The effect of landfill age on municipal leachate composition. Bioresour Technol 99:5981–5985

Kurniawan TA, Lo WH, Chan GY (2006) Physico-chemical treatments for removal of recalcitrant contaminants from landfill leachate. J Hazard Mater B 129:80–100

Kwak W, Rout PR, Lee E, Bae J (2020) Influence of hydraulic retention time and temperature on the performance of an anaerobic ammonium oxidation fluidized bed membrane bioreactor for low-strength ammonia wastewater treatment. Chem Eng J 386:123992
Lackner S, Gilbert EM, Vlaeminck SE, Joss A, Horn H, van Loosdrecht MCM (2014) Full-scale partial nitritation/anammox experiences – an application survey. Water Res 55:292–303
Li XR, Du B, Fu HX, Wang RF, Shi JH, Wang Y, Jetten MSM, Quan ZX (2009) The bacterial diversity in an anaerobic ammonium-oxidising (anammox) reactor community. Syst Appl Microbiol 32:278–289
Li Y, Huang Z, Ruan W, Ren H, Zhao M (2014) ANAMMOX performance, granulation, and microbial response under COD disturbance. J Chem Technol Biotechnol 90(1):139–148
Li J, Peng Y, Zhang L, Liu J, Wang X, Gao R, Pang L, Zhou Y (2019) Quantify the contribution of anammox for enhanced nitrogen removal through metagenomic analysis and mass balance in an anoxic moving bed biofilm reactor. Water Res 160:178–187
Liang Z, Liu JX (2007) Control factors of partial nitritation for landfill leachate treatment. J Environ Sci (China) 19:523–529
Liang Z, Liu J (2008) Landfill leachate treatment with a novel process: anaerobic ammonium oxidation (Anammox) combined with soil infiltration system. J Hazard Mater 151:202–212
Liao D, Li X, Yang Q, Zhao Z, Zeng G (2007) Enrichment and granulation of Anammox biomass started up with methanogenic granular sludge. World J Microbiol Biotechnol 23:1015–1020
Liao D, Li X, Yang Q, Zeng G, Guo L, Yue X (2008) Effect of inorganic carbon on anaerobic ammonium oxidation enriched in sequencing batch reactor. J Environ Sci (China) 20(8):940–944
Lin L, Chan GYS, Jiang BL, Lan CY (2007) Use of ammoniacal nitrogen tolerant microalgae in landfill leachate treatment. Waste Manag 27(10):1376–1382
Liu JB, Zhang HB, Zhang PY, Wu Y, Gou XY, Song YH, Tian ZY, Zen GM (2017) Two-stage anoxic/oxic combined membrane bioreactor system for landfill leachate treatment: pollutant removal performances and microbial community. Bioresour Technol 243:738–746
Long Y, Lao H, Hu L, Shen D (2008) Effects of *in situ* nitrogen removal on degradation/stabilization of MSW in bioreactor landfill. Bioresour Technol 99:2787–2794
Lopez H, Puig S, Ganigue R, Ruscalleda M, Balaguer MD, Colprim J (2008) Start-up and enrichment of a granular anammox SBR to treat high nitrogen load waste- waters. J Chem Technol Biotechnol 83:233–241
Lotti T, Kleerebezem R, Abelleira-Pereira JM, Abbas B, van Loosdrecht MCM (2015) Faster through training: the anammox case. Water Res 81:261–268
Loukidou MX, Zouboulis AI (2001) Comparison of two biological treatment process using attachedgrowth biomass for sanitary landfill leachate treatment. Environ Pollut 111:273–281
Luo JH, Qian GR, Liu JY, Xu ZP (2015) Anaerobic methanogenesis of fresh leachate from municipal solid waste: a brief review on current progress. Renew Sustain Energy Rev 49:21–28
Madigan MT (1995) Microbiology of nitrogen fixation by anoxygenic photosynthetic bacteria. Anoxygenic photosynthetic bacteria. Springer, New York, pp 915–928
Malamis S, Katsou E, Di Fabio S, Bolzonella D, Fatone F (2014) Biological nutrients removal from the supernatant originating from the anaerobic digestion of the organic fraction of municipal solid waste. Crit Rev Biotechnol 34(3):244–257
Martins CL, Fernandes H, Costa RHR (2013) Landfill leachate treatment as measured by nitrogen transformations in stabilization ponds. Bioresour Technol 147:562–568
Metcalf and Eddy (2003) Wastewater engineering treatment and reuse. The McGraw-Hill Companies, Inc
Miao L, Wang K, Wang S, Zhu R, Li B, Peng Y, Weng D (2014) Advanced nitrogen removal from landfill leachate using real-time controlled three-stage sequence batch reactor (SBR) system. Bioresour Technol 159:258–265
Miao L, Wang SY, Li BK, Cao TH, Xue TL, Peng YZ (2015) Advanced nitrogen removal via nitrite using stored polymers in a modified sequencing batch reactor treating landfill leachate. Bioresour Technol 192:354–360

Miao L, Wang SY, Cao TH, Peng YZ, Zhang M, Liu ZY (2016) Advanced nitrogen removal from landfill leachate via Anammox system based on Sequencing Biofilm Batch Reactor (SBBR): effective protection of biofilm. Bioresour Technol 220:8–16

Miao L, Yang GQ, Tao T, Peng YZ (2019) Recent advances in nitrogen removal from landfill leachate using biological treatments—a review. J Environ Manage 235:178–185

Molinuevo B, Garcia MC, Karakashev D, Angelidaki I (2009) Anammox for ammonia removal from pig manure effluents: effect of organic matter content on process performance. Bioresour Technol 100:2171–2175

Mulder A (2003) The quest for sustainable nitrogen removal technologies. Water Sci Technol 48:67–75

Mulder A, van de Graaf AA, Robertson LA, Kuenen JG (1995) Anaerobic ammonium oxidation discovered in a denitrifying fluidized bed reactor. FEMS Microbiol Ecol 16:177–183

Mustafa EM, Phang SM, Chu WL (2012) Use of an algal consortium of five algae in the treatment of landfill leachate using the high-rate algal pond system. J Appl Phycol 24:953–963

Ni SQ, Lee PH, Fessehaie A, Gao BY, Sung S (2010) Enrichment and biofilm formation of anammox bacteria in a non-woven membrane reactor. Bioresour Technol 101:1792–1799

Ni B, Ruscalleda M, Smets BF (2012) Evaluation on the microbial interactions of anaerobic ammonium oxidizers and heterotrophs in Anammox biofilm. Water Res 46:4645–4652

Nie W-B, Xie G-J, Ding J, Peng L, Lu Y, Tan X, Yue H, Liu B-F, Xing D-F, Meng J, Han H-J, Ren N-Q (2020) Operation strategies of n-DAMO and Anammox process based on microbial interactions for high rate nitrogen removal from landfill leachate. Environ Int 139:105596

Ozturk I, Altinbas M, Koyuncu I, Arikan O, Gomec-Yangin C (2003) Advanced physico-chemical treatment experiences on young municipal landfill leachates. Waste Manag 23:441–446

Paredes D, Kuschk P, Mbwette TSA, Stange F, Muller RA, Koser H (2007) New aspects of microbial nitrogen transformations in the context of wastewater treatment – a review. Eng Life Sci 7:13–25

Peng Y, Zhang SJ, Zeng W, Zheng SW, Mino T, Satoh H (2008) Organic removal by denitritation and methanogenesis and nitrogen removal by nitritation from landfill leachate. Water Res 42:883–892

Phan TN, Ha NB, Nguyen TTM, Bui XT, Nguyen PD (2014) Application of a partial nitritation and anammox system for the old landfill leachate treatment. Int Biodeter Biodegr 95:144–150

Puyol D, Carvajal-Arroyo JM, Sierra-Alvarez R, Field JA (2014) Nitrite (not free nitrous acid) is the main inhibitor of the anammox process at common pH conditions. Biotechnol Lett 36:547–551

Qi Z, Zhang XH, Boon N, Bossier P (2009) Probiotics in aquaculture of China-current state, problems and prospect. Aquaculture 290:15–21

Reginatto V, Teixera RM, Pereira F, Schmidell W, Furigo A, Menes R, Etchebehere C, Soares HM (2005) Anaerobic ammonium oxidation in bioreactor treating slaughterhouse wastewater. Braz J Chem Eng 22:593–600

Renou S, Givaudan JG, Poulain S, Dirassouyan F, Moulin P (2008) Landfill leachate treatment: review and opportunity. J Hazard Mater 150(3):468–493

Rozendal RA, Jeremiasse AW, Hamelers HVM, Buisman CJN (2008) Hydrogen production with a microbial biocathode. Environ Sci Technol 42:629–634

Ruscalleda M, Lopez H, Ganigue R, Puig S, Balaguer MD, Colprim J (2008) Heterotrophic denitrification on granular anammox SBR treating urban landfill leachate. Water Sci Technol 58:1749–1755

Sahinkaya E, Dursun N, Kilic A, Demirel D, Uyanik S, Cinar O (2011) Simultaneous heterotrophic and sulfur-oxidizing autotrophic denitrification process for drinking water treatment: control of sulfate production. Water Res 45(20):6661–6667

Scheu M, Bhattacharyya JK (1997) In: Coad A (ed) Reuse of decomposed waste, lessons from India in solid waste management. Department of International Development, UK Government

Shi Y, Hu SH, Lou JQ, Lu PL, Keller J, Yuan ZG (2013) Nitrogen removal from wastewater by coupling Anammox and methane-dependent denitrification in a membrane biofilm reactor. Environ Sci Technol 47:11577–11583

Shivaraman N, Shivaraman G (2003) Anammox – a novel microbial process for ammonium removal. Curr Sci 84:1507–1508

Shoener BD, Bradley IM, Cusick RD, Guest JS (2014) Energy positive domestic wastewater treatment: the roles of anaerobic and phototrophic technologies. Environ Sci: Processes Impacts 16:1204–1222

Show PL, Pal P, Leong HY, Juan JC, Ling TC (2019) A review on the advanced leachate treatment technologies and their performance comparison: an opportunity to keep the environment safe. Environ Monit Assess 191:227

Silpa K, Lisa CY, Perinaz B-T, Frank VW (2018) What a Waste 2.0: a global snapshot of solid waste management to 2050, urban development, World Bank Group, Washington, DC. https://openknowledge.worldbank.org/handle/10986/30317. Accessed 1 Aug 2021

Sri Shalini S, Joseph K (2012) Nitrogen management in landfill leachate: application of SHARON, ANAMMOX and combined SHARON-ANAMMOX process – a review. Waste Manag 32:2385–2400

Sri Shalini S, Joseph K (2018) Combined SHARON and ANAMMOX processes for ammoniacal nitrogen stabilization in landfill bioreactors. Bioresour Technol 250:723–732

Sri Shalini S, Joseph K, Wong JWC (2015) Aerobic and anaerobic ammonium oxidising bacterial enrichment from municipal solid waste. Int J Environ Sci Technol 12:1107–1122

Statom RA, Thyne GD, McCray JE (2004) Temporal changes in leachate chemistry of a municipal solid waste landfill cell in Florida, USA. Environ Geol 45:982–991

Strous M, Van Gerven E, Kuenen JG, Jetten MSM (1997) Effects of aerobic and microaerobic conditions on anaerobic ammonium-oxidising (anammox) sludge. Appl Environ Microbiol 63:2446–2448

Strous M, Heijnen JJ, Kuenen JG, Jetten MSM (1998) The sequencing batch reactor as a powerful tool for the study of slowly growing anaerobic ammonium-oxidizing microorganisms. Appl Microbiol Biotechnol 50:589–596

Strous M, Kuenen JG, Jetten M (1999) Key physiology of anaerobic ammonium oxidation. Appl Environ Microbiol 65:3248–3250

Strous M, Kuenen GJ, Fuerst JA, Wagner M, Jetten MSM (2002) The anammox case – a new experimental manifesto for microbiological eco-physiology. Antonie Van Leeuwenhoek 81:693–702

Sun H, Yang Q, Peng Y, Shi X, Wang S, Zhang S (2010) Advanced landfill leachate treatment using a two-stage UASB-SBR system at low temperature. J Environ Sci (China) 22(4):481–485

Tang CJ, Zheng P, Mahmood Q, Chen JW (2009) Startup and inhibition analysis of the Anammox process seeded with anaerobic granular sludge. J Ind Microbiol Biotechnol 36:1093–1100

Tang CJ, Zheng P, Zhang L, Chen JW, Mahmood Q, Chen XG, Hu BL, Wang CH, Yu Y (2010) Enrichment features of anammox consortia from methanogenic granules loaded with high organic and methanol contents. Chemosphere 79(6):613–619

Tatsi AA, Zouboulis AI (2002) A field investigation of the quantity and quality of leachate from a municipal solid waste landfill in a Mediterranean climate (Thessaloniki, Greece). Adv Environ Res 6:207–219

Tchobanoglous G, Burton FL, Stensel HD (2003) Wastewater engineering: treatment and reuse, 4th edn. McGraw-Hill Higher Education, Boston, 1819

Tengrui L, Harbawi AFA, Bo LM, Jun Z, Long XY (2007) Characteristics of nitrogen removal from old landfill leachate by sequence batch biofilm reactor. Am J Appl Sci 4:211–214

Terada A, Zhou S, Hosomi M (2011) Presence and detection of anaerobic ammonium-oxidizing (anammox) bacteria and appraisal of anammox process for high-strength nitrogenous wastewater treatment: a review. Clean Techn Environ Policy 13(6):759–781

Third KA, Paxman J, Schmid M, Strous M, Jetten MSM, Cord-Ruwisch R (2005) Enrichment of ANAMMOX from activated sludge and its application in the CANON process. Microb Ecol 459:236–244

Tran HT, Park YJ, Cho MK, Kim DJ, Ahn DH (2006) Anaerobic ammonium oxidation process in an upflow anaerobic sludge reactor with granular sludge selected from an anaerobic digester. Biotechnol Bioprocess Eng 11:199–204

Tsarpali V, Kamilari M, Dailianis S (2012) Seasonal alterations of landfill leachate composition and toxic potency in semi-arid regions. J Hazard Mater 233:163–171

Tsushima I, Kindaichi T, Okabe S (2007) Quantification of anaerobic ammonium oxidising bacteria in enrichment cultures by real-time PCR. Water Res 41:785–794

Wang T, Zhang H, Yang F, Liu S, Fu Z, Chen H (2009) Start-up of the Anammox process from the conventional activated sludge in a membrane bioreactor. Bioresour Technol 100(9):2501–2506

Wang Z, Peng YZ, Miao L, Cao TH, Zhang FZ, Wang SY, Han JH (2016) Continuous-flow combined process of nitritation and ANAMMOX for treatment of landfill leachate. Bioresour Technol 214:514–519

Weigelhofer G, Hein T, Bondar-Kunze E (2018) Phosphorus and nitrogen dynamics in riverine systems: human impacts and management options. In: Schmutz S, Sendzimir J (eds) Riverine ecosystem management. Aquatic Ecology Series 8, Springer, Cham

Winkler MKH, Straka L (2019) New directions in biological nitrogen removal and recovery from wastewater. Curr Opin Biotechnol 57:50–55

Winkler M, Coats ER, Brinkman CK (2011) Advancing post-anoxic denitrification for biological nutrient removal. Water Res 45:6119–6130

Wiszniowski J, Robert D, Surmacz-Gorska J, Miksch K, Weber JV (2006) Landfill leachate treatment methods: a review. Environ Chem Lett 4:51–61

Yang J, Zhang L, Fukuzaki Y, Hira D, Furukawa K (2010) High-rate nitrogen removal by the Anammox process with a sufficient inorganic carbon source. Bioresour Technol 101:9471–9478

Zhang H, Zhou S (2006) Treating leachate mixture with anaerobic ammonium oxidation technology. J Cent South Univ Technol 13:663–667

Zhang SJ, Peng YZ, Wang SY, Zheng SW, Guo J (2007) Organic matter and concentrated nitrogen removal by shortcut nitrification and denitrification from mature municipal landfill leachate. J Environ Sci (China) 19(6):647–651

Zhang L, Zheng P, Tang C, Jin R (2008) Anaerobic ammonium oxidation for treatment of ammonium-rich wastewaters. J Zhejiang Univ Sci B 9:416–426

Zhang T, Ding L, Ren H (2009) Pretreatment of ammonium removal from landfill leachate by chemical precipitation. J Hazard Mater 166(2–3):911–915

Zhang C, Lai C, Zeng GM, Huang DL, Yang CP, Wang Y, Zhou YY, Cheng M (2016) Efficacy of carbonaceous nanocomposites for sorbing ionisable antibiotic sulfamethazine from aqueous solution. Water Res 95:103–112

Zhao X, Zhou Y, Huang S, Qiu DY, Schideman L, Chai XL, Zhao YC (2014) Characterization of microalgae-bacteria consortium cultured in landfill leachate for carbon fixation and lipid production. Bioresour Technol 156:322–328

Zhu R, Wang S, Li J, Wang K, Miao L, Ma B, Peng Y (2013) Biological nitrogen removal from landfill leachate using anaerobic–aerobic process: Denitritation via organics in raw leachate and intracellular storage polymers of microorganisms. Bioresour Technol 128:401–408

# Chapter 10
# Valorization of Solid Waste from Landfill Activities

**M. S. S. R. Tejaswini, Pankaj Pathak, Sankar Ganesh Palani, and Prangya Ranjan Rout**

**Abstract** Open dumping is the most common way of managing solid wastes in developing countries. It causes environmental pollution and global climate change besides posing a threat to human health. Mismanagement of solid wastes results in emissions of toxic pollutants and greenhouse gases. The increase in population has projected ~3.4 billion tons of solid wastes generation by 2050. Also, developing nations face new challenges in managing solid wastes due to limited landfill space. The landfill wastes are heterogeneous in nature and reservoirs for precious and heavy metals, organics, plastics in the form of waste. It can be separated and converted as a resource and known as valorization resources through urban mining. Urban mining of various types of landfill wastes helps in the conservation of natural resources as well as increases the economic value of the disposed of wastes. It also ensures maximum recovery from the wastes that can be used as a secondary raw material for production purposes. Accordingly, this chapter determines environmental impacts associated with the landfill wastes and identifies different technologies that would help in the conversion of waste into resources through urban mining.

**Keywords** Landfill waste · Impact · Circular economy · Urban mining · Sustainability

---

M. S. S. R. Tejaswini
Department of Environmental Science, SRM University, Amaravati, Andhra Pradesh, India

P. Pathak (✉)
Department of Environmental Science, SRM University Andhra Pradesh, Guntur, Andhra Pradesh, India
e-mail: pankajpathak18@gmail.com

S. G. Palani
Environmental Biotechnology Laboratory, Department of Biological Sciences, BITS Pilani, Hyderabad Campus, Hyderabad, Telangana, India

P. R. Rout
Department of Biotechnology, Thapar Institute of Engineering & Technology, Patiala, Punjab, India



P. Pathak, S. G. Palani (eds.), *Circular Economy in Municipal Solid Waste Landfilling: Biomining & Leachate Treatment*, Radionuclides and Heavy Metals in the Environment, https://doi.org/10.1007/978-3-031-07785-2_10

## 10.1 Introduction

The incessant growth of solid wastes is a rising concern all over the world. Majority of the solid wastes generated from various sources viz., household, industries, shopping complexes, agriculture, are being directly open dumped into landfills in an unscientific manner. World Bank (2018) has projected that ~3.4 billion tons of solid waste would be generated by 2050 with increasing population. Open dumping of solid wastes in landfills are creating significant environmental and health risk as a result of toxic and greenhouses gas emitting from them. Moreover, countries are also lacking adequate amount of landfill space ($43 \times 10^7$ $m^3$) by 2050 to accommodate large heaps of waste in it (Pujara et al. 2020). Generally, improper collection and informal recycling are causing the mismanagement of solid wastes. Besides, the lack of efficient methods to treat solid waste is creating new challenges in developing nations (Pujara et al. 2020).

Solid wastes comprise biodegradable and nonbiodegradable recyclable and inert wastes. The biodegradable component of solid waste is usually processed through composting or anaerobic digestion, but handling and managing non-biodegradable inert wastes is becoming a challenge in today's society. The inert wastes are mainly produced from construction and demolition (C&D) activities, processing of glass, plastic packaging, electrical equipment, agricultural sector etc. About $12 \times 10^6$ tons of inert waste is generated annually from the construction and demolition sector in developing nations (Townsend 2017). On other hand, developed nations such as America and Europe are generating ~11.3 million tons and 12.3 million tons of e-wastes, respectively (Fu et al. 2019; Ramanayaka et al. 2020). Plastic inert wastes are breaks down into microplastics and harmful chemicals like heavy metals and polyhalogenated organics are leached out from e-waste that causes huge impacts to the environment (Chabhadiya et al. 2021a, b). However, due to the energy potential and presence of valuable metals, these inert wastes have been identified as a potential resource that can be recovered through the urban mining process (Chand et al. 2021). Urban mining maximizes the resource and economic worth of waste present in landfills by promoting resource conservation.

It has been reported that urban mining of e-waste can help in the optimum recovery of precious metals such cobalt, silver, gold etc., through different processes of metallurgical operations viz., pyro, hydro and bio-metallurgical operations (Pathak and Chabhadiya 2021; Pathak and Pathak 2021). Besides, advanced technology has been developed that helps to convert e-waste into nanoparticles and extract nanometals from wastes. Thus, it is beneficial to the ecology and economy (Tesfaye et al. 2017; Musariri et al. 2019). Urban mining of plastic wastes also helps in obtaining high value-added products through large-scale valorization. Valorization of plastic inert wastes into Refuse-Derived Fuel (RDF) and its utilization in road construction helps reduce raw materials, minimize environmental pollution, and create a pathway for closing the loop for discarded plastic wastes (Bao et al. 2020; Govani et al. 2020). Apart from this, plastics degradation is done by a catalytic depolymerization process that helps in achieving sustainable waste management (Chamas et al. 2020).

Furthermore, sustainable building materials such as green bricks, recycled concrete aggregates are produced from inert construction and demolishing wastes. These materials could be further utilized in landscaping, cementitious mortars, construction of pavements and new buildings.

The effective management of inert solid wastes from landfills involves applications of both advanced technologies and policy intervention in a sustainable manner (Pathak and Prangya 2021). Inert materials are imperative and significant part of solid wastes. However, if they are adequately valorized, they will help resource recovery and reduce the environmental burden (Pathak and Sinha 2021). Thus, this chapter explains about different types of landfill wastes and their environmental impacts. It also explains the valorization of inert wastes through urban mining.

## 10.2 Different Types of Landfill Waste

The global composition of different types of landfill waste is shown in Fig. 10.1. Globally, it has been reported that less than 20% of landfill wastes are recycled, approximately 11% of the wastes are incinerated, 5% of wastes are subjected to composting (World Bank 2018). On the other hand, a significant portion of solid wastes, almost 33%, goes for open dumping. In India, solid waste comprises 40–60% of organic waste, 20% recyclable waste and the remaining 30% in the form of inert waste and inorganic materials. The primary source of inert waste generation is construction and demolition activities, processing of glassware, and household sites. This chapter mainly explains construction and demolition (C&D) wastes, plastic, and e-wastes present in the landfill.

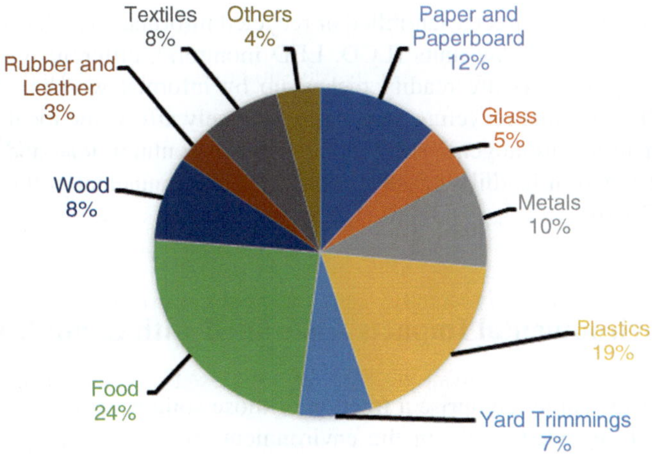


**Fig. 10.1** Global composition of landfill wastes

### *10.2.1 Construction & Demolition Wastes*

Developed nations like the United States of America and Australia produce 40–70 million tons of C&D waste annually (Townsend 2017). This type of waste comprises sand, gravel, concrete, limestone, metal, and wood. On the other hand, India and Africa, a developing nation, produce 10–15 million tons of C&D wastes annually. The composition of waste varies depending on construction activities, including demolition and renovation of old buildings, dismantling and expansion of roads. Geological parameters and natural calamities also play an essential role in contributing to the generation of this waste.

### *10.2.2 Inert Waste from MSW*

The inert materials from municipal solid waste (MSW) mainly comprise crushed glasses, ceramics, packaging bottles, jars, bulbs, windows, food wrappers, soiled cardboard, non-recyclable plastics (zip-lock bags, cheese wrappers, styrofoam), which often gets contaminated with other plastic and bio-medical waste, making challenging to manage the waste. Apart from this, landfill sites also contain a portion of household medical waste (discarded medicines, sharps, needles), and it remains on the landfill as inert waste. Landfill plastic waste mainly comes from MSW.

### *10.2.3 Metal Scraps from E-Waste*

In developing nations, less than 5% of e-waste is recycled formally, and the remaining 95% of e-waste is directly landfilled or recycled informally (Pathak et al. 2019). Cathode ray tubes, display units (LCD, LED monitors), different types of audio components, processors are readily picked up by informal workers (Pathak and Pathak 2021). Informal recycling of e-waste adversely affects the local workers by exposing them to carcinogenic substances such as strontium, lead and mercury—the e-waste found in landfills causing substantial environmental pollution (Pathak and Gupta 2020).

## 10.3 Environmental Impacts Associated with Landfill Waste

Landfill wastes mainly comprise a fraction of those solid wastes, which cannot be recycled and degraded easily in the environment. However, a major portion of wastes (biodegradable and nonbiodegradable) generated from various sources directly gets dumped in the landfills unscientifically, leading to environmental pollution and climate change. Moreover, organic waste on the landfill gets degraded and generates leachate that directly enters the ground and causes geoenvironmental problems. Apart from this, producing potent greenhouse gases like methane and carbon dioxide, significant amounts of volatile organic compounds, particulate matter, trace amounts of nitrogen, hydrogen, ammonia are also produced from landfill sites causing air pollution (Cremiato et al. 2018). These gases get often mix with an enormous amount of inert waste causes the firing in the landfill sites and neighboring habitats. The Deonar landfill in Mumbai, India, witness a huge fire in 2016. Almost 6000 million tons of waste per day were subjected to massive fire and showed insignificant and improper waste treatment facilities. It took more than a week to control the fire, besides residents suffered from breathing problems and skin rashes. Landfill sites are also known for attracting a wide range of vectors responsible for causing several diseases such as malaria, dengue, Jaundice, and chlorea (Pujara et al. 2020). The quality of leachate is affected by many factors like age of landfill, type of inert waste, precipitation etc. In general, as the age of landfill increases, chemical oxygen demand (COD) of the leachate decreases and the concentration of ammonia increases abruptly. The leachate with higher concentrations of ammonia when percolates through the landfill contaminate the groundwater and decrease the soil fertility, thereby affecting the growth of vegetation. Often, e-waste in developing countries is handled unsafely and in an unscientific manner by informal workers (Srivastava and Pathak 2020). The various health hazards associated with e-waste are summarized in Table 10.1.

## 10.4 Valorization of Landfill Waste

The process of converting wastes into resources or a more valuable product is known as valorization. Solid wastes from the landfill can be valorized into different products by recycling, reusing, urban mining, and landfill mining, as shown in Fig. 10.2. Sustainable landfill waste management involves the systematic collection and transportation of waste from the source, followed by the segregration and sorting of waste at the material recovery facilities (MRFs). Recyclable wastes are treated through physical, chemical, or tertiary recycling methods. On the other hand, biodegradable wastes are converted to either compost through aerobic or biogas through anaerobic digestion (Tejera et al. 2019). The remaining non-recyclable wastes are valorized through the process of urban mining. Urban mining maximizes the resource and economic worth of waste present in landfills by promoting resource conservation. It aims to convert waste materials into a resource that can be used as a secondary raw material for industrial production.

**Table 10.1** Impact on Human Health caused by leaching of toxic substances present in E-waste

| E-wastes components | Toxic metals | Limit ppm | Disease caused by the exposure to the above permissible limit |
|---|---|---|---|
| Ceramic capacitors, switches, batteries | Ag* | 5.0 | Excessive amount causing blue pigments on the body, damages brain, lung, liver, kidney |
| Gallium arsenide is used in light-emitting | As** | 5.0 | Chronic effect and causes skin disease and lung cancer and impaired nerve signaling |
| Electron tube, lubricant, fluorescent lamp, CRT gun | Ba** | <100 | Causes brain swelling, muscle weakness, damage to the heart |
| Power supply boxes, motherboard | Be** | 0.75 | Causes lung cancer, berylliosis, skin disease, carcinogens |
| PCBs, casing, PVC cables | Br** | 0.1 | Thyroid gland damage, hormonal issues, skin disorder, DNA damages, hearing loss |
| PCBs, battery, CRTs, semiconductors, infrared detectors, printer ink, toners | Cd** | 1.0 | Pose a risk of irreversible impacts on human health, particularly the kidney |
| Printed circuit boards (PCBs) | CN** | <0.5 | Cyanide poisoning, >2.5 ppm may cause to coma and death |
| Plastic computer hosing, cabling, hard discs, colorant in pigments | Cr(VI)** | 5.0 | Toxic in the environment, causing DNA damage and permanent eye impairment |
| Batteries, LCD, switches, backlight bulbs or lamps | Hg** | 0.2 | Damages brain, kidney and foetuses |
| Mobile, telephone, batteries | Li* | <10# | Diarrhoea, vomiting, drowsiness, muscular weakness |
| Batteries, semiconductor, CRT, PCB | Ni* | 20.0 | Causes allergic reaction, bronchitis, reduces lung function, lung cancers |
| Transistor, LED lead-acid battery, solder, CRT, PCBs, florescent tubes | Pb*** | 5.0 | Damages brain, nervous system, kidney, and reproductive system, causes acute and chronic effects on human health |
| CRT glass, plastic computer housing and a solder alloy | Sb** | <0.5 | Carcinogen, causing stomach pain, vomiting, diarrhoea and stomach ulcer |
| Fax machine, photoelectric cells | Se** | 1.0 | High concentration causes selenosis |
| CRT, batteries | Sr*** | 1.5 | Somatic as well the genetic changes due to this cancer in bone, nose, lungs, skin |
| Batteries, luminous substances | Zn** | 250.0 | Nausea, vomiting, pain, cramps and diarrhoea |
| Cooling units and insulation foam | CFCs** | <1.0 (8 h/day) | Impacts on the ozone layer which can lead to greater incidence of skin cancer |
| Transformer, capacitor, condensers | PCBs** | 5.0 | PCB causes cancer in animals and can lead to liver damage in human |
| Monitors, keyboard, cabling and plastic computer housing | PVC** | 0.03 | Hazardous and toxic air contaminants, release of HCl causes respiratory problems |

Source: Adapted from Pathak et al. (2017)

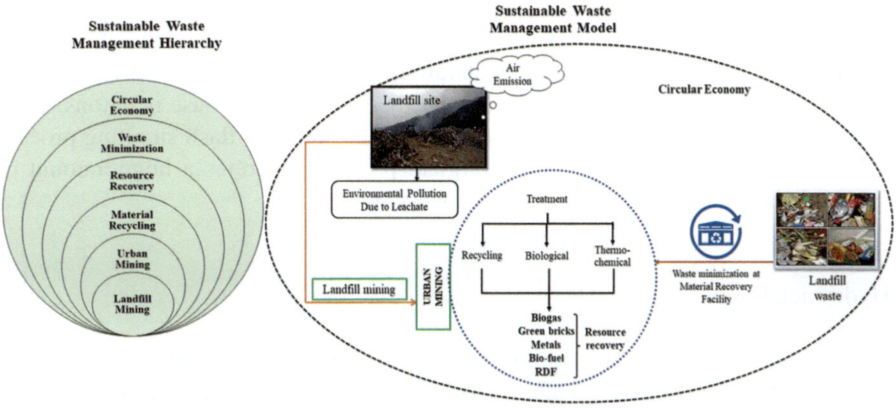


**Fig. 10.2** Sustainable model for the valorization of landfill waste

## *10.4.1 Urban Mining of E-Waste*

E-waste is a mixture of different types of metals, alloys, polymetallic compounds and is thus, considered as the best way of resource recovery than primary ores. In order to ensure maximum recovery of precious metals, the recycling process of the waste involves sequential steps of pretreatment, separation by physical and chemical methods, and purification by different metallurgical processes (pyro/ hydro/bio).

### 10.4.1.1 Physical Methods

Metallic and non-metallic components of e-waste are separated through physical separation methods, ensuring no loss of valuable substances. Dismantling is performed further through automatic or semi-automatic mechanical processes (Hsu et al. 2019). Desoldering is carried out, where the separate units are heated in a tin furnace at a high temperature. The remaining wastes are subjected to crushing, shredding and pulverization, which is done with the help of ball and disc milling (Tesfaye et al. 2017). This overall process of physical dismantling turns waste into powder. Magnetic separation (drum separators) separates ferrous metals from non-ferrous metals. In contrast, electrostatic separation helps in the potential recovery of copper and aluminum from cables and electrical wires (Lu and Xu 2016).

### 10.4.1.2 Metal Recovery Methods

Pyrometallurgy, hydrometallurgy and biometallurgy are the most commonly employed methods for recovering precious metals, as shown in the following.

### Pyrometallurgy

It involves series of steps such as incineration, smelting (1000 °C and 1300 °C) in blast furnaces, sintering, drossing followed by different gas-phase reactions. Lead and copper are separated from other metals by the bath and flash smelting process (Ma 2017). However, the pyrometallurgical process generates a huge amount of slag and also loses metals during the recovery process.

### Hydrometallurgy

It involves acid or alkaline leaching of e-waste, followed by solvent extraction (ion exchange and adsorption), which helps to extract metals (Agrawal et al. 2012; Khaliq et al. 2014). Conventionally, mineral acids viz., nitric, sulphuric, and hydrochloric acid are used in the leaching process based on the extraction efficiency of metals (Pathak et al. 2021). Organic acids such as citric acid, oxalic acid, and malic acid are green acids, commonly used to extract metals from e-waste (Chabhadiya et al. 2021a, b; Fu et al. 2019). Different organic and inorganic acids used for the recovery of metals have been summarized in Table 10.2.

**Table 10.2** Urban mining of precious and base metals using inorganic and organic acids

| Acid | Used E-waste | Recovered metal | References |
|---|---|---|---|
| 2.0 M, HCl | Printed circuit boards (PBCs) of mobile phones | Au (5.0%), Cu (94.91%) | Guo et al. (2016) and Yang et al. (2017) |
| 0.1 M, HCl | PBCs of mobile phones | Au (93.06%), Cu (0.58%) | Hsu et al. (2019) |
| 2–5 M $HNO_3$ | TV board scraps | Cu (88.5%–99.9%), Ag (14%–48%) | Tesfaye et al. (2017) |
| 0.5 M $H_2SO_4$ | Lithium-ion batteries | Mn (100%), Co (99%), Ni (85%) | Yang et al. (2017) |
| 5.5 M HCl | PBCs | Sn (95.97%) | Innocenzi et al. (2017) |
| DL-malic acid | LIBs | Li (96%), Ni (99%) | Khaliq et al. (2014) |
| Citric acid | LIBs | Co (95%), Li (97%) | Fu et al. (2019) |
| Benzenesulfonic acid | LIBs | Li(99.58%), Co (96.53%) | Musariri et al. (2019) |
| Oxalic acid | LEDs | Li (96%), Ni (99%) | Rocchetti et al. (2013) |

Source: Adapted from Ramanayaka et al. (2020)

Bio-metallurgy

Bio-leaching and bio-sorption are the two steps followed in biometallurgy. Ferrooxidans and thiooxidans are micro-organisms that can solubilize the metal sulfides from waste through direct or indirect mechanisms (Vakilchap et al. 2016). These methods are considered to be low-cost, environmentally friendly processes that can be carried out at room temperature. Pseudomonas putida is a type of bacteria that can leach approximately 48% of Au from e-waste. Other micro-organisms, such as Yarrowia lipolytica, have been discovered to carry out bio-leaching of gold nanoparticles (Ma et al. 2017).

### *10.4.2 Urban Mining of Construction Waste*

Construction and demolition wastes are usually managed through a process of waste minimization, reutilization and recycling. Reusing C&D waste ensures that a significant amount of the materials is converted as a resource (Townsend 2017). The materials that can be readily reused include wood, concrete, and steel, as shown in Fig. 10.3. Apart from this, reusing waste materials provides environmental benefits (reducing pollution) and economic benefits (Pathak and Srivastava 2021).

Further, it has been observed that after the demolition of old buildings and roads, concrete is frequently being disposed of in landfills. However, these materials can be recycled and used to make bricks, aggregates, paving blocks etc. The process of in-situ recycling involves crushing the waste material, followed by grinding in the milling machine. Catalyst is then added to the fine material in a pan mixer to produce green bricks (Li et al. 2017). Bricks of different sizes are dried in the sun for at least four weeks and later used for construction purposes. This type of process

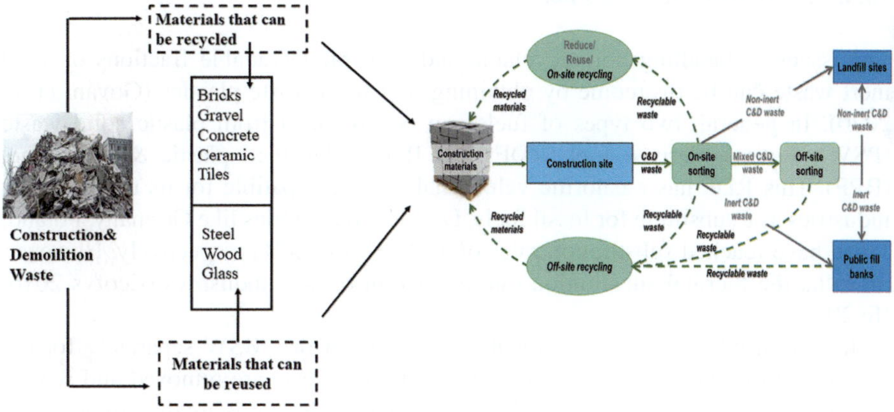


**Fig. 10.3** Valorization of construction and demolition wastes

**Table 10.3** Different types of CDW and their end products

| Material | Process | End-use/recycled product |
|---|---|---|
| Demolition waste | Crushed and sorted | Recycled aggregate |
| Clay bricks and roof tiles | Cleaned, crushed and pulverized | Reused for masonry<br>Mixed with lime to produce mortar |
| Reinforced concrete | Crushed, sorted and steel bars removed | Recycled concrete aggregate |
| Ceramic tiles | Cleaned and crushed | Flooring, cladding |
| Natural stone slabs | Sorted, crushed | Cladding, aggregate |
| Timber beams, doors etc. | Cleaned | Reused as beams, doors etc. |
| Mixed demolition waste (asphalt, bricks, concrete) | Crushed and hot mixed | Fill material |
| Asphalt paving | Crushed and cold mixed | Road construction excluding wearing course |
| Gypsum plasterboard | Cleaned, crushed, recycled | Reuse as boards, soil conditioner, new gypsum products. |

Source: Adapted from CPHEEO (2016)

involves saving transportation costs. The different types of C&D wastes and their recycled products are listed in Table 10.3.

Apart from the production of green bricks, C&D wastes are utilized for the production of recycled concrete aggregates (RCA). The shape of RCA produced depends upon the type of crusher used. RCAs generally have lower density, high absorption capacity, and tensile strength compared to conventional aggregates (Shi et al. 2016). The production process includes crushing, sorting, pre-sizing, screening, and elimination of contaminants.

## *10.4.3 Urban Mining of Inert Plastic Waste*

### 10.4.3.1 Refuse-Derived Fuel

The issues of handling non-recyclable and non- biodegradable fractions of solid inert waste can be overcome by obtaining fuel from waste plastics (Govani et al. 2020). In general, two types of fuels can be produced from plastic solid waste (PSW): Refuse- derived fuel (RDF) and Refuse-densified plastic & paper fuel (RPF). This RDF has a calorific value, making it compatible for its utilization in industries as a substitute for fossil fuels. Developed countries like Germany, Poland, Japan have reached substitution rates of 100%, 80%, 60%, respectively. However, in India, the thermal substitution rate is ~4% in cement industries (Ecofys 2016; Ifc 2017).

It is desirable to collect, segregate and transport dry MSW separately for the production of RDF. Recyclable fractions from dry waste are removed and sent to respective reprocessing facilities. The non-recyclable fractions are then subjected to screening, shredding, separation by air density, and blending as shown in following Fig. 10.4.

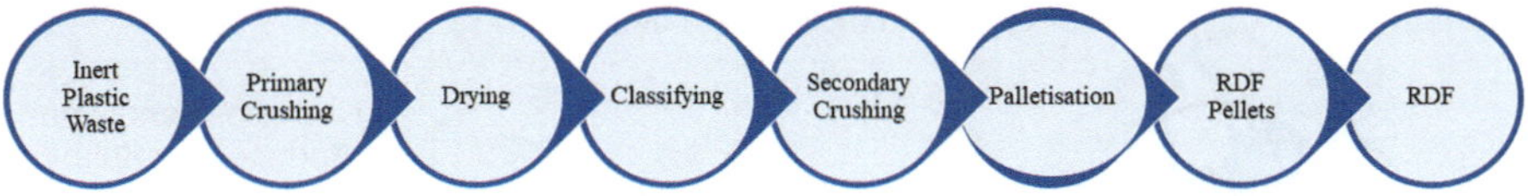


**Fig. 10.4** The RDF production process

#### 10.4.3.2 Bio-augmentation of Plastics

In most developing countries, landfilling, incineration, and recycling are commonly used to treat plastic wastes. However, landfilling and incineration of plastics have environmental hazards associated with them, which have been discussed in Sect. 10.2. Therefore, it was reported that bio-degradation of plastics can also be an efficient way of treating inert plastic wastes (Ho et al. 2018; Taniguchi et al. 2019). Literature studies showed that, in general, polyethylene is resistance towards degradation owing to its characteristics such as long carbon chains, high hydrophobic level, and mechanical properties (Tejaswini et al. 2022). Therefore, to make polyethylene susceptible to microbial attack, modification is required at the crystalline level with the addition of two additives, starch, and pro-oxidant. The susceptibility towards microbes can be increased by exposing it to ultraviolet (UV) radiation, pretreating it with nitric acid for six days at 80 °C, which helps activate inert materials. When polyethylene is blended with starch, the material becomes hydrophilic, which is easily catalyzed by amylase enzymes that provide the pathway for hydrobiodegradation by micro-organisms, whereas pro-oxidant (transition metal) additive helps in carrying out oxidative degradation of polyethylene.

It has been observed that when disposable LDPE pieces were blended with 6% starch and buried in the soil, fungal strains such as Aspergillus terreus and Penicillum sp. could form biofilm on the surface of LDPE and decrease its hydrophobicity after ten months. The surface properties (weak and eroded) changed along with the 0.2% weight loss in 10 years when LDPE was inoculated with Fusarium species in the soil. However, whitening of the degraded area, considerable weight loss, formation of small holes in PE films were observed after 32 years in aerobic soil.

Further, Taniguchi et al. (2019) observed that enzymes like PETase and MHETase, isolated from Ideonella sakaiensis are able to carry out PET hydrolysis together at 30 °C for 96 h, into terephthalic acid and ethylene glycol. Degradation of PET can be increased by incorporating aliphatic monomers such as polymethylene adipate terephthalate (PTMAT) and polybutylene adipate terephthalate (PBAT), that provide weak ester linkages to the polymer thereby, increasing the susceptibility towards hydrolysis by micro-organisms. Nowadays, synthetic biology is applied to obtain microbial cell factories by applying recombinant DNA technology, which could depolymerize plastic wastes more efficiently, as shown in Fig. 10.5. Furthermore, these depolymerized products are utilized in the production of secondary products.

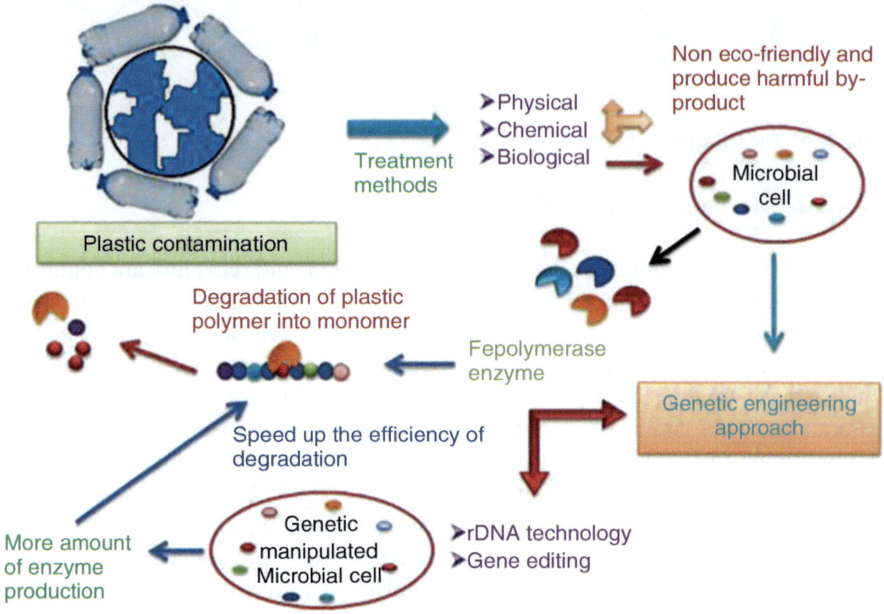


**Fig. 10.5** Methods to increase bio-degradation of plastic waste. (Adapted from Jaiswal et al. 2020)

#### 10.4.3.3 Plastic to Road Construction

Utilization of plastic for construction purposes is another sustainable way of managing post-discarded plastic wastes, as shown in Fig. 10.6. Polystyrene, Polypropylene, Polyethylene are commonly used for making roads (Govani et al. 2020). Plastic wastes are generally shredded and passed through 4.75 mm sieve, ensuring better spread and coating with aggregate. After shredding into the required size, plastic wastes are mixed with hot stone at a temperature between 150 °C to 170 °C. At this temperature, bitumen gets diffused through molten plastic layers either dry or wet processes (Chabhadiya et al. 2021a, b). It has been observed that making roads with plastic wastes reduces the cost of construction and maintenance.

## 10.5 Conclusions

This chapter emphasized various environmental impacts associated with the open dumping of inert wastes in landfills. It causes serious health issues and poses a threat to the environment if not managed properly. If inert solid wastes are identified as a potential resource, they can be exploited for several environmental and economic benefits. Urban mining from e-waste helps in recovery of precious metals and valuable products besides providing revenue generations and green jobs in the

**Fig. 10.6** Road Constructions Process with Bitumen Plastic Mix Aggregates

future. On the other hand, the best method of handling construction and demolition wastes is by reutilization and reusing it from the source of generation, thereby reducing transportation costs. Plastic inert wastes are managed sustainably by converting them into refuse-derived fuel and their utilization for construction purposes. However, inert solid wastes in the landfills can be managed effectively with advancements in technologies and the proper legislation rules and regulations in developing nations.

## References

Agrawal A, Pankaj P, Mishra D, Sahu KK (2012) Solvent mediated interaction for the selective recovery of cadmium from Ni-Cd spent battery. J Mol Liq 173:77–84

Bao Z, Lee WMW, Lu W (2020) Implementing on-site construction waste recycling in Hong Kong: barriers and facilitators. Sci Total Environ 747:10141091

Chabhadiya K, Pujara Y, Patel HT, Govani G, Pathak P, Mashru D (2021a) Different treatment technologies used for municipal solid waste management in India: material & energy recovery, waste management policies and practices in BRICS nations. CRC Press of Taylor & Francis group. eBook ISBN9781003007579

Chabhadiya K, Srivastava RR, Pankaj P (2021b) Two-step leaching process and kinetics for an eco-friendly recycling of critical metals from spent Li-ion batteries. J Environ Chem Eng 9(3):105232

Chamas A, Hyunjin M, Jiajia, Yang Q, Tarnuma T, Jang JH, Mahdi O, Susannah L, Scott, Suh S (2020) Degradation rates of plastics in the environment. ACS Sustain Chem Eng 8(9):3494–3511

Chand S, Rout PR, Pathak P (2021) Basic concepts, potentials and challenges of urban mining sustainable approaches of urban mining for waste management and resource recovery. CRC Press of Taylor & Francis Group, in print

Cremiato R, Mastellone ML, Tagliaferri C, Zaccariello L, Lettieri P (2018) Environmental impact of municipal solid waste management using life cycle assessment: the effect of anaerobic digestion, material recovery and secondary fuels production. Renew Energy 124:180–188

Ecofys (2016) Market opportunities for use of alternative fuels in cement plants across the EU: assessment of drivers and barriers for increased fossil fuel substitution in three EU member states: Greece, Poland and Germany. Available from : https://coprocessamento.org.br/wp-content/uploads/2019/09/Ecofys_Report_Market_Opportunities_Coprocessing_20160501.pdf. Accessed 17 Aug 2021

Fu Y, He Y, Qu L, Feng Y, Li J, Liu J et al (2019) Enhancement in leaching process of lithium and cobalt from spent lithium-ion batteries using benzenesulfonic acid system. Waste Manag 88:191–199

Govani J, Patel H, Pujara Y, Chabhadiya K, Pathak P (2020) Technological upgradation and advancement for treatment of plastic waste in developing countries. In: Hussain C (ed) Handbook of environmental materials management. Springer, Cham. ISBN 978-3-319-58538-3

Guo Y, Li F, Zhu H, Li G, Huang J, He W (2016) Leaching lithium from the anode electrode materials of spent lithium-ion batteries by hydrochloric acid (HCl). Waste Manag 51:227–233. https://doi.org/10.1016/j.wasman.2015.11.036

Ho BT, Roberts TK, Lucas S (2018) An overview on biodegradation of polystyrene and modified polystyrene: the microbial approach. Crit Rev Biotechnol 38:308–320. https://doi.org/10.1080/07388551.2017.1355293

Hsu E, Barmak K, West AC, Park AHA (2019) Advancements in the treatment and processing of electronic waste with sustainability: a review of metal extraction and recovery technologies. Green Chem 21(5):919–936

IFC (2017) Increasing the use of alternative fuels at cement plants: international best practices, report. Accessed 30 Aug 2021

Innocenzi V, De Michelis I, Veglio F (2017) Design and construction of an industrial mobile plant for WEEE treatment: investigation on the treatment of fluorescent powders and economic evaluation compared to other e-wastes. J Taiwan Inst Chem Eng 80:769–778. https://doi.org/10.1016/j.jtice.2017.09.019

Jaiswal S, Babita S, Pratyoosh S (2020) Integrated approaches in microbial degradation of plastics. Environ Technol Innov 17:100567

Khaliq A, Rhamdhani M, Brooks G, Masood S (2014) Metal extraction processes for electronic waste and existing industrial routes: a review and australian perspective. Resources 3(1):152–179. https://doi.org/10.3390/resources3010152

Li Y, Zhou H, Su L, Hou H, Dang L (2017) Investigation into the application of construction and demolition waste in urban roads. Adv Mat Sci Eng 2017

Lu Y, Xu Z (2016) Precious metals recovery from waste printed circuit boards: a review for current status and perspective. Resour Conserv Recycl 113:28–39

Ma Y, Tang H, Lin X, Jiang T, Li G, Li Z et al (2017) Pyrometallurgical recovery of platinum group metals from spent catalysts. JOM 69(9):1553–1562

Manual on Municipal Solid Waste Management, Central Public Health and Environmental Engineering Organisation (2016) Available from http://cpheeo.gov.in. Accessed 17 Aug 2021

Musariri B, Akdogan G, Dorfling C, Bradshaw S (2019) Evaluating organic acids as alternative leaching reagents for metal recovery from lithium-ion batteries. Miner Eng 137:108–117. https://doi.org/10.1016/j.mineng.2019.03.027

Pathak P, Chabhadiya K (2021) Recycling of rechargeable batteries: a sustainable tool for urban mining. Handbook of solid waste management. Springer, Singapore. https://doi.org/10.1007/978-981-15-7525-974-1

Pathak P, Gupta DK (2020) Strontium contamination in the environment. Volume series 88, handbook of environmental chemistry. Springer Nature Switzerland. ISBN 978-3-030-15314-4

Pathak VM, Navneet (2017) Review on the current status of polymer degradation: a microbial approach. Bioresour Bioprocess 4:15

Pathak P, Pathak S (2021) Electronic waste: an emerging contaminant in the geo-environment, emerging contaminants in the environment. Elsevier, in print

Pathak P, Prangya R (2021) Urban mining for waste management and resource recovery: sustainable approaches. CRC Press, Taylor & Francis. in press

Pathak P, Sinha S (2021) Valorisation of waste tires into fuel and energy, advanced technology for the conversion of waste into fuels and chemicals. Elsevier. 978-0-323-90150-5
Pathak P, Srivastava RR (2021) Alternate energy resources: a way to sustainable modern society, volume series 99, Book Series: Handbook of Environmental Chemistry. Springer Nature Switzerland. ISBN 978-3-030-57922-7
Pathak P, Rajiv RS, Ojasvi (2019) Chapter 5 – environmental management of E-waste. Electronic Waste Management and Treatment Technology, pp 103–132
Pathak P, Singh VK, Chabhadiya K (2021) Sequential leaching of strategic metals from LNCM cathode material using organic and mineral acid lixiviant. J of Metals 73:1386–1394
Pujara Y, Govani J, Patel HT, Chabhadiya K, Vaishnav K, Pathak P (2020) Waste-to-energy: suitable approaches for developing countries. Alternative Energy Resources – The Way to a Sustainable Modern Society. The Handbook of Environmental Chemistry. Springer, Berlin, Heidelberg
Ramanayaka S, Keerthanan S, Vithanage M (2020) Urban mining of E-waste: treasure hunting for precious nanometals. Handbook of Electronic Waste Management: International Best Practices and Case Studies, pp 19–54. https://doi.org/10.1016/B978-0-12-817030-4.00023-1
Rocchetti L, Veglio F, Kopacek B, Beolchini F (2013) Environmental impact assessment of hydrometallurgical processes for metal recovery from WEEE residues using a portable prototype plant. Environ Sci Technol 47(3):1581–1588
Shi C, Li Y, Zhang J, Li W, Chong L, Xie Z (2016) Performance enhancement of recycled concrete aggregateea review. J Clean Prod 112:466–472
Solid Waste Management (2018) World Bank. Available on http://www.worldbank.org/en/topic/urbandevelopment/brief/solid-waste-management. Accessed 13 Aug 2021
Srivastava RR, Pathak P (2020) Chapter 4 – Policy issues for efficient management of E-waste in developing countries. Handbook of Electronic Waste management, pp 81–99, Elsevier, ISBN 9780128170304
Taniguchi I, Yoshida S, Hiraga K, Miyamoto K, Kimura Y, Oda K (2019) Biodegradation of PET: current status and application aspects. ACS Catal 9:4089–4105. https://doi.org/10.1021/acscatal.8b05171
Tejaswini MSSR, Pathak P, Ramkrishna S, Ganesh SP (2022) A comprehensive review on integrative approach for sustainable management of plastic waste and its associated externalities. Sci Total Environ 825:153973
Tejera J, Ruben M, Daphne H, Iñigo U, Carlos N, Ángeles B (2019) Treatment of a mature landfill leachate: comparison between homogeneous and heterogeneous photo-Fenton with different pretreatments. Water 11(9):1849
Tesfaye F, Lindberg D, Hamuyuni J, Taskinen P, Hupa L (2017) Improving urban mining practices for optimal recovery of resources from e-waste. Miner Eng 111:209–221
Townsend GT (2017) Benefits of construction and demolition debris recycling in the United States. Available from https://cdrecycling.org/. Accessed 3 Sept 2021
Vakilchap F, Mousavi SM, Shojaosadati SA (2016) Role of Aspergillus Niger in recovery enhancement of valuable metals from produced red mud in Bayer process. Bioresour Technol 218:991–998
Yang Y, Xu S, He Y (2017) Lithium recycling and cathode material regeneration from acid leach liquor of spent lithium-ion battery via facile co-extraction and coprecipitation processes. Waste Manag 64:219–227. https://doi.org/10.1016/j.wasman.2017.03.018

# Chapter 11
# Biomined and Fresh Municipal Solid Waste as Sources of Refuse Derived Fuel

**Atun Roy Choudhury, Lakshmi Prasad Boyina, D. Laxman Kumar, Neha Singh, Sankar Ganesh Palani, Mohammad Mehdizadeh, M. V. Praveen Kumar, A. Leelavathi, B. Koteswara Rao, S. U. Abitha Begum, and Kayibanda Patrick**

**Abstract** With scarce land availability and ever-soaring quantity of municipal solid waste (MSW), recovery and recycling of the waste stream can only be the savior. An integrated system that facilitates recovery in the form of nutrients, materials, and energy is the need of the hour. Conventional energy sources are getting scarcer with aggressive consumption in industries such as thermal power plants, metal and cement industries etc. While the combustible fraction of MSW colloquially known as refuse-derived fuel (RDF) can supplement the demand up to a considerable extent. Additionally, the application of RDF significantly lowers the operational expenditure and carbon footprint. Source-specific characterization, quality assessment, and promotion of RDF as a promising fuel for cement manufacturing plants

---

A. R. Choudhury
Chadwick's FSM Laboratory, Shed no. 37, Banka BioLoo Limited, Hyderabad, India

Department of Biological Sciences, Birla Institute of Technology and Science, Hyderabad, Telangana, India
e-mail: atnroy10@gmail.com

L. P. Boyina · B. K. Rao
Institute of Science and Technology, Jawaharlal Nehru Technological University, Hyderabad, India

Department of Environment & Sustainable Development, Environment Protection Training and Research Institute (EPTRI), Hyderabad, India

D. L. Kumar
Institute of Science and Technology, Jawaharlal Nehru Technological University, Hyderabad, India

N. Singh
Chadwick's FSM Laboratory, Shed no. 37, Banka BioLoo Limited, Hyderabad, India

S. G. Palani
Environmental Biotechnology Laboratory, Department of Biological Sciences, Birla Institute of Technology and Science, Pilani, Hyderabad Campus, Hyderabad, Telangana, India



P. Pathak, S. G. Palani (eds.), *Circular Economy in Municipal Solid Waste Landfilling: Biomining & Leachate Treatment*, Radionuclides and Heavy Metals in the Environment, https://doi.org/10.1007/978-3-031-07785-2_11

were the primary objectives of the study. The investigation revealed the waste characteristics have been significantly varied between the source of generation, transfer station, and the MSW processing and disposal facility. The calorific value (CV) and moisture content (MC) of the mixed RDF consortium were the primary parameters of interest. CV and MC delineated an inversely proportional correlation. The CV value of 3200 cal/gm at source depleted to 1300 cal/gm in segregated combustible fraction, while the MC value has elevated from 14% to 33%. The findings direct better source segregation and transport of combustible fraction can enhance the RDF quality and encourage cement industries for better adaptation of RDF as a potential alternate.

**Keywords** Municipal solid waste · Refuse derived fuel · Collection and transportation · Calorific value · Moisture content

## 11.1 Introduction

With the soar in the quantity of municipal solid waste (MSW) generation and contrarily significant depletion in the entire quantum of reserved fossil fuel resources, an urge to utilize the incinerable as an alternate fuel is needed of the hour. In India, the current urban MSW production rate is 109,598 tonnes per day (or 0.34 kg/capita/day) and is assumed to reach 376,639 tonnes per day (or 0.7 kg/capita/day) by 2025 (Hoornweg and Bhada-Tata 2012). Despite many Technologies adopted for the treatment still there is a lacuna and many of the municipalities are still in the nascent stage.

Mainer nations across the globe still finds it difficult to deal with the ultimate disposal and get rid of the secondary pollution-related issues with residual MSW. The rapid increase in MSW generation, declination in the availability of the raw materials, and shrinking landfill spaces are demanding alternate strategies and their implementation. Complying with stringent norms and proper waste management made us to rely on waste to energy value chain. With continuous decrease in natural resources and raw materials additional limelight is shifted towards the

M. Mehdizadeh (✉)
Department of Agronomy and Plant Breeding, Faculty of Agriculture and Natural Resources, University of Mohaghegh Ardabili, Ardabil, Iran
e-mail: mehdizadeh.mohammad@uma.ac.ir

M. V. P. Kumar · A. Leelavathi
Department of Environment & Sustainable Development, Environment Protection Training and Research Institute (EPTRI), Hyderabad, India

S. U. A. Begum
Department of Civil Engineering, Faculty of Engineering and Technology, Annamalai University, Annamalai Nagar, Tamil Nadu, India

K. Patrick
Department of Civil Engineering, Pan African University Institute for Basic Sciences, Technology and Innovation, Nairobi, Kenya

recycling and reuse rather than incineration of the waste material. In this perspective alternate fuel such as RDF is one of the promising candidates to compensate the demand of all major fuel consuming industries. Generally, the organic fraction of MSW (biowaste) is digested as compost while the residual of the pre-sorting operation is turned into RDF through further processing. Sapuay (2016) has stated that the resource recovery from waste is not only limited to recyclable materials such as glass, paper, wood, metals cans, etc. it as well involves recovery of residual waste. RDF recovered from MSW is an energy-rich waste that should not be disposed of in landfills in order to preserve the air volume, facilitate resource-recovery, extend economic viability, and contribute to the circular economy. RDF as a raw material consists of combustible solid waste fraction such as plastics, wood, pulp, cotton, cloth etc. which can be further processed as pellets or fluff. This processed RDF in the form of pellets or fluff would serve as a fuel for cement industries, waste to energy plants, thermal power plants, etc. The present work assessed the scenario of Hyderabad city to establish an understanding of material flow balance between total MSW and RDF generation. The data can be considered as the baseline to ascertain the highest quantity of industrial energy compensation for a particular location irrespective of the heterogeneity of MSW.

Presently, approx. 6000–7000 tons of MSW is getting generated within the premises of Greater Hyderabad Municipal Corporation (GHMC), out of which nearly 30–35% of the waste weighing around 2000–2200 tons has higher calorific value and the same is processed as RDF.

The basic idea of the study was to evaluate RDF production, its end use, and its role in the waste to energy cycle. It also attempts to figure out the main obstacles that affect its current and future scenario. The work tries to investigate the laggings in the existing system and also suggested technical up-gradation for further improvement towards the enhancement of RDF yield and quality. Mechanized source segregation at the initial stage will maximize the production and minimizes the human intervention. Segregation of nonreusable, nonrecyclable with high calorific value will further expand its production and increases the efficiency of the recycling plants at the same time. To make the chain of waste to energy more productive and profitable RDF quality should be compliant with the necessary regulatory requirements. The same could be achieved through a multistage waste segregation process, efficient technical knowledge, and pertinent mechanical intervention. Non-discriminated attention towards RDF consumption can certainly strengthen the concept of the circular economy. Parallelly, it will enhance the acceptability of RDF as the primary energy source, rather than an auxiliary fuel (Longo et al. 2020).

Among all the processing and disposal technologies, RDF manufacturing is considered the most efficient technique for handling MSW and substitutes conventional fossil fuels (Reza et al. 2013; Jayawardhana et al. 2016). Refuse Derived Fuel (RDF) is an alternate fuel that helps eliminate the non-combustible fraction, moisture content, that could be traded and co-processed in the industries for power generation. It is majorly used in the cement industries globally. The paper deals with assessing RDF production at various compositions, which helps compensate for the energy demand produced in the industries.

## 11.2 Basics of RDF

RDF is defined as the alternate fuel derived by assorting the high CV fractions of MSW such as most of the inorganics like cardboard, wood, paper, rexine, leather, plastic etc. processed by sequential micronization, moisture removal, and purification operations.

RDF has primarily been utilized in energy synthesis-oriented applications. Presently, it is the sole choice of substrate for the incineration-based waste to energy plants (WtEP). Ancillary applications include kiln fuel for cement and thermal power plants (Sharma et al. 2018).

### *11.2.1 RDF Composition*

RDF typically comprises of the residual dry combustible fraction of the MSW incorporates the following (Fig. 11.1).

### *11.2.2 Qualitative Assessment of RDF*

Characterization operation for RDF was performed by analyzing the moisture and ash content (proximate analysis), elemental analysis (ultimate analysis), and CV (higher heating value). The findings for segregated combustible fraction (SCF) are portrayed in Table 11.1 (Nobre et al. 2021).

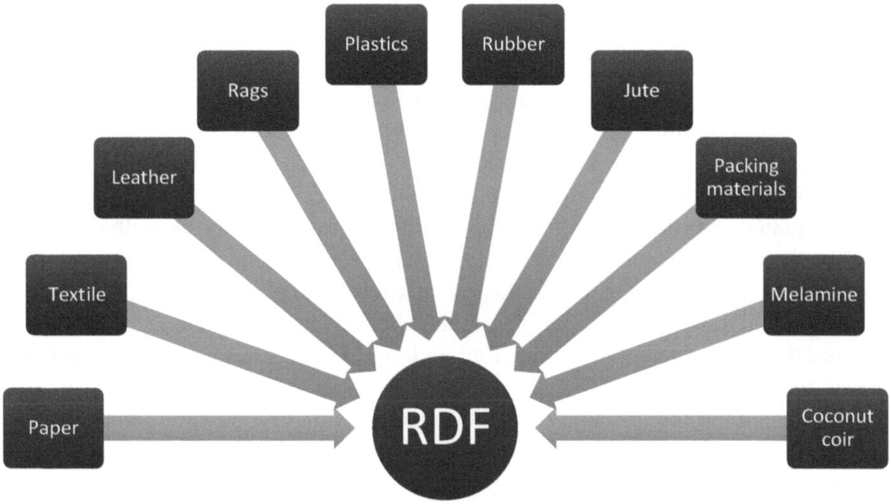


**Fig. 11.1** Basic composition of RDF

**Table 11.1** Results of proximate and ultimate analysis of SCF

| No. | Parameter | Composition (%) |
|---|---|---|
| 1 | Ash content | 28.0 |
| 2 | Moisture content | 32.28 |
| 3 | Carbon | 13.12 |
| 4 | Hydrogen | 2.01 |
| 5 | Oxygen | 8.95 |
| 6 | Nitrogen | 0.51 |
| 7 | Sulfur | 0.1 |
| 8 | Chlorine (Cl) | 0.23 |

### *11.2.3 Classification of RDF*

RDF is classified into seven prominent grades based on parameters such as CV, moisture & ash content, chloride and sulphur content etc. by the American Society for Testing and Materials (ASTM). The details are delineated in Fig. 11.2.

Presently, it is strictly prohibited for any MSW processing and disposal site to dispose of any waste with CV more than 1500 cal/gm in the sanitary landfills. As per the protocol, the same should be processed as RDF and utilized in WtEPs or cement industries. On an average, the footprint of combustible fraction lies about 20% against the comprehensive MSW generation in India.

Other than the classification established by ASTM (Fig. 11.2), an interim categorization developed by the Ministry of Housing and Urban Affairs (MoHUA) is the basis of differentiation between different grades of RDFs in India. Scientifically, RDF was categorized into three basic grades (Table 11.2) by MoHUA citing different ranges of heating value, moisture, salt, and ash content.

Further citing various factors of environmental and bio-safety, MoHUA also indulged in making a guideline for a list of items to be prohibited in RDF streams are portrayed in Fig. 11.3.

### *11.2.4 MSW Sampling Procedure for RDF Making*

The sampling of the waste is the primary most crucial component of the study; the detailed sampling procedure is delineated below. At the waste deposition area, the waste heap was classified into three layers according to height (top, middle and lower layer) and five different points were opted to ensure the collection of a representative sample. Hence from all the five different points and three layers, 15 different samples were drawn. These randomly collected samples were blended in a quadrant method. The collected samples were separated into four quadrants and in which the first and fourth quadrants were mixed while the second and third are

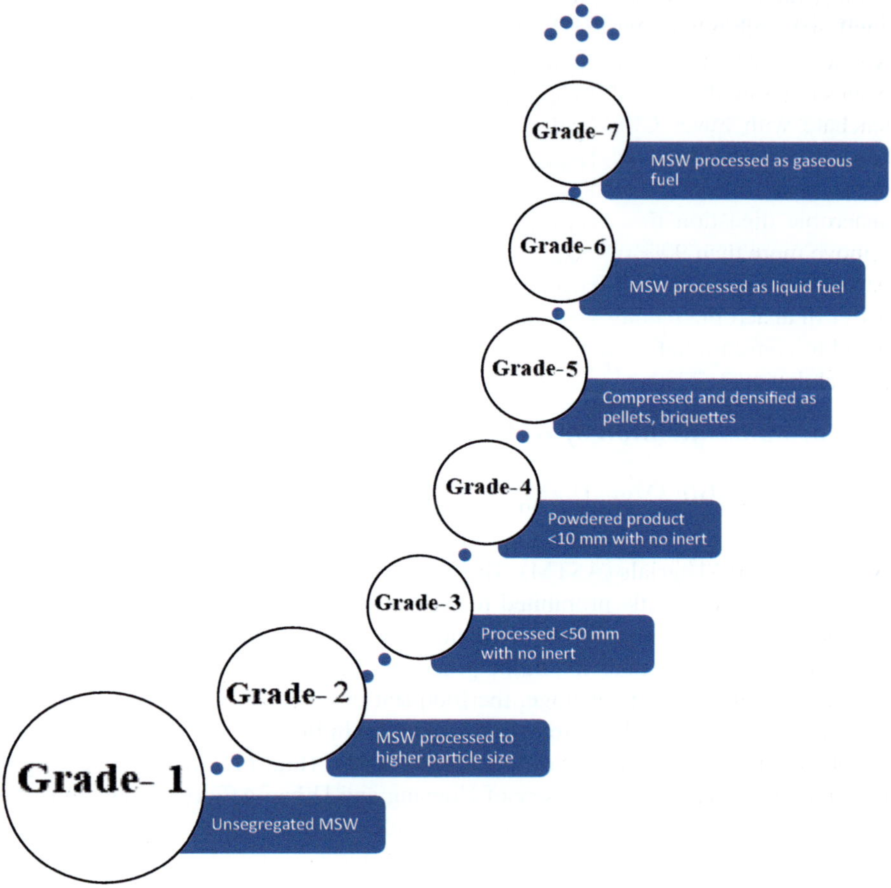


**Fig. 11.2** Different grades of RDF as per the ASTM Standard

**Table 11.2** Different grades and quality comparison of RDF as per MoUHA

| Parameters | RDF-Grade-III | RDF-Grade-II | RDF-Grade-I |
|---|---|---|---|
| Size | 20 mm < particle size <50 mm | | |
| Ash | <15 | <10% | <10% |
| Moisture | <20% | <15% | <10% |
| Chlorine | <1.0% | <0.7 | <0.5 |
| Sulfur | <2% | <1.5 | <1.5 |
| Net Calorific Value (NCV) | >3000 Cal/gm | >3750 Cal/gm | >4500 Cal/gm |

eliminated. The procedure was repeated until our required 10 kg of MSW sample was obtained. Then the samples were spread and the components were weighed individually, followed by finding the percentage of the MSW which should not exceed the initial value. The person handling the sample should know about the

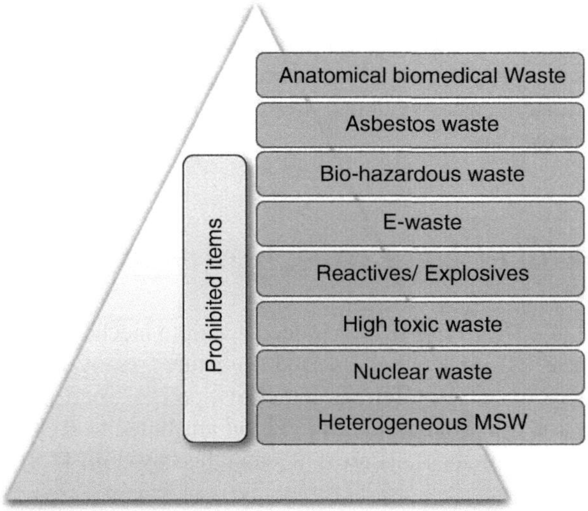


**Fig. 11.3** List of prohibited items in RDF

degradable, partially degradable, and non-degradable materials, which should be tabulated for future use. This is considered as the visual inspection of the sampling.

### *11.2.5 Brief on Conventional RDF Making Process*

The proposed RDF is produced after the segregation of the suitable residues. Then the segregated wastes are dried over the sun or by hot air passing or through the oven at 105 °C and screened for separation of residues based on sizes. Then it is followed by size reduction and magnetic separation. The reduced residues are made in three ways according to their particle size distribution.

The larger-sized particles are made into briquettes which includes materials like MSW reject (plastic bags, papers & all the other materials excluding the PVC & textiles) that is then mixed with additives like (coir dust, rice husk, and water hyacinth) with a binder named cassava starch, all the bound materials are then pressed in a uniaxial piston machine for fabrication and is dried before use. While the medium-sized particles are processed as fluff and the small residues become pellets.

The density of the briquettes is determined. The moisture content of the briquette is determined by drying the known weight of the sample in an oven at 105 ± 5 °C for an hour. The volatile and ash content were calculated according to the American Standards for Testing and Material (ASTM). On the other hand, chlorine in RDF samples was determined by ASTM Method E776–87 (2004), Fixed carbon, hydrogen, nitrogen, and sulfur were also determined using CHNS elemental analyzer. Oxygen was determined by subtracting the percentages (estimation by difference) of carbon, hydrogen, nitrogen and sulfur from the whole rate (100%). ASTM D440–86 method was used to determine Impact Resistance Index (IRI). Gross Calorific values were experimentally measured using the bomb calorimetry method. Data for comparisons were obtained from the Energy research Centre for the Netherland ECN database.

### *11.2.6 Detailed RDF Processing Mechanism*

The production of RDF from MSW is indeed a simple mechanism that relies on the primary principle of size reduction and moisture removal, post segregation. Separation of ferrous and non-ferrous materials, process inert, and segregation of light and heavy fraction are auxiliary operation attributed to RDF processing. The sequential unit operational details are delineated hereby (Fig. 11.4).

### *11.2.7 Unit Operations*

The combustible fraction gets separated initially from the compostable fraction by 70–100 mm mesh size trommel at the pre-segregation area. The particle size of SCF varies between 70 and 120 mm.

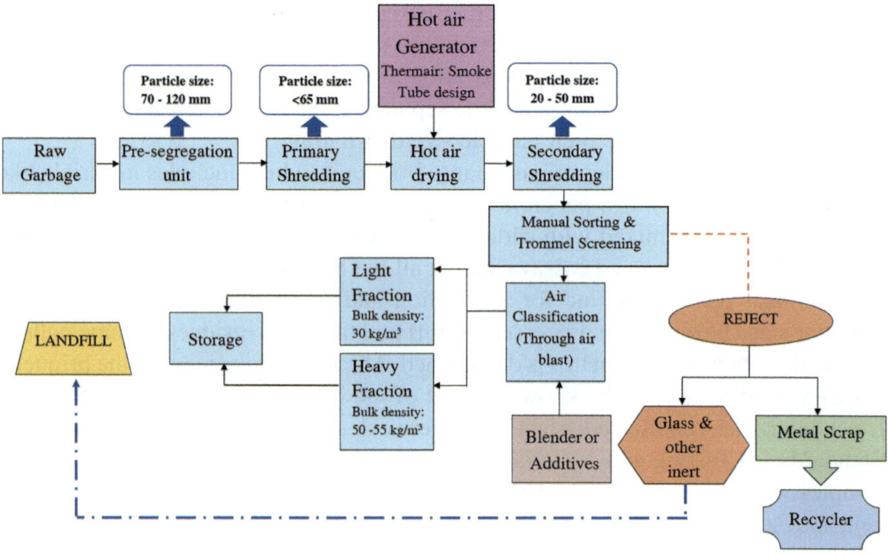


Fig. 11.4 RDF production process flow diagram

### *11.2.8 Primary Shredding and Air Drying*

SCF was subjected to a twin-blade crocodile crusher and the particle was chopped off below 60 mm. The constant mechanical action and friction emerge in the form of temperature elevation up to 90 °C. The higher temperature successively helps in moisture removal as an ancillary benefit. The shredded material was thereafter further desiccated using an in-line static drier assembly. Dehydrated hot air swiftly diminishes the inherent moisture in shredded SCF using the dual effect of radiation and convection.

### *11.2.9 Sorting*

The precision required for feedstock sorting is derived from several factors such as source, the end use of the product, and the operations and technology involved.

Generally, in an MSW processing & disposal (P&D) facility sorting is performed either manually or mechanically. The manual process is sluggish and risky. While mechanical processes offer various levels of accuracy ranging from trommels to infrared enabled artificial intelligence-based systems.

- Unsegregated waste was fed onto a slow-moving (~5 m per min) conveyor belt.
- Items not suitable for the trommel were removed manually, such as glass bottles, any domestic hazardous waste such as paints, creams, CFL bulbs, tube lights etc. and stored in puncture-resistant containers.
- During the feeding process, the heap height was always ensured less than 15 cm on the moving conveyor belt to optimize the manual segregation.
- All the ferrous metals present in the waste stream were segregated by a suspended magnetic separator prior to the manual segregation operation. The velocity of the conveyor belt was minimized and the heap height was also restricted to obtain the maximum segregation efficacy.
- Post-separation of ferrous metals, non-ferrous metals such as aluminum, copper, alloy and others are segregated from the stream using eddy current machine

### *11.2.10 Secondary Shredding & Screening*

Post sorting, the entire chunk was further micronized to a particle size of 50 mm or so using a secondary jaw cruncher and fine screens were engaged to strictly restrict the particle size below 50 mm for regulatory compliance.

### *11.2.11 Ballistic Separation*

A ballistic separator was commissioned for the ultimate removal of heavy inert and grits from the RDF stream making use of varied density and elasticity of components.

### *11.2.12 Air Classifiers*

The device was used for the separation of light fraction (i.e. mostly paper, plastics, and multi-layer packing materials) and heavy fractions (i.e. jute, leather, rexine, coconut, rubber, rags etc.) of RDF based on weight and density of the material.

## 11.3 Results of Microexperiment

Primarily, municipal solid waste is highly heterogeneous in nature and the characteristics differ based on factors such as geographic location, topography, economy, and certain special occasions. The present study encountered a huge variation in the characteristics of MSW between the point of generation and the P&D facility. Due to the non-segregated collection practice, CV of the conjugated fraction starts depleting with hauling time and distance. Further, the mixing of organic waste resulted in a higher ash value. Composite grab samples were collected from the waste stream in a sequential manner and characterized and analyzed for profound understanding (Dianda and Mahidin 2018; Chiou and Chen 2021). The findings are symmetrized in Fig. 11.5.

Figure 11.5 explicitly depicts as the hierarchy progresses; the proportion of components also largely gets varied. The concentration of organics sharply descends with process hierarchy and grade III typically doesn't contain any organics. While the quantity of certain fractions such as paper, plastic, and other combustibles get drastically increased depending on the process efficacy.

Further, the present work also undertook the analytical study of MSW and various RDF grades (Gallardo et al. 2014). Both proximate and ultimate analysis was performed for each of the samples collected over the tenure of 1 year from various locations and the consolidated report is portrayed in Figs. 11.6 and 11.7.

## 11.4 Discussions

Various analyses and tests have been done with the RDF samples. The analysis like physical, proximate and ultimate analysis has been done. The proximate analysis like moisture content, ash content and the volatile matter has been done. The

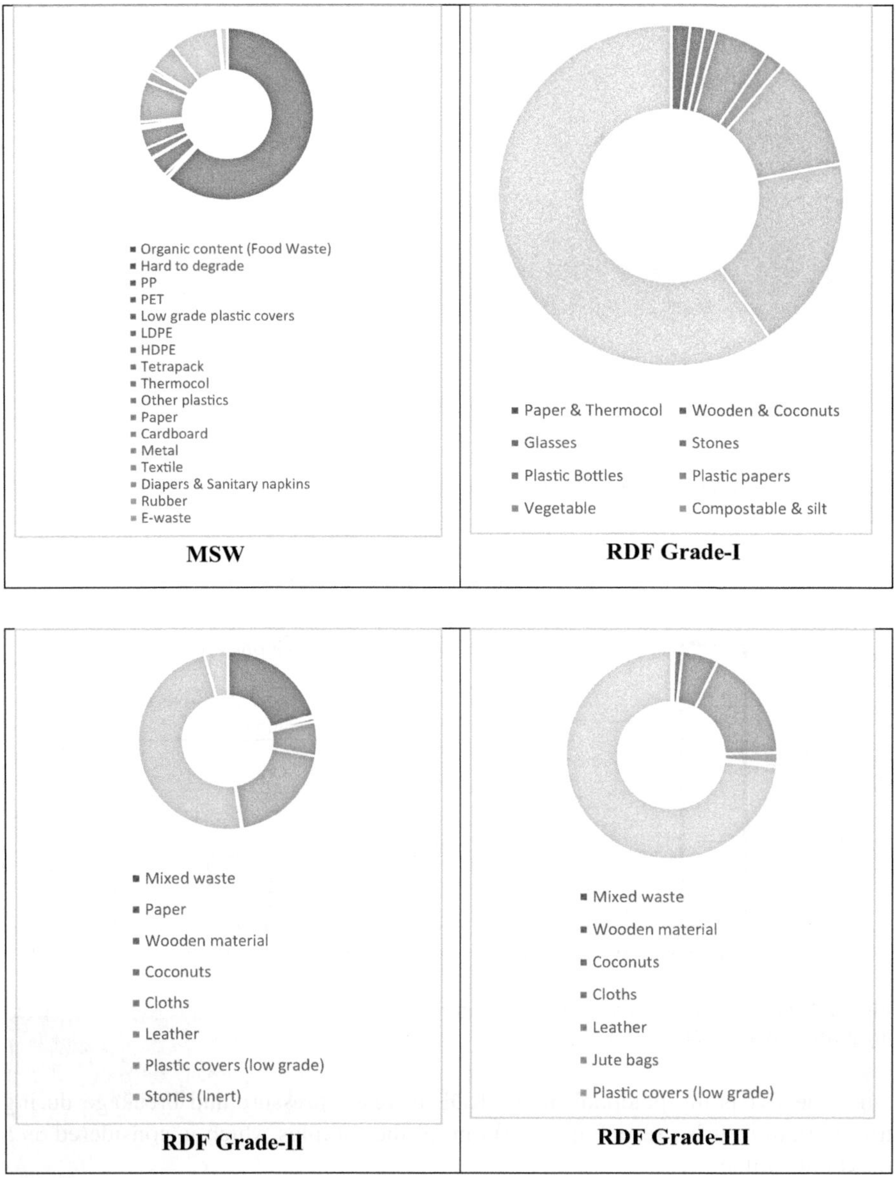


**Fig. 11.5** Compositional Analysis of MSW stream as per the hierarchy of the process

ultimate analysis like carbon, hydrogen, oxygen and calorific value, and some minor components like sulfur, nitrogen and chlorine has also been done. Physical analysis like testing its size, density and IRI component has been taken into account. The produced RDF briquette was 60 mm in length and 50 mm in diameter with a hole of 15 mm at the center. The density of the product ranged between 0.558 and 0.733 g/

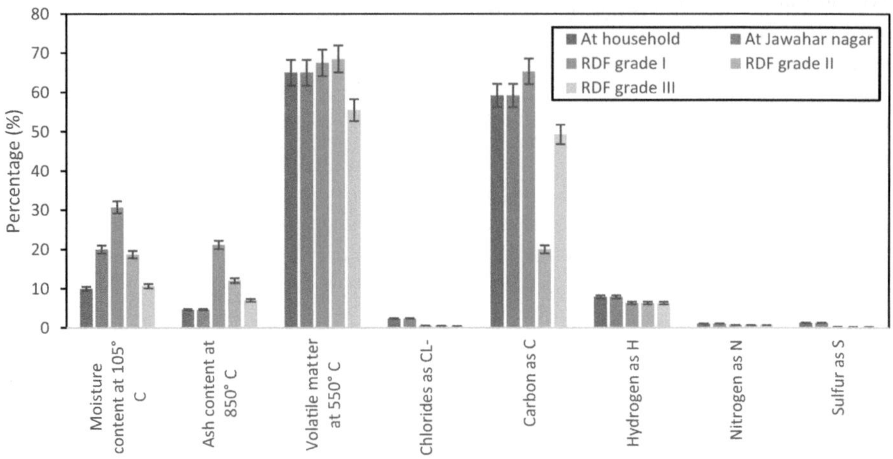


**Fig. 11.6** Proximate and Ultimate analysis report of various MSW stream as per the hierarchy of the RDF manufacturing process

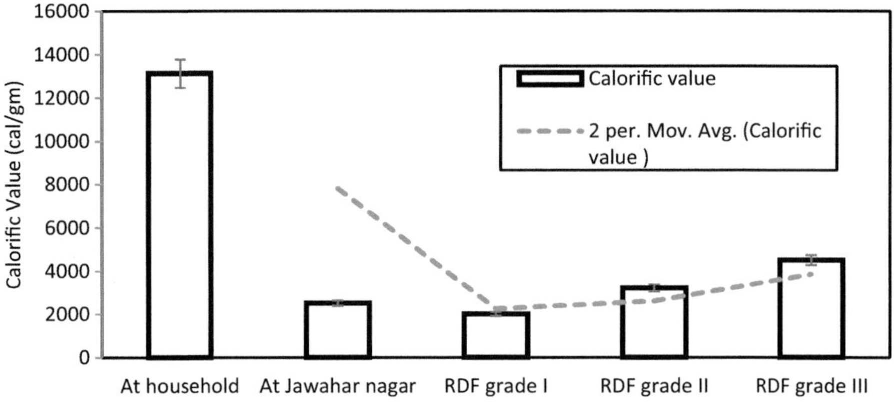


**Fig. 11.7** Variation in CV of various MSW samples as per the hierarchy of the RDF manufacturing process. (Cheela et al. 2021)

$cm^3$. The IRI is the potential of the RDF to resist pressure and breakage during transportation. It is present about 200 in all the samples which is considered as a maximum value.

Proximate analysis of the RDF sample consists of moisture content, loss on ignition, ash content, and calorific value. These all parameters play a vital role in characterizing any RDF sample. A potential RDF sample has a lower moisture content, high loss on ignition, and calorific value of approximately 2500 cal/gm. The lower the moisture content, the higher the calorific value of the samples. The moisture content of the sample 3 was 30.7% as it majorly consists of compostable matter. In the case of ash content, samples 3, 4, 5 had 21.2, 12.1, 7.1% were slightly higher when compared to the other two samples and could be a result of the binders too. In the case of volatile matter, the content was the same in all the samples.

Ultimate analysis of the samples involves the analysis of the various major and minor components. The significant components like carbon, oxygen and hydrogen are responsible for the heating values of the RDF samples. In the present study hydrogen, carbon and oxygen had values ranging from 47.71 to 57.45% wt, 6.7 to 7.70% wt, and 30.30 to 36.41% wt, respectively. The parameters are important as the environment and technical aspects are being assessed based on the chemical composition of RDF.

Minor component (sulfur, nitrogen and chlorine) is responsible for the formation of harmful gases when combusting the RDF for example nitrogen for constructing nitrogen oxides. Lower values of 0.3–0.5% wt lead to the lower formation of harmful products (Zhao et al. 2016). In this study nitrogen content in RDF processed varied from 0.21% to 0.61% by weight, which agrees and is within the range reported by Akdağ et al. (2016). Sulfur values varied from 0.42% to 0.69% wt and mean value of 0.51% wt. According to Rezaei et al. 2020 0.50% wt is the maximum value set as the standard for sulfur in RDF. On the other hand, chorine had values ranging from 0.38% to 0.49% wt which is below 1% wt a European Union standard (Chen et al. 2010).

High and Low heating values (HHV and LHV) in the presents study were ranging from 5085 to 6474 kcal $kg^{-1}$ and 44,427 to 5709.5 kcal $kg^{-1}$ which compares well with Indian imported coal which has calorific values ranging from 6200 to 6500 kcal $kg^{-1}$ according to Indian Ministry of Coal (Wagland et al. 2011; Kuspangaliyeva et al. 2021).

### *11.4.1 Possible Utilization of RDF*

The cement industry is the primary consumer of the major chunk of RDF. Presently, the bulk cement manufacturers in the country such as Ultratech cement, Bharati cement, Ramco cement etc. have already switched over to alternate energies to curtail their dependency on coal and other fossil fuels. Recently, in the guidelines on the usage of RDF in various Industries (2018), Central Public Health and Environmental Engineering Organisation (CPHEEO), the Government of India has recorded its commitment towards the replacement of at least 25% of conventional fuel for cement industries by 2025 with RDF and alternate fuel (AF) from hazardous waste management facilities (Wolny-Koładka and Żukowski 2019). Co-processing of RDF and AF in cement kiln is gaining rampant popularity in the country due to facts such as inexpensiveness, diminishing carbon footprint, local availability in cases etc. Whereas, some of the retarding factors in this business include, middleman system, quality issue, high hauling distance in cases, nonavailability, inadequate tipping fee, awareness and knowledge etc. The other most prevalent application involves consumption as feedstock in WtEPs. Incineration-based WtEPs are primarily relying on RDF and these self-sufficient carbon-neutral giants are no doubt going to rule the waste management industry in forthcoming decades. Unfortunately, at present, a couple of crucial troubleshooting needs to be performed before mass implementation of the technology as a wholesome waste management solution. The typical complications include mammoth capital expenditure, import of equipment and machinery, lack of skilled manpower, leachate handling and management, local environmental pollution from maltreated exhaust gases in cases etc. The least preferred application of RDF is restricted to research and trials in thermal power plants and brick manufacturing units. Elevated salinity and the presence of non-recyclable plastics are the primary hindrances against using RDF in the prior-mentioned industries (Infiesta et al. 2019).

Intrinsically, the CPHEEO RDF usage guideline, 2018 advised a mandatory bare minimum replacement of conventional fuel by 5% for all the fossil fuel consuming industries located within the 100 km radius of MSW P&D facility within the first 6 months from the issuance of the guideline. Whereas, the cement industries were put up with a stringent task of mending their existing system to enable consumption of RDF and ensure a minimum of 15% replacement within 6 months from the date of amendment of the guideline.

### *11.4.2 Economic Viability of RDF Manufacturing*

The heating value of RDF is a variable of the grade. Better the grade higher the CV. So, the commercial evaluation of RDF is presently performed considering CV as a baseline and it is expressed as Indian rupee (INR) per 1000 Cal/gm. RDF with CV more than 2500 Cal/gm is agreed for a commercial rate of 40 paise per unit of measurement. Further, the comprehensive cost can be derived by multiplying the unit rate with the overall weightage of the shipment. The particulars are portrayed in Fig. 11.8.

The above charges delineated in the figure above are exclusive of shipping charges. Further, the charge for transport is classified into two categories by CPHEEO as up to 100 km and more than it. The rate is adjudicated as Rs. 5 and 3/km for the former and latter categories.

### *11.4.3 Biomining and RDF Recovery*

Recovery of RDF from legacy dump through biomining process is a simple intervention. The existing dump is shaped in the form of a trapezoidal heaps, composted, and successively screened through 20 mm, 12 mm, and 4 mm mesh size trommels. The reject of each trommel gets assorted together and processed further to get rid of the inert and metals. Generally, the RDF sourced from legacy waste does not include fibrous materials such as coconut coir. It results in a moderate increment of the CV. However, the elevated level of the soil (mostly present in conjugation with the

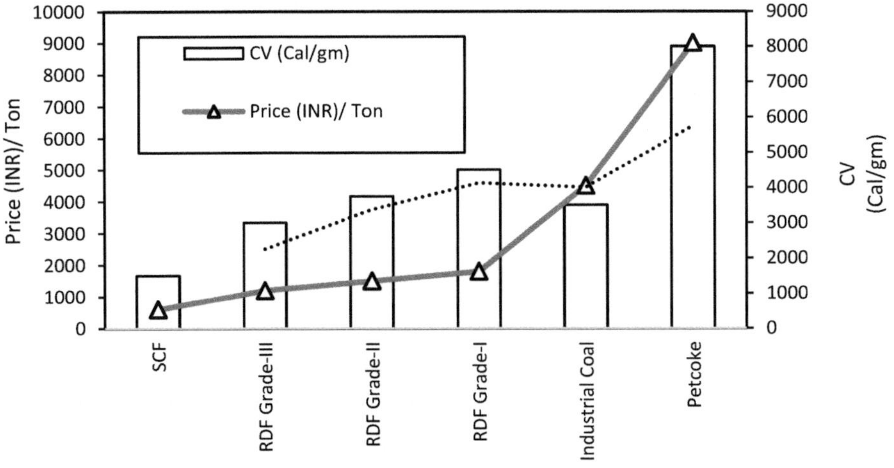


**Fig. 11.8** Variation in prices of different grades of RDF in comparison of industrial coal and pet coke

old plastics) impacts ash generation and further disposal. Detailed lab-scale experimentation was performed to ascertain the feasibility of RDF recovery from the legacy dump of Hyderabad city at Jawaharnagar. The details of the same are delineated below.

Jawaharnagar has initially designated as the dumpsite for Hyderabad city and catered as such between 2000 and 2012. The mountain of dump engulfed a large footprint of approx. 300 acres and later in the year 2016 a world-class MSW P&D facility was set up as a joint venture between the government and third-party service provider. Citing the quality and quantity of the dump, consolidation and capping was the agreed technology adopted for Jawaharnagar. While the present study involved a bitty fraction of the legacy dump and explored the possible recovery opportunities through biomining. The process of biomining was performed in three sequential phases. The non-stabilized waste chunk was processed through biological stabilization during the preliminary stabilization phase. The legacy dump was transformed into a trapezoidal heap of windrow and bio-culture of effective microorganisms were sprayed in multiple layers citing the heap height as a variable. The heap height was restricted up to 3 m to ensure optimal oxygen supply. The addition of bio-culture and deodorizer lowered the vector nuisance and pathogen breeding while accelerated the decomposition process (Chakrabarti and Dubey 2015; Kurakalva et al. 2016; Kaksonen et al. 2017). Frequent turnings during the primary phase ensured adequate moisture removal and an aerobic environment. Further, the biochemical process yielded an exothermic reaction and escalation in temperature resulted in volume and ancillary moisture removal. The bioremediation process over 30 days yielded a mature mass with zero odor, gas, and leachate issue.

**Table 11.3** Characteristics of bio-mined RDF from legacy dumping

| Parameters | RDF-Grade-III |
|---|---|
| Ash | 33.2% |
| Moisture | 22% |
| Chlorine | 1.8% |
| Sulfur | 0.9% |
| Gross Calorific Value (GCV) | 1800 cal/gm |

The remediated mass during the second phase was segregated from the inert and bulkier chunks using 20 mm and 12 mm mesh size trommels respectively. The separation process ensured the segregation of combustibles and composted matter.

During the third phase, the segregated inert such as stones and soil was processed as fine aggregate. The soundness of the alternate aggregate was compromised. While, the assorted combustible fraction was further polished through multiple stages of treatment such as shredding, air drying, metal separation etc. to ensure superior quality. Unlike the regular, the RDF processed through biomining lacks a couple of fundamental components such as paper, jute, coir etc. The CV of the bio-mined RDF didn't exceed 2200 cal/gm. The comprehensive analytical data of bio-mined RDF is delineated in Table 11.3.

Significantly, the quality of bio-mined RDF is not on-par with the regulatory reequipments possibly due to the humid Hyderabad climate. Uncontrolled soaking for about last two decades made it further difficult for the existing process to tackle the elevated humidity.

## 11.5 Conclusions

RDF has a great potential to compensate the overall or partial energy demand of the nearby fuel consuming industries if processed in a holistic manner. The existing elevated salt concentration in RDF is a factor of worry for a wide variety of industries such as thermal power plants and other corrosion sensitive industries. The study interpreted salt concentration is a variable of conjugation of organics. The issue can be possibly compensated by a proportionate amalgamation of saw dust with RDF. The multi-dimensional benefit of saw dust mixing include lowering of overall particle size, moisture and salt content, elevation of CV etc. Further, interpretations as per the process hierarchy are delineated below.

- RDF quality largely varies with the geographic locations, consumption pattern, local economy, etc.
- The primary segregation at the source will enhance the quality of the RDF.
- Observed the highest and greatest CV as well as the Economic value of the RDF at the source level since it is devoid of contamination at the household level.

- Adoption of Decentralized waste management facilities at transfer stations can potentially improve the quality of RDF.
- Co-operation from cement plants and payment of necessary wages shall motivate the service provider to yield better quality of RDF.

## References

Akdağ AS, Atımtay A, Sanin FD (2016) Comparison of fuel value and combustion characteristics of two different RDF samples. Waste Manag 47(Pt B):217–224. https://doi.org/10.1016/j.wasman.2015.08.037

Chakrabarti M, Dubey A (2015) Remediation techniques, for open dump sites, used for the disposal of municipal solid waste in India. J Basic Appl Eng Res 2:1510–1513

Cheela VRS, John M, Dubey B (2021) Quantitative determination of energy potential of refuse derived fuel from the waste recovered from Indian landfill. Sustain Environ Res 31:24. https://doi.org/10.1186/s42834-021-00097-5

Chen D, Guan Z, Liu G, Zhou G, Zhu T (2010) Recycling combustibles from aged municipal solid wastes (MSW) to improve fresh MSW incineration in Shanghai: investigation of necessity and feasibility. Front Environ Sci Eng China 4:235–243. https://doi.org/10.1007/s11783-010-0016-5

Chiou I-J, Chen C-H (2021) Municipal solid waste landfill age and refuse-derived fuel. Waste Manag Res 39:601–606. https://doi.org/10.1177/0734242X20961832

Dianda P, Mahidin ME (2018) Production and characterization refuse derived fuel (RDF) from high organic and moisture contents of municipal solid waste (MSW). IOP Conf Ser Mater Sci Eng 334. https://doi.org/10.1088/1757-899X/334/1/012035

Gallardo A, Carlos M, Bovea MD, Colomer FJ, Albarrán F (2014) Analysis of refuse-derived fuel from the municipal solid waste reject fraction and its compliance with quality standards. J Clean Prod 83:118–125. https://doi.org/10.1016/j.jclepro.2014.07.085

Guidelines on Usage of Refuse Derived Fuel in Various Industries (2018) Central Public Health and Environmental Engineering Organisation (Cpheeo), Ministry of Housing and Urban Affairs, Government of India

Hoornweg D, Bhada-Tata P (2012) What a waste: a global review of solid waste management. Urban development series; knowledge papers no. 15. World Bank, Washington, DC. © World Bank. https://openknowledge.worldbank.org/handle/10986/17388. License: CC BY 3.0 IGO

Infiesta LR, Ferreira CRN, Trovó AG, Borges VL, Carvalho SR (2019) Design of an industrial solid waste processing line to produce refuse-derived fuel. J Environ Manag 236:715–719. https://doi.org/10.1016/j.jenvman.2019.02.017

Jayawardhana Y, Kumarathilaka P, Herath I, Vithanage M (2016) Municipal solid waste biochar for prevention of pollution from landfill leachate. In: MNV P, Shih K (eds) Environmental materials and waste. Academic Press, London, pp 117–148. https://doi.org/10.1016/B978-0-12-803837-6.00006-8

Kaksonen AH, Boxall NJ, Bohu T, Usher K, Morris C, Wong PY, Cheng KY (2017) Recent advances in biomining and microbial characterisation. Solid State Phenom 262:33–37. https://doi.org/10.4028/www.scientific.net/ssp.262.33

Kurakalva RM, Aradhi KK, Mallela KY, Venkatayogi S (2016) Assessment of groundwater quality in and around the Jawaharnagar municipal solid waste dumping site at Greater Hyderabad, Southern India. Procedia Environ Sci 5:328–336. https://doi.org/10.1016/j.proenv.2016.07.013

Kuspangaliyeva B, Suleimenova B, Shah D, Sarbassov Y (2021) Thermogravimetric study of refuse derived fuel produced from municipal solid waste of Kazakhstan. Appl Sci 11:1219. https://doi.org/10.3390/app11031219

Longo S, Cellura M, Girardi P (2020) Life cycle assessment of electricity production from refuse derived fuel: a case study in Italy. Sci Total Environ 738:139719. https://doi.org/10.1016/j.scitotenv.2020.139719
Nobre C, Alves O, Durao L, Sen A, Vilarinho C, Gonclaves M (2021) Characterisation of Hydrochar and process water from the hydrothermal carbonization of Refuse Derived fuel. Waste Manag 120:303–313. https://doi.org/10.1016/j.wasman.2020.11.040
Reza B, Soltani A, Ruparathna R, Sadiq R, Hewage K (2013) Environmental and economic aspects of production and utilization of RDF as alternative fuel in cement plants: a case study of Metro Vancouver Waste Management. Resour Conserv Recycl 81:105–114. https://doi.org/10.1016/j.resconrec.2013.10.009
Rezaei H, Yazdan Panah F, Lim CJ, Sokhansanj S (2020) Pelletization of Refuse-Derived Fuel with varying compositions of plastic, paper, organic and wood. Sustainability 12:4645. https://doi.org/10.3390/su12114645
Sapuay GP (2016) Resource recovery through RDF: current trends in solid waste management in the Philippines. Procedia Environ Sci 35:464–473. https://doi.org/10.1016/j.proenv.2016.07.030
Sharma A, Gupta AK, Ganguly R (2018) Impact of open dumping of municipal solid waste on soil properties in mountainous region. J Rock Mech Geotech Eng 10:725–739. https://doi.org/10.1016/j.jrmge.2017.12.009
Wagland ST, Kilgallon P, Coveney R, Garg A, Smith R, Longhurst PJ, Pollard SJT, Simms N (2011) Comparison of coal/solid recovered fuel (SRF) with coal/refuse derived fuel (RDF) in a fluidised bed reactor. Waste Manag 31:1176–1183. https://doi.org/10.1016/j.wasman.2011.01.001
Wolny-Koładka KA, Żukowski W (2019) Mixed municipal solid waste hygienisation for refuse-derived fuel production by ozonation in the novel configuration using fluidized bed and horizontal reactor. Waste Biomass Valoriz 10:575–583. https://doi.org/10.1007/s12649-017-0087-7
Zhao L, Giannis A, Lam W, Lin S, Yin K, Yuan G, Wang J (2016) Characterization of Singapore RDF resources and analysis of their heating value. Sustain Environ Res 26:51–54. https://doi.org/10.1016/j.serj.2015.09.003

# Chapter 12
# Recovery of Metal Values by Treating the Municipal Solid Waste Incineration Ashes

**Sadia Ilyas, Vinay K. Singh, Hyunjung Kim, and Rajiv R. Srivastava**

**Abstract** In a rapid development of urban societies, municipal solid waste (MSW) has emerged as one of the biggest challenges to environmental sustainability. To cope with an improper disposal of MSW, the incineration technique is widely adopted to continuously receive and burn the waste that significantly reduces the refuse mass and volume. However, the secondary waste generated during incineration in the form of fly ash and bottom residues still needs to be managed. Approximately 3–4 wt.% of the waste input belongs to fly ash (of size ≤1 mm), while most solids (in between 20 and 25 wt.% of the input mass) refer to the bottom ash (of size ranging from several μm to m). This chapter aims to identify the best practice technologies for (heavy/valuable) metals recovery from the secondary waste after incineration of MSW in order to avoid their landfilling. Within this scope, different cases are presented to discuss the recycling of metal values from the MSW incineration refuse.

**Keywords** Municipal solid waste · Waste disposal · Metal recovery · Environmental challenge · Incinerated ashes

S. Ilyas · H. Kim
Department of Earth Resources & Environmental Engineering, Hanyang University, Seoul, South Korea
e-mail: sadiailyas1@yahoo.com

V. K. Singh
Faculty of Science, Department of Chemistry, The Maharaja Sayajirao University of Baroda, Vadodara, Gujarat, India

R. R. Srivastava (✉)
Center for Advanced Chemistry, Institute of Research and Development, Duy Tan University, Da Nang, Vietnam
e-mail: r2.srivastava@gmail.com



P. Pathak, S. G. Palani (eds.), *Circular Economy in Municipal Solid Waste Landfilling: Biomining & Leachate Treatment*, Radionuclides and Heavy Metals in the Environment, https://doi.org/10.1007/978-3-031-07785-2_12

## 12.1 Introduction

The anthropologic activities on a daily basis are generating a large amount of waste. The type of waste primarily produced by domestic households, including some industrial wastes of similar nature, is commonly named as the "Municipal Solid Waste" (MSW). Basically, they are collected by municipals and/or dumped at the places designated by the municipalities. MSW consists of daily used items, like product packaging, plastic and glass bottles, grass clippings, furniture, clothing, food-waste, papers, appliances, and sometimes batteries and metal wires as well. In a rapid development of urban societies, MSW has emerged as one of the biggest challenges due to the potential contamination risks of water resources, air, and farmland (Pujara et al. 2019). The traditional approach of landfilling disposal is therefore not suitable in terms of environmental and material sustainability. A-half of the worldwide production of about 2.01 billion tons of MSW/year (The World Bank 2021) is composted, recycled, and ends up in the overcapacity landfilling.

To cope with an improper disposal of MSW (as depicted in Table 12.1), incineration technique is widely adopted to reduce the refuse mass and volume as well. After MSW incineration (MSWI) treatment, the volume of waste can be significantly reduced by approximately 85–90%, the mass by 60–90%, and the organic matter by nearly 100% (Fruergaard and Astrup 2011; Leckner 2015). It is a waste-to-energy process that involving the burning of MSW at high temperature and leaving two waste streams in the form of fly ash and bottom ash which still require to be managed.

It has been observed that the incineration proportion increased with increasing generation of MSW. In between 2006 and 2018, the amount of MSWI rapidly increased from 11.4 to 101.8 million tonnes (Mt), i.e., nearly an increase of nine-folds. The US generated 742 kg MSW/capita, significantly higher than Australia 558 kg MSW/capita and the European Union (EU) 482 kg MSW/capita in 2017. Japan is the most frequent user of MSWI (78.5%) along with the next highest use of incineration in the UK (38.4%) and the EU (27.4%).

**Table 12.1** The potential issues with the landfilling discharge of MSW

| Environmental | Financial | Recycling |
|---|---|---|
| Long-term spillage of heavy metals and salts from MSW disposal sites<br>Utilization of high quantity of cement for stabilization | High-cost of MSW incineration ashes treatment and disposal | Low recycling rates of secondary raw materials from MSW incineration ashes (i.e., metals and minerals) |

Adapted with permission from Bakalár et al. (2021), copyright Elsevier

## 12.2 Characterization of Incinerated MSW

There are two main fraction by-products yielding from MSWI, namely the incinerated bottom ash (IBA) and incinerated fly ash (IFA) (Quina et al. 2014; Verbinnen et al. 2017). Approximately 3–4 wt.% of the waste input belongs to fly ash (of size ≤1 mm), while most solids (in between 20 and 25 wt.% of the input mass) refer to the bottom ash (of size ranging from several μm to m), depending on different MSWI furnaces.

IFA is extremely irregular in shape that contains substantial number of irregular particles (Mao et al. 2020), which has been classified as a hazardous waste due to the high contents of heavy metals like mercury, lead, molybdenum, cadmium, selenium, nickel, etc. and other contaminants like dioxins, furans. IBA can be widely used as a secondary raw material either in the construction area considering its low concentrations of contaminants or metals recovery due to concentrated metal alloys (del Valle-Zermeño et al. 2017; Dou et al. 2017; Silva et al. 2017; Yin et al. 2018). The chemical analysis of both types of incineration ash is summarized in Table 12.2. As can be seen from Table 12.1, both types of incineration ash contain a significant number of heavy metals/compounds. Prudent and adequate treatments for IFA and IBA are required in accordance with strict environmental regulations regarding the collection, storage, and disposal of MSW-incineration ashes. The ample contents of

**Table 12.2** Elemental composition of fly ash and incineration fly ash (Astrup et al. 2016; Šyc et al. 2020)

| Elements | Concentration in fly ash (mg/kg) | Concentration in bottom ash |
|---|---|---|
| Al | 14,000–79,000 | 4% |
| Ca | 8600–170,000 | 16.5% |
| Fe | 3100–150,000 | 1.3% |
| K | 660–16,000 | 3.1% |
| Mg | 240–26,000 | 1.4% |
| Mn | 7.7–3200 | 829 mg/kg |
| Na | 2200–42,000 | 3.7% |
| P | 440–10,500 | 0.9% |
| Si | 4300–308,000 | 8.1% |
| As | 0.12–190 | 71 mg/kg |
| Cd | 0.3–70 | 129 mg/kg |
| Cu | 190–25,000 | 707 mg/kg |
| Cr | 20–3400 | 520 mg/kg |
| Mo | 2.5–280 | 19 mg/kg |
| Ni | 7.0–4300 | 100 mg/kg |
| Pb | 74–14,000 | 1350 mg/kg |
| Sn | 2.0–470 | 65.3 mg/kg |
| V | 16–120 | 30 mg/kg |
| Zn | 10–20,000 | 1300–15,600 |

reactive silica, alumina, and lime in incineration ashes set forth its potential pozzolanic properties and small particles guaranteed the sufficient reactive surface area for different chemical reaction (Ferraro et al. 2019; Zhou et al. 2017) that can be utilized for cement clinker manufacturing, preparation of lightweight aggregates, and the potential metals' recovery. This chapter is focused on the recovery of valuable metals from MSW-incinerated ashes.

## 12.3 Treatment of Incinerated Fly Ash

Three major techniques for metals recovery are being applied that including thermal separation, aqueous processing (chemical and bio leaching), and electrochemical processing (Geng et al. 2020a). Basically, a thermal separation technique utilizes the volatility of targeted metals at high temperatures that amasses in the gas treatment system and generates metal enriched material that can be processed for metals' extraction (Kurashima et al. 2019). The additives like $MgCl_2$ and HCl has been found beneficial to transform the less volatile metal/compounds into more volatile metal chlorides (Yu et al. 2016). For example, Cu, Pb, and Zn were effectively volatilized (in between 80% and 95%) by adding $MgCl_2$ at 900 °C, which were only 60% of Cu, 87% of Pb, and 40% of Zn. In other methods, Geng et al. (2020a, b) investigated co-reduction of IFA with red mud, enhancing metal reduction and altering the melting point of $CaO$-$SiO_2$-$Al_2O_3$ system. At a mass ratio for IFA-to-red mud of 3:7, the recovery was found to be 76% Fe, 74% Cu, 83% Ni, and 58% Cr. Similarly, Lane et al. (2020) separated Zn, Cu, Pb, Sn, and Sb from MSW-IFA using selective volatilization and condensation processes. Metal concentrates were produced by heating IFA under a reducing atmosphere with 10% $H_2$ and 90% $N_2$ at condensing temperatures 900 °C and 1100 °C. The thermodynamic equilibrium predicted the volatility of Zn, Pb, and Cu in the reducing atmosphere in the presence of air (as shown in Fig. 12.1).

The extent of volatilisation determined experimentally for both the reducing and oxidising atmosphere revealed that the concentrate produced at 900 °C consists of ~60% Zn and ~5% Pb, however, Bi, Sn, and Sb were largely concentrated in comparison to the untreated IFA showing a great potential for the developed process to separate multiple metal values. Despite those further refinements to the process are needed to separate minor impurities of Na, Cd, and K, etc. It was also observed that the reducing gas atmosphere increased Zn-volatility albeit suppressed Cu-volatility and Pb-volatility.

The high temperatures required for thermal separation process is a highly energy consuming and high-cost process. Hence, aqueous processing using chemical and bacterial activity is alternatively explored by several researchers. In a chemical leaching encompasses first-step with HCl and second-step with NaCl could recover about 90% Pb and Cd, as well as 80% Cu and Zn (Weibel et al. 2018). Loginova et al. (2019) studied a 3-step combinational treatment of IFA using ethylenediaminetetraacetic acid (EDTA) and gluconate that improved the leaching efficiency of

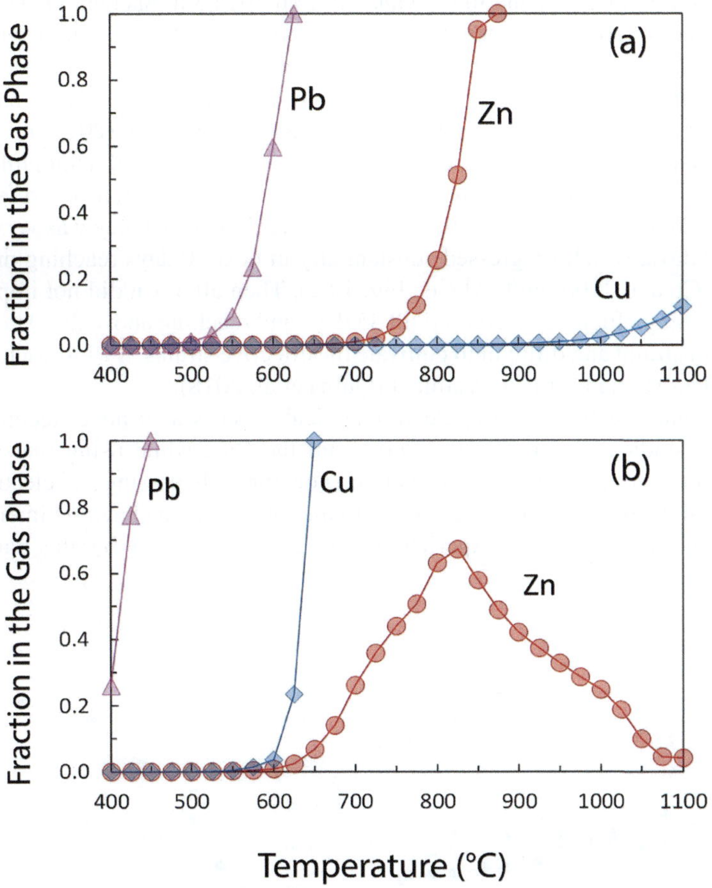


**Fig. 12.1** The thermodynamic fractions of Zn, Cu, and Pb in the gas phase with changing temperatures at 10% $H_2$/90% $N_2$ (**a**) and air (**b**). (Adapted with permission from Lane et al. 2020, copyright Elsevier)

metals (Cd, Cu, and Zn) about 10–1000-folds higher than water-based leaching. Tang et al. (2018, 2019) suggested the integrated aqueous processing for an effective recovery of Cu and Zn via HCl leaching followed by solvent extraction techniques. The IFA feed containing 5.4 g/kg Cu and 5.8 g/kg Zn was preferably leached in HCl solution over $HNO_3$ (Tang and Steenari 2015, 2016) at a pulp density of 25% at ambient temperature for 4 h yielding 100% Cu and 75% Zn. Subsequently, using 0.08 mol/L LIX 860N-I for Cu-extraction and 0.06 mol/L Cyanex 572 for Zn-extraction at an organic-to-aqueous ratio of 1:4, an overall recovery of 95% Cu and 61% Zn could be obtained. A pilot-scale study was conducted by Schlumberger et al. (2007) for Zn-recovery using leaching-solvent extraction method. A pH-controlled leaching was performed at around 4.0 value and pulp density of 33% IFA to generate the leach liquor, which was treated for mercury and iron removal prior to >99% Zn-extraction with 30% Cyanex 272 (prepared in Shellsol D 70). Finally, Zn-enriched solution of concentration ~190 g/L could be recovered which was taken for electrowinning of 99.99% pure Zn-metal.

Additionally, bioleaching is another stream of aqueous processing that can be applied for metal recovery from incinerated waste. Funari et al. (2019) investigated a controlled experiment using a culture with dominant strains of acidophilic bacteria, mainly *Acidothiobacillus thiooxidans* and *Acidothiobacillus ferrooxidans*. With respect to time, it was observed that Cu and Al leaching efficiency was very low for initial 10 days, which progressed substantially in next 10 days reaching more than 30 mg/L Cu and 2000 mg/L Al (see Fig. 12.2). Then after, Cu did not improve but Al bioleaching efficiency increased till 35 days and reaching above 2500 mg/L. Due to time constraint and difficult to cultivate the effective strains, the bioleaching technique is less studied for IFA treatment (Quina et al. 2018).

Other than the bioleaching, electrochemical process as a novel technique has also been studied for metals' recovery from the IFA, while using the oxidation (anode) and reduction (cathode) phenomenon through passing an electric field. Zhang et al. (2017) used 3D-electrode with the sample region having dimensions of 8 × 8 cm, in which, the experiment was greatly affected by the nature and

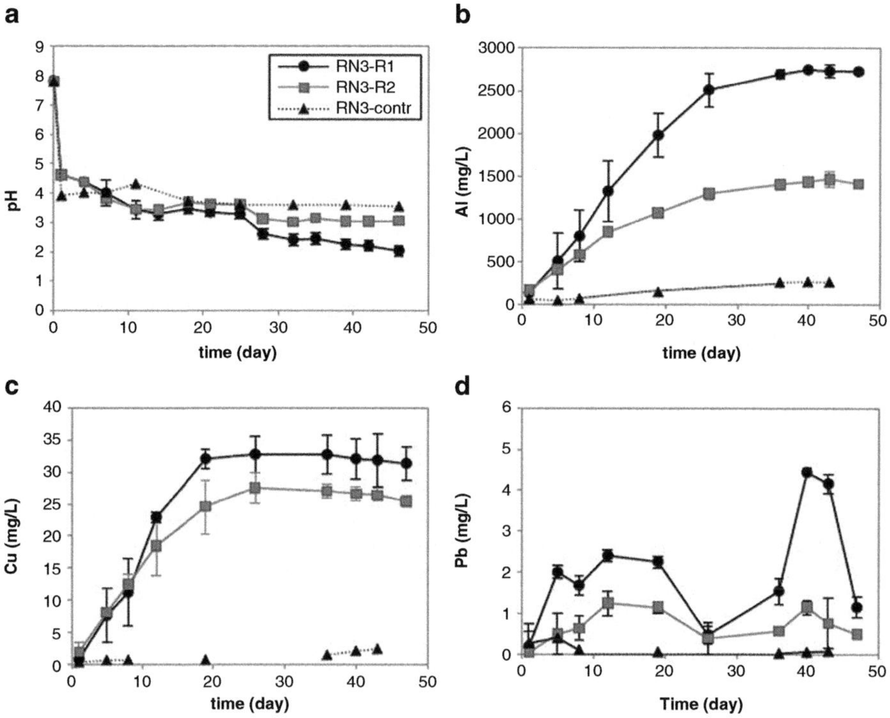


**Fig. 12.2** Change in pH with different types of culture (**a**) removal of Al (**b**) Cu (**c**) and Pb (**d**) with respect to time during bioleaching of MSW-IFA. (Modified from Funari et al. 2019)

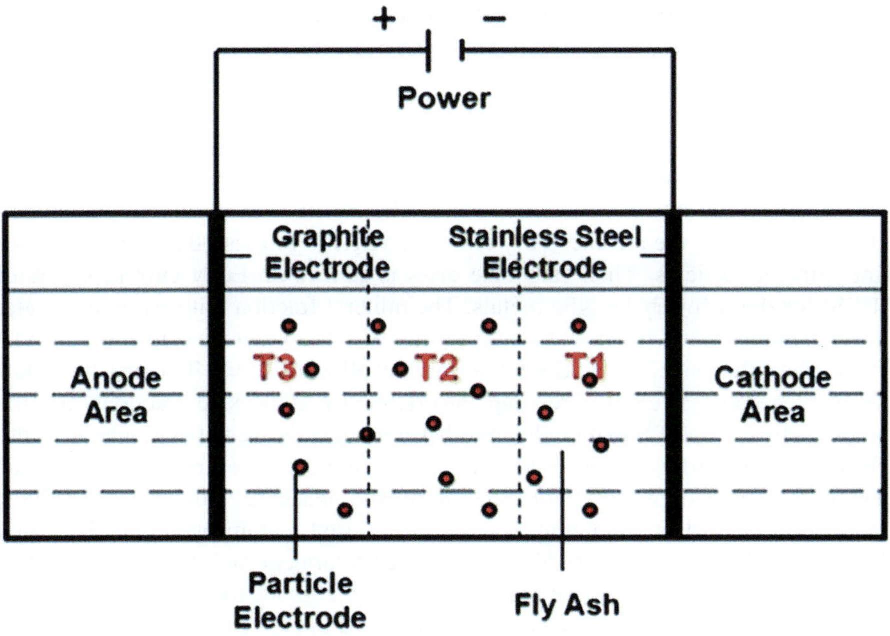


**Fig. 12.3** The experimental schematic of electrochemical processing of MSW-IFA. (Adapted from Zhang et al. 2017, copyright Royal Society of Chemistry)

concentration of the particle electrode (activated carbon and graphite). The experimental schematic is shown in Fig. 12.3, wherein the sample region was divided into three sections: T1 (near the cathode channel), T2 (middle area) and T3 (near the anode tank). The optimal dimension (6 × 6 mm) to achieve high heavy metals' removal was determined on the basis of following three factors: (i) the particle electrode ratio, (ii) the voltage gradient and (iii) the repair time by applying a voltage gradient of 9 V for 5 days at a dosing ratio of 5%. Huang et al. (2018a) observed that a 3D-electrode significantly improved the metals depletion as compared to the traditional 2D-electrolyzer following the recovery rate of Zn, Pb, Cd, and Cu increased by 44%, 41%, 17%, and 46%, respectively. The acidification and ultrasonic pretreatment further enhanced the metals' recovery (Huang et al. 2018b).

## 12.4 Treatment of Incinerated Bottom Ash

Until now, the most of IBA treatment is dependent upon the separation of magnetic and non-magnetic metal particles using the magnetic and/or Eddy current separation. In this regard, the treatment plants mainly located in European countries deals with wet/dry separation techniques. With an aim to separate metal values and then washing of soluble salts and metal compounds from the IBA mineral fraction. The leachable heavy metals in fine fractions undergoes to washing treatment. A major disadvantage of this route has been identified as the sludge production (ca. 50,000 tons) with a large content of heavy metals that must requires a further treatment (Born 2018a, b). The input and output stream materials balance of Alkmaar plant is shown in Table 12.3. The Heros plant (The Netherlands) treats 650,000 t of IBA/year using a similar type of approach. After ageing and the separation of Fe-scraps, the IBA is sieved into nine different sizes. The fractions of particles smaller than 10 mm are sent to the advanced dry residue (ADR) to separate the light and sticky fine mineral particles. Then each size class is sent to an Eddy current separator (ECS) for the recovery he NFe metals. The mineral fraction with particles greater than 4.0 mm is water- washed at a pulp density of 25% to remove sulphate, chloride.

On the other side, a dry discharge system allows better efficiency in metals recovery because it eases screening into defined particle size fractions and the potential treatment of fine particles. Furthermore, metals are not agglomerated into clusters by wet particles reduces the water consumption, lower transportation costs, less contents of organic carbon, and improved leaching efficiency of metals. However, due to dust formation, it requires a closed system operation (Böni and Morf 2018; Quicker et al. 2015). A centralized treatment facility (MSWI KEZO) built in Hinwil (Zurich) with a designed capacity of 200,000 tons IBA/year commissioned in 2017, of which the schematic is depicted in Fig. 12.4. Wherein, after the magnetic separation of coarse metallic scraps, the IBA is divided into four sizes belonging (A) >80 mm, (B) 30–80 mm, (C) 12–30 mm, and (D) <12 mm. The fraction-A is sent to a handpicking recovery of metals and separating large particles to resend for crushing to be mixed with particles size of <80 mm. The fraction-B is sent to magnetic separation and crushing to particles size of <30 mm. The fraction-C is sent to magnetic separation → glass separator → two stainless steel separators → ECS → crusher to <12 mm. The fraction-D is sent to two magnetic separators and two high-frequency ECS, resulting in separation of light fraction (usually Al) over the heavy fraction (usually Cu and precious metals) using the densimetric tables. Consequently, about 10.08% iron, 4.45% as non-ferrous metals, and 1.07% glass are obtained.

**Table 12.3** Annual mass balance at the Alkmaar plant

| Items | | Weight in tons |
|---|---|---|
| Input mass | MSW-IBA | 235,866 |
| Output mass | Sludge | 49,745 |
| | Sand | 99,745 |
| | Granulate | 81,610 |
| | Iron scrap | 15,965 |
| | Stainless steel | 2590 |
| | Unburnt material | 4168 |
| | Non-ferrous metals | 8321 |
| | Heavy non-ferrous metals (<4 mm particle size) | 800 |

Born (2018a, b); adapted with permission from Šyc et al. (2020), copyright Elsevier

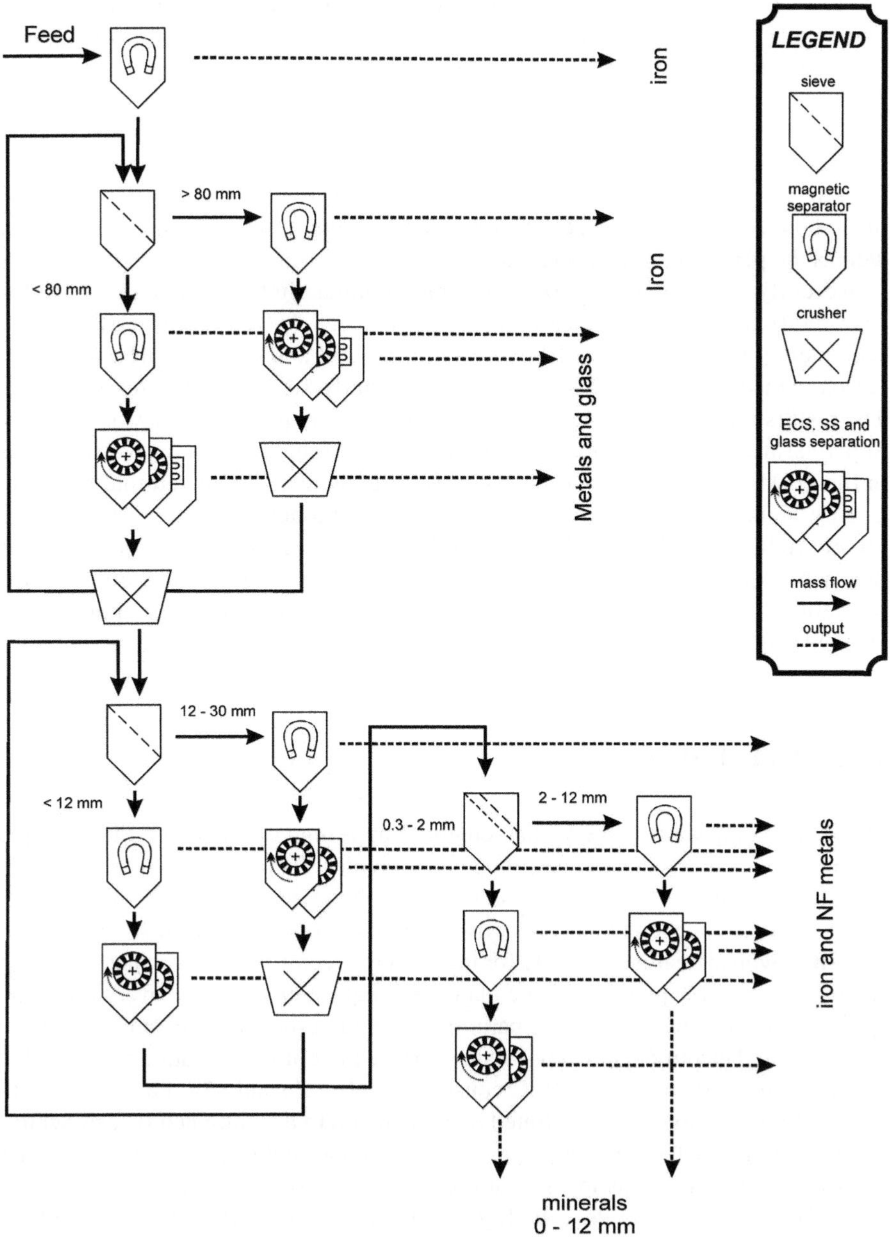


**Fig. 12.4** Scheme of the dry bottom ash treatment plant in Hinwil (CH). (Böni and Morf 2018; adapted with permission from Šyc et al. 2020, copyright Elsevier)

Besides the commissioned plants, several research items have been reported time-to-time on the treatment and recovery of metals from MSW-IBA. Sun and Yi (2020) used a simple water washing technique for heavy metals' (Pb, Zn, Ni, Cd,

Cu, and Cr) removal from with the in-situ change in pH evolution. It was observed that the pH of wash liquor increases steeply until the initial 3 h of washing, which could be corroborated to the dissolution of quicklime and portlandite (see Fig. 12.5). Then after, decline in pH with increasing of washing duration was ascribed to the consumption of hydroxyl ions via precipitation of metal hydroxides. In wash liquor, the dissolution of Pb, Zn, and Ni showed a similar trend with pH and time, however, Cd, Cu, and Cr increased with prolonged washing. Hence, the optimum washing time should be determined accordingly based on the most concerned metal(s), as well as the pH evaluation during washing.

Recently, manganese-oxidizing aerobic granular sludge (Mn-AGS) is used to treat the bottom ash leachates that containing high As(III) and Cu(II). Using a sequencing batch reactor (SBR), metals in influent and effluent, pH and chemical oxygen demand (COD) were monitored daily for 91 days refer Fig. 12.6. The results indicated that As(III) and Cu(II) removal performance was sensitive to pH variation. Metals removed from IBA containing high As(V) was obtained in the Fe/Mn oxide-bound phase. The bacterial analysis indicated that the community richness was only slightly changed at moderate acidic pH (4.0) but greatly reduced at pH ~6.0 (He et al. 2020). Earlier, Aouad et al. (2008) showed interactions between IBA and *Pseudomonas aeruginosa* bacteria exhibiting the formation of biofilm on mineral surfaces, possibly protecting them to be leached out.

## 12.5 Conclusions

Municipal solid waste (MSW) has emerged as a biggest challenge due to rapid urbanization and associated anthropogenic activities. Incineration has been widely adopted as a suitable treatment to reduce the volume and mass of the huge waste. This activity also generates secondary waste in the form of fly ash (IFA) and bottom ash (IBA) whose treatment mainly for the resource recovery of heavy metals therein can be a sustainable practice. Three major techniques for metals recovery are being applied that including thermal separation, aqueous processing (chemical and bio leaching), and electrochemical processing. The extent of volatilisation revealed that the concentrate produced at 900 °C consists of ~60% Zn and ~5% Pb, however, Bi, Sn, and Sb were largely concentrated in comparison to the untreated IFA showing a great potential for the developed process to separate multiple metal values. It has been found that the reducing gas atmosphere increased Zn-volatility albeit suppressed Cu-volatility and Pb-volatility. Metal recovery from IBA (both wet and dry mass) has remained dependent mainly on the dry mechanical processes albeit wet treatment process has also been applied at higher scale. Metals' recovery can be carried out directly at the incineration facility plant or at the centralized facility that receives the material from different plants. In order to treat both types of ash waste,

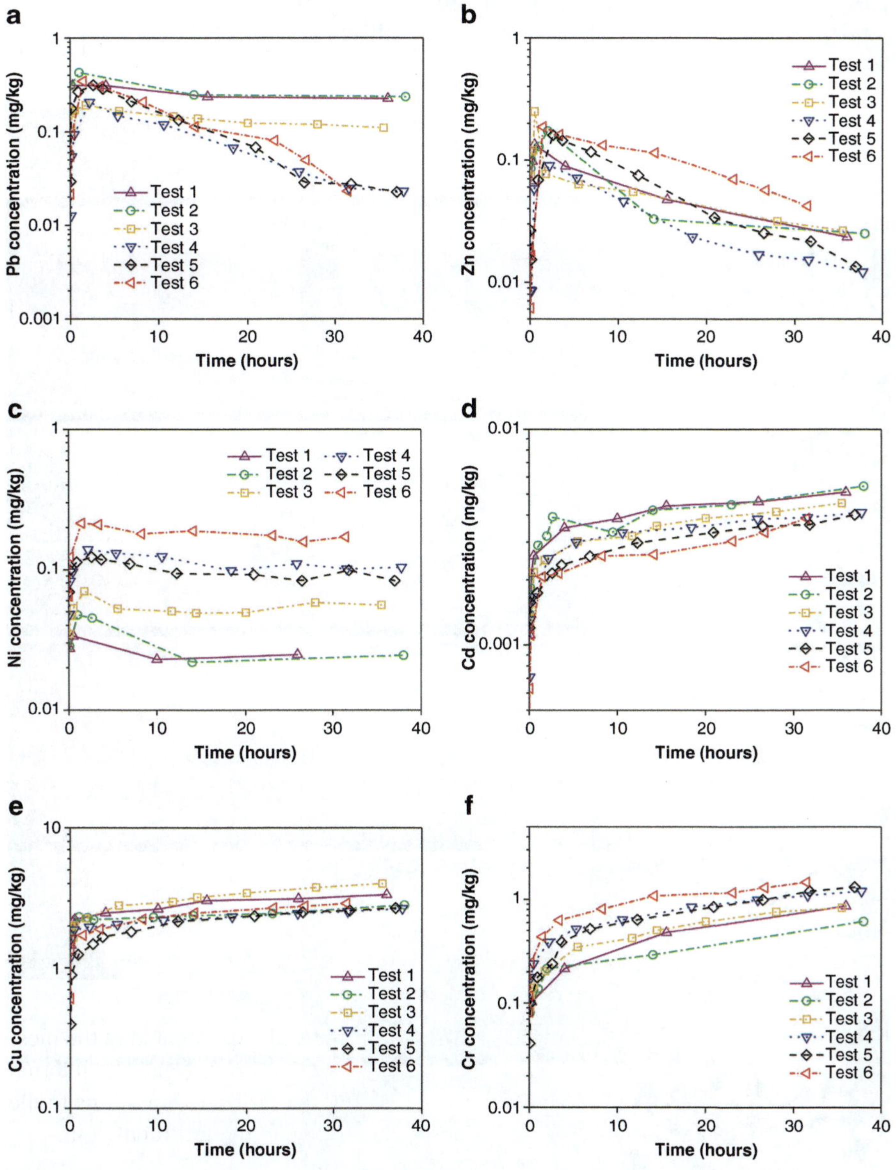


**Fig. 12.5** Concentrations of heavy metals versus washing time: (**a**) Pb, (**b**) Zn, (**c**) Ni, (**d**) Cd, (**e**) Cu, and (**f**) Cr. (Adapted with permission from Sun and Yi 2020, copyright Elsevier)

the bioleaching is still applicable on lab scale while the electrochemical treatment has potential to be applied in future at a higher scale either alone or in combination with the mechanical treatment processing.

**Fig. 12.6** Macroscopic and microscopic photographs of granules at different time. The sludge was sampled on days 1 (**a**), 9 (**b**), 23 (**c**), 31 (**d**), 52 (**e**), and 91 (**f**). (Adapted with permission from He et al. 2020, copyright Elsevier)

**Acknowledgements** This work was supported by the Brain Pool Program through the National Research Foundation of Korea (NRF) funded by the Ministry of Science and ICT (Grant No. 2021H1D3A2A01100016) and Basic Science Research Program through the National Research Foundation of Korea (NRF) funded by the Ministry of Education (Project No. 2020R1I1A1A01074249).

## References

Aouad G, Crovisier J-L, Damidot D, Stille P, Hutchens E, Mutterer J, Meyer J-M, Geoffroy VA (2008) Interactions between municipal solid waste incineration bottom ash and bacteria (*Pseudomonas aeruginosa*). Sci Total Environ 393:385–393

Astrup T, Muntoni A, Polettini A, Pomi R, Van Gerven T, Van Zomeren A (2016) Treatment and reuse of incineration bottom ash. In: Prasad, MNV, Shih K (eds) Evironmental materials and waste resource recovery and pollution prevention. Academic Press, Elsevier, Oxford, pp 607–645. https://publications.tno.nl/publication/34627970/5np6r7/b16004.pdf

Bakalár T, Pavolová H, Hajduová Z, Lacko R, Kyšeľa K (2021) Metal recovery from municipal solid waste incineration fly ash as a tool of circular economy. J Clean Prod 302:126977

Böni D, Morf LS (2018) Thermo-recycling – efficient recovery of valuable materials from dry Bottom ash. In: Holm O, Thomé-Kozmiensky E (eds) Removal, treatment and utilisation of waste incineration bottom ash. Thomé-Kozmiensky Verlag GmbH, Nietwerder, pp 25–37

Born JP (2018a) Mining IBA for precious metals. VDI Conf Met Miner Recover. From IBA, Düsseldorf, 8 November 2018

Born JP (2018b) Mining incinerator bottom ash for heavy non-ferrous metals and precious metal. In: Holm O, Thome-Kozmiensky E (eds) Removal, treatment and utilisation of waste incineration bottom ash. TK Verlag, Neuruppin, pp 11–24

del Valle-Zermeño R, Gómez-Manrique J, Giro-Paloma J, Formosa J, Chimenos JM (2017) Material characterization of the MSWI bottom ash as a function of particle size. Effects of glass recycling over time. Sci Total Environ 581–582:897–905

Dou X, Ren F, Nguyen MQ, Ahamed A, Yin K, Chan WP, Chang VWC (2017) Review of MSWI bottom ash utilization from perspectives of collective characterization, treatment and existing application. Renew Sust Energ Rev 79:24–38

Ferraro A, Farina I, Race M, Colangelo F, Cioffi R, Fabbricino M (2019) Pre-treatments of MSWI fly-ashes: a comprehensive review to determine optimal conditions for their reuse and/or environmentally sustainable disposal. Rev Environ Sci Biotechnol 18:453–471

Fruergaard T, Astrup T (2011) Optimal utilization of waste-to-energy in an LCA perspective. Waste Manag 31:572–582

Funari V, Gomes HI, Cappelletti M, Fedi S, Dinelli E, Rogerson M, Rovere M (2019) Optimization routes for the bioleaching of MSWI fly and bottom ashes using microorganisms collected from a natural system. Waste Biomass Valoriz 10(12):3833–3842

Geng C, Liu J, Wu S, Jia Y, Du B, Yu S (2020a) Novel method for comprehensive utilization of MSWI fly ash through co-reduction with red mud to prepare crude alloy and cleaned slag. J Hazard Mater 384:121315

Geng C, Chen C, Shi X, Wu S, Jia Y, Du B, Liu J (2020b) Recovery of metals from municipal solid waste incineration fly ash and red mud via a co-reduction process. Resour Conserv Recycl 154:104600

He Z, Zhang Q, Wei Z, Zhu Y, Pan X (2020) Simultaneous removal of As(III) and Cu(II) from real bottom ash leachates by manganese-oxidizing aerobic granular sludge: performance and mechanisms. Sci Total Environ 700:134510

Huang T, Liu L, Zhou L, Yang K (2018a) Operating optimization for the heavy metal removal from the municipal solid waste incineration fly ashes in the three-dimensional electrokinetics. Chemosphere 204:294–302
Huang T, Zhou L, Liu L, Xia M (2018b) Ultrasound-enhanced electrokinetic remediation for removal of Zn, Pb, Cu and Cd in municipal solid waste incineration fly ashes. Waste Manag 75:226–235
Kurashima K, Matsuda K, Kumagai S, Kameda T, Saito Y, Yoshioka T (2019) A combined kinetic and thermodynamic approach for interpreting the complex interactions during chloride volatilization of heavy metals in municipal solid waste fly ash. Waste Manag 87:204–217
Lane DJ, Hartikainen A, Sippula O, Lähde A, Mesceriakovas A, Peräniemi S, Jokiniemi J (2020) Thermal separation of zinc and other valuable elements from municipal solid waste incineration fly ash. J Clean Prod 253:120014
Leckner B (2015) Process aspects in combustion and gasification waste-to-energy (WtE) units. Waste Manag 37:13–25
Loginova E, Proskurnin MA, Brouwers HJ (2019) Municipal solid waste incineration (MSWI) fly ash composition analysis: a case study of combined chelatant-based washing treatment efficiency. J Environ Manag 235:480–488
Mao Y, Wu H, Wang W, Jia M, Che X (2020) Pretreatment of municipal solid waste incineration fly ash and preparation of solid waste source sulphoaluminate cementitious material. J Hazard Mater 385:121580
Pujara Y, Pathak P, Sharma A, Govani J (2019) Review on Indian Municipal Solid Waste Management practices for reduction of environmental impacts to achieve sustainable development goals. J Environ Manag 248:109238
Quicker P, Stockschläder J, Zayat-Vogel B, Pretz T, Garth A, Koralewska R, Malek S, Gellermann C, Brämer T, Gabor E (2015) Wertstoffpotenziale von trocken und nass ausgetragenen Abfallverbrennungsaschen. In: Thomé-Kozmiensky KJ (eds) Mineralische Nebenprodukte Und Abfälle 2, Aschen, Schlacken, Stäube Und Baurestmassen. TK Verlag Karl Thomé-Kozmiensky, Neuruppin, Germany, pp 117–135
Quina MJ, Bordado JM, Quinta-Ferreira RM (2014) Recycling of air pollution control residues from municipal solid waste incineration into lightweight aggregates. Waste Manag 34:430–438
Quina MJ, Bontempi E, Bogush A, Schlumberger S, Weibel G, Braga R, Funari V, Hyks J, Rasmussen E, Lederer J (2018) Technologies for the management of MSW incineration ashes from gas cleaning: new perspectives on recovery of secondary raw materials and circular economy. Sci Total Environ 635:526–542
Schlumberger S, Schuster M, Ringmann S, Koralewska R (2007) Recovery of high purity zinc from filter ash produced during the thermal treatment of waste and inerting of residual materials. Waste Manag Res 25:547–555
Silva RV, de Brito J, Lynn CJ, Dhir RK (2017) Use of municipal solid waste incineration bottom ashes in alkali-activated materials, ceramics and granular applications: a review. Waste Manag 68:207–220
Sun X, Yi Y (2020) pH evolution during water washing of incineration bottom ash and its effect on removal of heavy metals. Waste Manag 104:213–219
Šyc M, Simon FG, Hykš J, Braga R, Biganzoli L, Costa G, Grosso M (2020) Metal recovery from incineration bottom ash: state-of-the-art and recent developments. J Hazard Mater 393:122433
Tang J, Steenari BM (2015) Solvent extraction separation of copper and zinc from MSWI fly ash leachates. Waste Manag 44:147–154
Tang J, Steenari B-M (2016) Leaching optimization of municipal solid waste incineration ash for resource recovery: a case study of Cu, Zn, Pb and Cd. Waste Manag 48:315–322
Tang J, Su M, Zhang H, Xiao T, Liu Y, Liu Y, Steenari BM (2018) Assessment of copper and zinc recovery from MSWI fly ash in Guangzhou based on a hydrometallurgical process. Waste Manag 76:225–233

Tang J, Su M, Wu Q, Wei L, Wang N, Xiao E, Xiao T (2019) Highly efficient recovery and clean-up of four heavy metals from MSWI fly ash by integrating leaching, selective extraction and adsorption. J Clean Prod 234:139–149

The World Bank (2021). Available at: https://datatopics.worldbank.org/what-a-waste/trends_in_solid_waste_management.html. Accessed 28 September 2021

Verbinnen B, Billen P, Van Caneghem J, Vandecasteele C (2017) Recycling of MSWI bottom ash: a review of chemical barriers, engineering applications and treatment technologies. Waste Biomass Valoriz 8:1453–1466

Weibel G, Eggenberger U, Kulik DA, Hummel W, Schlumberger S, Klink W, Fisch M, Mäder UK (2018) Extraction of heavy metals from MSWI fly ash using hydrochloric acid and sodium chloride solution. Waste Manag 76:457–471

Yin K, Ahamed A, Lisak G (2018) Environmental perspectives of recycling various combustion ashes in cement production — a review. Waste Manag 78:401–416

Yu J, Sun L, Ma C, Qiao Y, Xiang J, Hu S, Yao H (2016) Mechanism on heavy metals vaporization from municipal solid waste fly ash by $MgCl2{\cdot}6H2O$. Waste Manag 49:124–130

Zhang Y, Huang T, Huang X, Faheem M, Yu L, Jiao B, Yin G, Shiau Y, Li D (2017) Study on electro-kinetic remediation of heavy metals in municipal solid waste incineration fly ash with a three-dimensional electrode. RSC Adv 7:27846–27852

Zhou X, Zhou M, Wu X, Han Y, Geng J, Wang T, Wan S, Hou H (2017) Reductive solidification/stabilization of chromate in municipal solid waste incineration fly ash by ascorbic acid and blast furnace slag. Chemosphere 182:76–84

GPSR Compliance

*The European Union's (EU) General Product Safety Regulation (GPSR) is a set of rules that requires consumer products to be safe and our obligations to ensure this.*

*If you have any concerns about our products, you can contact us on ProductSafety@springernature.com*

In case Publisher is established outside the EU, the EU authorized representative is:

Springer Nature Customer Service Center GmbH
Europaplatz 3
69115 Heidelberg, Germany

**Batch number: 10726135**

Printed by Printforce, the Netherlands